PROGRESS IN BRAIN RESEARCH

VOLUME 89

PROTEIN KINASE C AND ITS BRAIN SUBSTRATES:
Role in Neuronal Growth and Plasticity

Recent Volumes in PROGRESS IN BRAIN RESEARCH

PROGRESS IN BRAIN RESEARCH

VOLUME 89

PROTEIN KINASE C
AND ITS BRAIN SUBSTRATES:

Role in Neuronal Growth and Plasticity

Proceedings of the Third International Meeting on Brain Phosphoproteins,
held at Zeist (The Netherlands), 24–26 August, 1990

EDITED BY

W.H. GISPEN and A. ROUTTENBERG

Rudolf Magnus Institute, Vondellaan 6, 3521 GD Utrecht (The Netherlands) and
Northwestern University, Cresap Neuroscience Laboratory, 2021 Sheridan Road, Evanston, IL 60208 (U.S.A.)

ELSEVIER
AMSTERDAM - LONDON - NEW YORK - TOKYO
1991

© 1991, Elsevier Science Publishers B.V.

ISBN 0-444-81436-1 (volume)
ISBN 0-444-80104-9 (series)

This book is printed on acid-free paper.

Published by:
Elsevier Science Publishers B.V.
P.O. Box 211
1000 AE Amsterdam
The Netherlands

Sole distributors for the U.S.A. and Canada:
Elsevier Science Publishing Company, Inc.
655 Avenue of the Americas
New York, NY 10010
U.S.A.

Printed in The Netherlands

List of Contributors

D.L. Alkon, Section of Neural Systems, Laboratory of Molecular and Cellular Neurobiology, National Institutes of Health, Bethesda, MD 20892, U.S.A.

Y. Asaoka, Departments of Biochemistry and Pharmacology, Kobe University School of Medicine, Kobe 650, Japan, and Biosignal Reseach Center, Kobe University, Kobe 657, Japan

D. Au, Department of Pharmacology, University of Washington, Seattle, WA 98195, U.S.A.

L.I. Benowitz, Department of Neurosurgery, Children's Hospital and Harvard Medical School, Boston, MA, U.S.A.

N. Berry, Departments of Biochemistry and Pharmacology, Kobe University School of Medicine, Kobe 650, Japan, and Biosignal Reseach Center, Kobe University, Kobe 657, Japan

S. Biffo, Dipartimento di Biologia Animale, Università di Torino, Albertina 17, Turin, Italy

S. Bock, Department of Cell Biology, Vanderbilt Medical School, Nashville, TN 37232, U.S.A.

D.A. Brickey, Department of Molecular Physiology and Biophysics, Vanderbilt University, Nashville, TN 37232-0615, U.S.A.

E.R. Chapman, Department of Pharmacology, University of Washington, Seattle, WA 98195, U.S.A.

K.-H. Chen, Section on Metabolic Regulation, Endocrinology and Reproduction Research Branch, National Institute of Child Health and Human Development, National Institutes of Health, Bethesda, MD 20892, U.S.A.

P.J. Coggins, Departments of Medical Physiology/Medical Biochemistry, University of Calgary Health Sciences Centre, 3330 Hospital Drive, N.W. Calgary, Alberta, Canada T2N 4N1

R.J. Colbran, Department of Molecular Physiology and Biophysics, Vanderbilt University, Nashville, TN 37232-0615, U.S.A.

P.D. Coleman, Department of Neurobiology and Anatomy, School of Medicine and Dentistry, University of Rochester, 601 Elmwood Avenue, Rochester, NY 14642, U.S.A.

B. Costello, Department of Cell Biology, Vanderbilt Medical School, Nashville, TN 37232, U.S.A.

P.N.E. De Graan, Division of Molecular Neurobiology, Institute of Molecular Biology and Medical Biotechnology, Rudolf Magnus Institute, University of Utrecht, Padualaan 8, 3584 CH Utrecht, The Netherlands

L.V. Dekker, Division of Molecular Neurobiology, Rudolf Magnus Institute and Institute of Molecular Biology and Medical Biotechnology, University of Utrecht, Padualaan 8, 3584 CH Utrecht, The Netherlands

L.A. Dokas, Departments of Neurology, and Biochemistry and Molecular Biology, Medical College of Ohio, Toledo, OH 43699-0008, U.S.A.

S.K. Doster, Department of Anatomy and Neurobiology, Washington University School of Medicine, 660 South Euclid Avenue, St. Louis, MO 63110, U.S.A.

M.C. Fishman, Developmental Biology Laboratory, Massachusetts General Hospital, Boston, MA 02114, U.S.A.

D.G. Flood, Department of Neurobiology and Anatomy, and Department of Neurology, School of Medicine and Dentistry, University of Rochester, 601 Elmwood Avenue, Rochester, NY 14642, U.S.A.

Y.L. Fong, Department of Molecular Physiology and Biophysics, Vanderbilt University, Nashville, TN 37232-0615, U.S.A.

J.A. Freeman, Department of Cell Biology, Vanderbilt Medical School, Nashville, TN 37232, U.S.A.

K. Fukunaga, Department of Molecular Physiology and Biophysics, Vanderbilt University, Nashville, TN 37232-0615, U.S.A.

W.H. Gispen, Rudolf Magnus Instituut, Vondellaan 6, 3521 GD Utrecht, The Netherlands

C.A. Gonçalves, Departamento de Bioquimica, Instituto de Biociencias, Universidade Federal do Rio Grande do Sul, 90.050 Porto Alegre, Brazil

M. Grillo, Department of Neurosciences, Roche Institute of Molecular Biology, Nutley, NJ 07110, U.S.A.

Y.-f. Han, Department of Biochemistry and Molecular Biology, Medical College of Ohio, Toledo, OH 43699-0008, U.S.A.

F.L. Huang, Section on Metabolic Regulation, Endocrinology and Reproduction Research Branch, National Institute of Child Health and Human Development, National Institutes of Health, Bethesda, MD 20892, U.S.A.

K.-P. Huang, Section on Metabolic Regulation, Endocrinology and Reproduction Research Branch, National Institute of Child Health and Human Development, National Institutes of Health, Bethesda, MD 20892, U.S.A.

U. Kikkawa, Departments of Biochemistry and Pharmacology, Kobe University School of Medicine, Kobe 650, Japan, and Biosignal Reseach Center, Kobe University, Kobe 657, Japan

A. Kishimoto, Departments of Biochemistry and Pharmacology, Kobe University School of Medicine, Kobe 650, Japan, and Biosignal Reseach Center, Kobe University, Kobe 657, Japan

H. Koide, Departments of Biochemistry and Pharmacology, Kobe University School of Medicine, Kobe 650, Japan, and Biosignal Reseach Center, Kobe University, Kobe 657, Japan

A. Kose, Departments of Biochemistry and Pharmacology, Kobe University School of Medicine, Kobe 650, Japan, and Biosignal Reseach Center, Kobe University, Kobe 657, Japan

R. Leal, Departamento de Bioquimica, Instituto de Biociencias, Universidade Federal do Rio Grande do Sul, 90.050 Porto Alegre, Brazil

D.S. Lester, Section of Neural Systems, Laboratory of Molecular and Cellular Neurobiology, National Institutes of Health, Bethesda, MD 20892, U.S.A.

L.-H. Lin, Department of Cell Biology, Vanderbilt Medical School, Nashville, TN 37232, U.S.A.

A.M. Lozano, Toronto Western Hospital, Toronto, Ont. M5T 258, Canada

C.W. Mahoney, Section on Metabolic Regulation, Endocrinology and Reproduction Research Branch, National Institute of Child Health and Human Development, National Institutes of Health, Bethesda, MD 20892, U.S.A.

R. Malinow, Department of Physiology and Biophysics, University of Iowa, Iowa City, IA 52242, U.S.A.

F.L. Margolis, Department of Neurosciences, Roche Institute of Molecular Biology, Nutley, NJ 07110, U.S.A.

A. Meymandi, Department of Cell Biology, Vanderbilt Medical School, Nashville, TN 37232, U.S.A.

T.A. Nicolson, Department of Pharmacology, University of Washington, Seattle, WA 98195, U.S.A.

Y. Nishizuka, Departments of Biochemistry and Pharmacology, Kobe University School of Medicine, Kobe 650, Japan, and Biosignal Reseach Center, Kobe University, Kobe 657, Japan

J.J. Norden, Department of Cell Biology, Vanderbilt Medical School, Nashville, TN 37232, U.S.A.

T. Oda, Departments of Biochemistry and Pharmacology, Kobe University School of Medicine, Kobe 650, Japan, and Biosignal Reseach Center, Kobe University, Kobe 657, Japan

A.B. Oestreicher, Division of Molecular Neurobiology, Institute of Molecular Biology and Medical Biotechnology, Rudolf Magnus Institute, Utrecht, The Netherlands

K. Ogita, Departments of Biochemistry and Pharmacology, Kobe University School of Medicine, Kobe 650, Japan, and Biosignal Reseach Center, Kobe University, Kobe 657, Japan

N.I. Perrone-Bizzozero, Department of Biochemistry, University of New Mexico, Mexico City, NM, U.S.A.

M.R. Pisano, Department of Biochemistry and Molecular Biology, Medical College of Ohio, Toledo, OH 43699-0008, U.S.A.

E. Rocha, Departamento de Bioquimica, Instituto de Biociencias, Universidade Federal do Rio Grande do Sul, 90.050 Porto Alegre, Brazil

R. Rodnight, Departamento de Bioquimica, Instituto de Biociencias, Universidade Federal do Rio Grande do Sul, 90.050 Porto Alegre, Brazil

K.E. Rogers, Department of Neurobiology and Anatomy, School of Medicine and Dentistry, University of Rochester, 601 Elmwood Avenue, Rochester, NY 14642, U.S.A.

A. Routtenberg, Northwestern University, Cresap Neuroscience Laboratory, 2021 Sheridan Road, Evanston, IL 60208, U.S.A.

D.P. Rich, Department of Molecular Physiology and Biophysics, Vanderbilt University, Nashville, TN 37232-0615, U.S.A.

N. Saito, Departments of Biochemistry and Pharmacology, Kobe University School of Medicine, Kobe 650, Japan, and Biosignal Reseach Center, Kobe University, Kobe 657, Japan

C.G. Salbego, Departamento de Bioquimica, Instituto de Biociencias, Universidade Federal do Rio Grande do Sul, 90.050 Porto Alegre, Brazil

P. Schotman, Laboratory for Physiological Chemistry, University of Utrecht, Padualaan 8, 3584 CH Utrecht, The Netherlands

L.H. Schrama, Laboratory for Physiological Chemistry, University of Utrecht, Padualaan 8, 3584 CH Utrecht, The Netherlands

S.M. Schuh, Department of Anatomy and Neurobiology, Washington University School of Medicine, 660 South Euclid Avenue, St. Louis, MO 63110, U.S.A.

M.S. Shearman, Departments of Biochemistry and Pharmacology, Kobe University School of Medicine, Kobe 650, Japan, and Biosignal Reseach Center, Kobe University, Kobe 657, Japan

T. Shinomura, Departments of Biochemistry and Pharmacology, Kobe University School of Medicine, Kobe 650, Japan, and Biosignal Reseach Center, Kobe University, Kobe 657, Japan

K. Smith, Department of Molecular Physiology and Biophysics, Vanderbilt University, Nashville, TN 37232-0615, U.S.A.

T.R. Soderling, Department of Molecular Physiology and Biophysics, Vanderbilt University, Nashville, TN 37232-0615, U.S.A.

S. Spencer, Department of Anatomy and Neurobiology, Washington University School of Medicine, 660 South Euclid Avenue, St. Louis, MO 63110, U.S.A.

D.R. Storm, Department of Pharmacology, University of Washington, Seattle, WA 98195, U.S.A.

C. Tanaka, Departments of Biochemistry and Pharmacology, Kobe University School of Medicine, Kobe 650, Japan, and Biosignal Reseach Center, Kobe University, Kobe 657, Japan

R.W. Tsien, Department of Molecular and Cellular Physiology, Beckman Center, Stanford University Medical Center, Stanford, CA 94305, U.S.A.

D. Valenzuela, Developmental Biology Laboratory, Massachusetts General Hospital, Boston, MA 02114, U.S.A.

J. Verhaagen, Division of Molecular Biology, Rudolf Magnus Institute for Pharmacology and Institute of Molecular Biology and Medical Biotechnology, Utrecht, The Netherlands

M.B. Willard, Department of Anatomy and Neurobiology, Washington University School of Medicine, 660 South Euclid Avenue, St. Louis, MO 63110, U.S.A.

S.T. Wofchuk, Departamento de Bioquimica, Instituto de Biociencias, Universidade Federal do Rio Grande do Sul, 90.050 Porto Alegre, Brazil

H. Zwiers, Departments of Medical Physiology/Medical Biochemistry, University of Calgary Health Sciences Centre, 3330 Hospital Drive, N.W. Calgary, Alberta, Canada T2N 4N1

Preface

This volume summarizes the presentations made at the Third International Meeting on Brain Phosphoproteins. The two previous meetings took place in Utrecht, but this one was convened in nearby Zeist. The pleasant surroundings and cordial hospitality of the local committee paved the way for both an enjoyable and productive conference.

The first conference took place in 1981, the second in 1985 and this one in 1990. While there have been several different participants in each conference we have retained the primary focus on the function of those nervous system proteins that are kinases or substrates for kinases.

We wish to thank the participants for their enthusiastic participation at the workshop and their co-operation in the final steps of preparing this volume. With gratitude we acknowledge Pierre de Graan and Loes Schrama for their help in organizing this meeting.

The meeting would not have been possible without the support from agencies in both The Netherlands and Italy. Under the auspices of the Pharmacologisch Studiefonds Utrecht, The Netherlands, the University of Utrecht, The Netherlands, Elsevier Science Publishers B.V., Amsterdam, The Netherlands, and Fidia Research Laboratories, Abano Terme, Italy, we were able to invite the current leaders in the field to contribute to this volume.

With the cloning of protein kinase C and one of its substrates B-50/F1/GAP-43 in 1986–1987, the opportunity became available for the study of the function of these molecules using the precise tools of recombinant DNA technology.

This volume reveals the first wave of research that emerges from the cloning of the cDNA for both kinase and substrate. Of particular note is that several laboratories have begun to describe the organization of the entire gene for substrate and kinase. This will no doubt lead to our understanding of the exact regulatory mechanisms involved in synthesis of mRNA and protein. Furthermore, cell transfection and in vivo mutagenesis have already been used to describe growth functions of B-50/F1/GAP-43.

There was an air of excitement at this meeting as we listened to the discoveries reported, the convergent lines of evidence from the different laboratories and yes, as well, the clear disagreements. Even in this latter case, there was the evident feeling that these were the issues that could be resolved, certainly before our next meeting. Indeed, the level of expectancy in terms of what will be forthcoming was another theme which, while not discussed explicitly, was implicitly realized by all participants.

One focus of this volume was on protein kinase C and its substrates. The plenary lecture delivered by Professor Nishizuka set the tone for the entire meeting. His incisive presentation provided a basis for discussion of many of the contributions which followed.

A final word, somewhat self-congratulatory we fear, but nonetheless worth noting. When we began these meetings, it was our conviction that brain phosphoproteins would be an important focus in the study of brain. The second meeting did nothing to discourage this view as we moved from a period of characterizing spots on a gel to thinking about what they might mean physiologically. Now, we have new tools, and a considerably broader base derived from diverse fields: behavioral psychology, neuro-physiology, pharmacology, cellular and molecular biology. Each, from its unique perspective, provides us with a view of brain phosphoprotein function from a unique perspective. Perhaps even more reassuring is the vigor with which it is being pursued, the rapid acceleration in our knowledge is clearly evident. So with some pride we point to our tentative beginnings, the growth of this field and the excitement which we feel now as the discoveries of the future seem so imminent.

January 1991   Willem Hendrik Gispen (Utrecht, The Netherlands)
                                                    Aryeh Routtenberg (Evanston, IL, U.S.A.)

Contents

Section III – Structure and Characteristics of Protein Kinases

Section IV – Plasticity and Function of the PKC Substrate B-50/F1/GAP-43/ Neuromodulin

SECTION I

Structure and Characteristics of the Growth and Plasticity Associated PKC Substrate B-50/F1/GAP-43/Neuromodulin

W.H. Gispen and A. Routtenberg (Eds.)
Progress in Brain Research, Vol. 89
© 1991 Elsevier Science Publishers B.V.

CHAPTER 1

B-50: structure, processing and interaction with ACTH

Henk Zwiers and Philip J. Coggins *

Departments of Medical Physiology / Medical Biochemistry, University of Calgary Health Sciences Centre, 3330 Hospital Drive, N.W. Calgary, Alberta, Canada T2N 4N1

Introduction

Information concerning the phosphoprotein B-50 is enough to fill this and many other books. This was not the case in 1974, nor did our preliminary results suggest that such would ever be the case. Consequently, it may not be inappropriate to devote a few introductory paragraphs to the historical perspective preceding this exciting and challenging era in the study of B-50. In a desperate move to avoid disaster and save the career of their pupil, certain PhD supervisors (unnamed, to protect the innocent) decided that one of us (HZ) should study the possible effect of neuroactive peptides on protein phosphorylation in synaptic plasma membranes. The original plan was simply to find a neurochemical correlate to the effects of neuropeptides in learning and memory. At that

time, behavioural effects of peptides were scarce, often unreliable, and difficult to reproduce and interpret. From a neurochemical perspective, the situation was even worse. Many a fresh and enthusiastic young researcher had seen his/her enthusiasm wane while trying to locate central neuropeptide receptors, especially for ACTH and related peptides; at that time one of the hottest and most likely candidates for a role in learning and memory. Despite this inauspicious start, we entered a period of challenging research that many students and colleagues enjoyed in the years thereafter. And now to the details.

We started with a strategy that others had applied to the study of cAMP dependent phosphorylation of synaptic membrane proteins, which was simply to in vitro phosphorylate purified membranes and separate the labelled proteins by SDS-PAGE, followed by autoradiography. To our great surprise and delight the first autoradiograms showed a vast array of labelled protein bands in the control samples; the technique worked in our hands. Repeating the experiment in the presence of $ACTH_{1-24}$, we immediately recognized that endogenous phosphorylation was equally robust for most of the proteins, and yet, a small number of low molecular weight bands were

---

* *Present address:* Department of Biomedical Science, Division of Neuroscience, McMaster University, Hamilton, Ontario, Canada L8N 3Z5.
Correspondence: Dr. Henk Zwiers, Departments of Medical Physiology/Medical Biochemistry, University of Calgary Health Sciences Centre, 3330 Hospital Drive, N.W. Calgary, Alberta, Canada T2N 4N1. Telephone: (403) 220-6111 Telefax: (403) 270-0979.

not labelled at all. One band in particular got our attention. With the highest molecular weight of the affected radiolabelled bands, this protein did not appear to be a proteolysis product of a larger protein. By dividing the gel into four alphabetically labelled regions, with further numerical subdivisions, this ACTH$_{1-24}$ sensitive band now had a name: B-50 (the first of many that were to follow). The apparent molecular weight on 11% SDS-PAGE gels was about 48 kDa (Fig. 1) and a short time later Skene and Willard would discover the same protein, for completely different reasons, and call it GAP-43, based on its apparent molecular weight. In the years that followed its discovery, an increasing number of students and postdocs in the Division of Neurobiology (Institute of Molecular Biology, University of Utrecht) amassed an enormous amount of information about B-50. A few of the earlier milestones included: B-50 was an exclusively neuronal protein; phosphorylation was strongly correlated to excessive grooming activity; B-50 kinase was identical to protein kinase C (PKC); the protein was developmentally regulated; and finally, B-50 was an inhibitor of PIP-kinase and therefore might regulate signal transduction via the polyphosphoinositide pathway. We also developed two procedures to extract and purify the protein from synaptic membranes. This turned up an unexpected bonus (our intent was primarily to obtain pure protein for sequence analysis) since one of the purification methods yielded a complex of co-purifying proteins that included B-50, PKC, a lipid kinase, and a protease that specifically cleaved B-50 to the smaller protein B-60, while the N-terminal fraction was an inhibitor of PKC-mediated B-50 phosphorylation. In this chapter, we will review some of our more recent work on the structure of the protein and its post-translational processing. In addition, we will revisit the interactions of B-50 and ACTH. The peptide continues to give us surprising information about

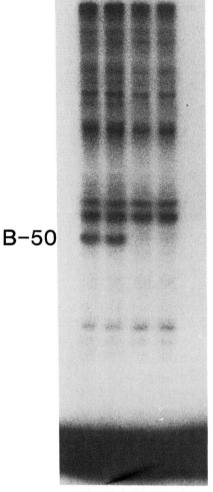

Fig. 1. Autoradiogram of ^{32}P-labelled SPM proteins, produced in 1975, showing the main reason why a band called B-50 was chosen for further study: a marked, specific inhibition by the neuroactive peptide ACTH$_{1-24}$.

the protein and we remain loyal to it after more than 15 years.

As the neurochemical data continued to accumulate it became apparent that functional correlates (e.g. to behaviour and phosphoinositide metabolism) were linked to the phosphorylation state of B-50. Therefore, it became increasingly important to isolate the protein from the numerous others present in s.p.m. This way, we intended to obtain the full primary sequence, define the phosphorylation site(s), and obtain a more accurate indication of the effects of phosphorylation in defined neurochemical and behavioural assays.

Our view of the purification

The methods commonly used to purify B-50 essentially fall into two categories: mild detergent, or alkali/heat extraction of a semi-purified synaptosomal membrane fraction (Zwiers et al., 1980a; Oestreicher et al., 1984; McMaster et al., 1988). Both methods have been routinely combined with ammonium sulphate precipitation and isoelectric focussing. The choice of protocol depends on the nature of subsequent experiments. For sequence analysis, secondary structure determination and analysis of putative functional domains within the protein, combined alkali extraction with heat and ammonium sulphate precipitation is the preferred method for producing large quantities of highly purified protein (routinely 10 mg from 1 kg of brain tissue) (McMaster et al., 1988; see Fig. 2 for analysis by SDS-PAGE of B-50 at three different stages of this purification procedure). Our initial attempts to sequence the IEF purified B-50 were not successful, apparently as a consequence of a blocked N-terminus. However, subsequent investigation revealed that the apparent homogeneous band on SDS-PAGE following IEF (Fig. 2) was in fact a heterogeneous mixture of intact and truncated forms of B-50, whose molecular weights were indistinguishable by SDS-PAGE. Despite the relatively high molecular weight of B-50, reverse phase HPLC on C18 columns proved to be a successful method of obtaining homogeneous forms of B-50, some of which are open to chemical sequencing.

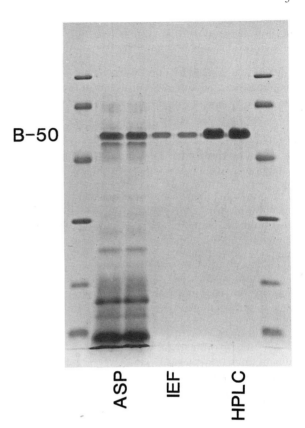

B-50

ASP   IEF   HPLC

Fig. 2. Protein pattern after Fast green staining of SDS-gel, showing bovine B-50 at three different stages of purity. The ammoniumsulphate fraction precipitating between 57 and 82% saturation (ASP) is enriched in B-50. One homogeneous protein is visible after flat bed isoelectric focusing (IEF), however, silver staining would reveal some contaminants in this fraction (not shown). Complete purification of B-50, of a sequenceable grade, is achieved by reverse phase HPLC.

The alternative route of purification involves extraction with Triton X-100. The yields of pure B-50 are rather low (about 1 μg per g brain tissue) but, as mentioned before, the advantage of this procedure is that B-50 seemingly copurifies with its kinase (Zwiers et al., 1980b, 1982) and a lipid kinase (Jolles et al., 1980). This purification strategy can also be utilized to investigate alkali-sensitive thioester bonds which may be involved in attaching B-50 to the plasma membrane (Skene and Virag, 1989). In order to separate the phosphorylated form of B-50 (p-B-50) from its dephosphorylated form (dp-B-50), affinity column chromatography on calmodulin-Sepharose (CaM-Sepharose) in the presence of the detergent Lubrol-PX (Andreasen et al., 1983) can be used. Interestingly, the B-50 batches that we have obtained so far consist for the major part of dp-B-50 (> 80% of the total). Therefore, the major form of B-50 in vivo may be in the dephosphorylated form. However, the question as to exactly where and when B-50 is phosphorylated will almost certainly be answered in greater detail in the near future, as monoclonal antibodies specific for each form of B-50 become available.

Several groups have reported on the unusual amino acid composition (Zwiers et al., 1980a; Benowitz et al., 1987; Wakim et al., 1987) of the protein, and subsequently the primary sequence, from a variety of species. In bovine B-50 (Andreasen et al., 1983; Wakim et al., 1987), the values are essentially identical to other species, nonpolar amino acids account for 42% of the residues, 9% and 22% of the total are proline and alanine respectively, the acidic amino acids glutamic acid and aspartic acid together comprise 22%, basic amino acids 14.5% (hence the very low isoelectric point), but the protein contains no tyrosine or tryptophan residues. The anomalous migration of B-50 on SDS-PAGE which results in a wide range of apparent molecular weights, and very different from the actual molecular weight found by various groups, is probably due to high negative charge at its surface, thereby preventing SDS binding (see also Benowitz et al., 1987).

Primary structure

As indicated in the previous section, direct chemical sequencing of the protein could only be achieved after further purification using reverse phase HPLC. At the risk of confusing the reader, the B-50 peaks were designated as: 1 (left), m

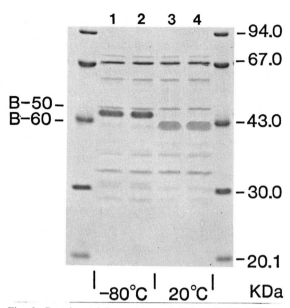

Fig. 3. Protein staining pattern of Triton X-100/KCl extracted B-50 complex from rat brain membranes, showing that upon incubation at 20°C B-50 is completely and highly specifically converted to a slightly lower molecular weight entity, called B-60. The positions of B-50 and B-60 are indicated. Molecular weight marker proteins were run on the same gel.

(mid), and r (right) B-50, with reference to their HPLC elution times (McMaster et al., 1988). The major forms of the protein (m and r) appeared to be N-terminally blocked while the minor form, l-B-50, yielded a 24-residue N-terminal sequence. In addition to B-50, a variable amount of B-60, the major degradation product of B-50 (see Introduction and Zwiers et al., 1980b, 1982; Fig. 3) was obtained. SDS-PAGE and HPLC demonstrated that high levels of B-60 could be detected when frozen brain tissue was used, however, our experience is that the shorter the time between defrosting of tissue and obtaining the total particulate fraction, the less B-60 is recovered in the ammonium sulphate precipitate fraction. Thus, the amount of B-60 produced can be minimized, resulting in a more efficient isolation of B-50. Research currently underway indicates that B-60 is not a purification artifact since it can also be detected in freshly prepared intact synaptosomes (see below). Chemical sequencing of apparently homogeneous, HPLC purified B-60 gave more than one N-terminal sequence. However, further purification using a different HPLC solvent system (HFBA instead of TFA) separated multiple forms of B-60, the major one being B-50$_{41-226}$ (rat), while the others were slightly shorter, based on amino acid composition and sequencing. In fact, peptide mapping by Staphylococcus aureus protease followed by analysis of fragments on HPLC and amino acid sequencing, suggested that

all the B-50 and B-60 forms that could be isolated had the same C-termini, and were therefore N-terminally truncated forms of intact B-50. This was immediately confirmed when the predicted amino acid sequence of a GAP-43 CDNA clone nucleotide sequencing became available (Basi et al., 1987; Karns et al., 1987; Nielander et al., 1988). Hence, l-B-50 is B-50$_{5-226}$, and the major B-60 form is B-50$_{41-226}$.

Correlation of amino acid analysis, SAP fragment data, and the predicted nucleotide sequence provided evidence that the N-terminus of

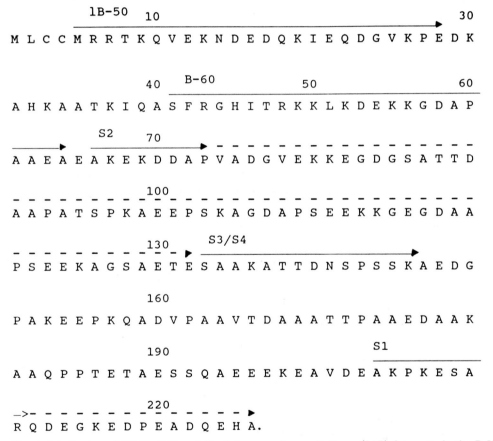

Fig. 4. Localization of l-B-50, B-60 and Staphylococcus Aureus Protease (SAP) fragments in the B-50 sequence. Solid lines represent amino acid sequences obtained from Edman degradation. Broken lines represent the probable size of S_1 and S_2 based on correlation of amino acid composition with the predicted sequence. The B-60 sequenced was the major form isolated. S_3/S_4 extend to the C-terminus of B-60, as also evidenced by crossreactivity with C-terminal antibodies on Western blots.

m-B-50 (and also r-B-50) starts at position 1 (Met) of the predicted sequence. Two major SAP digest fragments, M_W approx. 28 and 14 kDa, respectively, most likely represent the N-terminal part of the protein. A 25 kDa (SDS-PAGE) fragment S_3/S_4) which is easy to overlook, because it does not stain with Fast Green, and only poorly with silver stains, was sequencable and subsequently identified as B-50$_{131-226}$ (because of its N-terminal sequence and its cross reactivity with antibodies directed against a (synthetic) carboxy terminal peptide from rat B-50 (B-50$_{214-226}$; see also Coggins and Zwiers, 1989)). Two small SAP fragments (S_1 and S_2) were also sequenced and corresponded to B-50$_{204-226}$ and B-50$_{66-132}$, respectively (see also Fig. 4).

The information obtained by isolation and sequencing of B-50 and various protease derived fragments not only confirmed the predicted sequence of the protein, but was very useful to us in mapping the exact site of phosphorylation in B-50 (see below). As soon as sequence information became available to us a computer search was initiated. We, and others have found no homology with any other sequenced protein. However, while the whole protein has low sequence homology, certain short sequences are homologous to short regions of B-50: neurofilament protein (LaBate and Skene, 1989), a casein kinase consensus sequence in the c-*myc* proto-oncogene product *myc*, and certain proopiocortin derived peptides, in particular ACTH (McMaster et al., 1988; Zwiers and Coggins, 1990). Also of interest is the presence of several internally homologous sequences within the protein (McMaster et al., 1988) and a distinctive arrangement of certain residues. Hence, in rat B-50, 4 out of 5 arginine residues are in the extreme N-terminus and positioned in either the membrane attachment domain (Skene and Virag, 1989; Zuber et al., 1989a) or the CaM binding domain (Alexander et al., 1988; Coggins and Zwiers, 1990a). Three out of 4

paired lysine residues are contained in an internally homologous peptide motif, while the remaining pair is positioned in the proposed CaM domain. In line with the absence of a characteristic signal sequence, there is no experimental evidence to suggest that B-50 is glycosylated (Zwiers et al., 1985; Masure et al., 1986; Benowitz et al., 1987).

Secondary structure

We have examined the secondary structure of B-50 by one- and two-dimensional proton NMR spectroscopy. Batches of 5 mg of bovine B-50 were purified and subjected to repeated freeze-drying and resuspension in ^2H_2O. ^1H NMR spectra were recorded at various pH values (2.5–9.0) and at different temperatures. The results (Fig. 5 is a representative spectrum) suggested that virtually all amino acid residues appeared at their respective random coil positions and we observed no chemical shift dispersion characteristic of a folded protein (Coggins et al., 1989). This is in partial agreement with data from Masure et al. (1986), who concluded, from circular dichroism spectroscopy, that B-50 contains 1% x-helix, 78% random coil and 21% β-sheet. B-50 contains only one Phe residue, yet, interestingly, all the B-50 preparations used so far, have contained peaks with chemical shift characteristic of other aromatic compounds. These compounds are probably not located in the extreme N-terminus of B-50 since they are also present in B-60 (B-50$_{41-226}$). Further analysis is underway to identify these peaks, but preliminary evidence suggests that these aromatic protons are not dissimilar to those found in the polyisoprenoid farnesol. While most other proton peaks behaved in a predictable manner, we observed that the three His residues showed a rather complex pH titration behavior.

Evidence for several slowly interconverting species was obtained, suggesting an interaction of the His residues with other parts of the B-50 molecule.

Phosphorylation

From our earliest encounter with B-50, we have been interested in the regulation of its phospho-

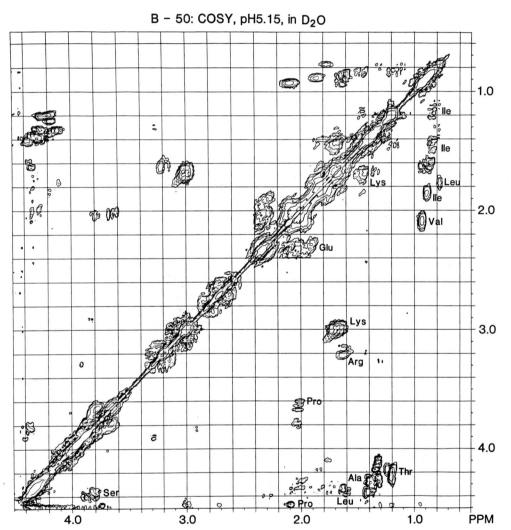

B – 50: COSY, pH5.15, in D₂O

Fig. 5. Two-dimensional 400 MHz ^{1}H-NMR spectrum of bovine B-50 (0.4 mM) in deuterium oxide (25°C, pH 5.15). One of the possibilities to study the structure of proteins has been the application of high-resolution NMR techniques. We used proton chemical shift correlation analysis (COSY correlated spectroscopy) to study the secondary structure of B-50. The technique allows for the detection of intraprotein amino acid interactions (protons connected through chemical bonds) resulting from organized structures, like α-helix, β-sheet. In case identifiable amino acids have strong crossreactions, their protons would show up as much larger, but flatter, "hills" in this 3D-picture. Therefore, this representative figure in fact shows that virtually all amino acids appeared at their random coil positions, with no shifts characteristic for a folded protein.

rylation, addressing the question of why ACTH had such a dramatic and specific effect on the phosphorylation of this protein. Was it because of some manner of structural similarity between the site(s) of phosphorylation of B-50 and ACTH? In addition, we have postulated a function for phosphorylation of B-50, i.e., phospho-B-50 reacted differently than dephospho-B-50 in the regulation of polyphosphoinositide metabolism (Van Dongen et al., 1985). To address these basic questions, knowledge of the exact site of phosphorylation was essential. Based on exhaustive phosphorylation (using kinase C) and dephosphorylation (using *E. coli* alkaline phosphatase) we were able to show a shift in the positions of 4 B-50 spots, following isoelectric focussing over a very narrow pH range. At that time, we concluded that these 4 spots might represent multi-phosphorylated forms of B-50, although our enzymatic treatments never resulted in the expected shift to only one IEF spot (Zwiers et al., 1985). These results, however, may be explained in part by subsequent studies (McMaster et al., 1988) that demonstrated multiple forms of B-50 following HPLC, which reflected heterogeneity of purified B-50. There-

fore, it now appears that phosphorylation was one more factor contributing to heterogeneity on IEF gels.

Recently, with the amino acid sequence available to us, we initiated a study to reveal the exact site(s) of phosphorylation. Of the 28 putative phosphorylation sites in B-50 (Ser and Thr residues, Fig. 6), the 14 Thr residues can be eliminated on the basis of phosphoamino acid analysis of ^{32}P-labelled protein, indicating that serine is the only phosphorylated amino acid (there are no tyrosines). SAP digestion of kinase C phosphorylated B-50 gave rise to a collection of radiolabelled B-50 fragments none of which appeared to have the same HPLC retention time as our three characterized SAP fragments; S_1, S_2 and S_3/S_4 (Figs. 4 and 6; McMaster et al., 1988). S_1 corresponds to B-50$_{204-226}$ whereas S_2 is B-50$_{66-132}$; together they contain 7 Ser residues. Therefore, residues 85, 96, 103, 110, 121, 127 and 209 were eliminated as putative PKC-mediated phosphorylation sites. Since the S_3/S_4 complex (in fact B-50$_{133-226}$) was not phosphorylated, we were able to conclude by exclusion, that Ser$_{41}$ was the single kinase C mediated phosphorylation site

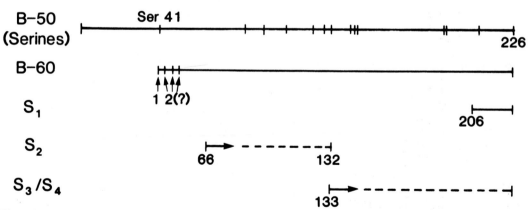

Fig. 6. Schematic representation of the B-50 molecule showing the position of the 14 Ser residues, being potential sites for kinase C phosphorylation. Further shown are the sequences of characterized SAP fragments (S_1, S_2 and S_3/S_4), and the known (B-60$_1$) or proposed (B-60$_2$) N-terminus of major and minor forms of B-60. Solid line represents sequence data, broken line amino acid analysis, or Western blots for S_3/S_4.

in B-50. Indirect confirmation of this finding was provided by differential phosphorylation of major and minor forms of B-60 that have their N-termini at, or C-terminal to, the Ser_{41} residue and are the major products of specific B-50 proteolysis. Only those forms of B-60 that contained the Ser_{41} residue incorporated phosphate label (Fig. 6). While there is no agreed consensus sequence for PKC substrates, Kemp and co-workers (House et al., 1987) have reported that PKC appears to recognize a group of basic residues adjacent to (preferably C-terminal to) a putative phosphorylation site, which in turn should be preceded by a neutral aliphatic amino acid residue. Given these criteria, the Ser_{41} residue is almost unique within the B-50 sequence, although serines at positions 110 and 120 in rat B-50 also qualify on the basis of sequence homology to the PKC phosphorylation site in the α-subunit of the nicotinic acetylcholine receptor (Steinbach and Zempel, 1987), but do not appear to be phosphorylated, at least in vitro. Interestingly, Pisano et al. (1988) reported that B-50 is also a substrate for casein kinase II (CKII), incorporating a maximum of 1 mol phosphate per mol protein. Most likely this site is distinct from the PKC site and could be either Ser-145 or Ser-192 based on CKII's recognition of acidic residues C-terminal to the phosphorylation site (especially in the N + 3 position (Kuenzal et al., 1987)).

Another interesting aspect of Ser_{41} phosphorylation relates to the calmodulin (CaM)-binding properties of the protein, studied in detail by Storm's group (this volume). They found that B-50 binding to CaM has the highly unusual feature of occurring in the absence of calcium (Andreasen et al., 1983). In addition, the binding has an absolute requirement for dp-B-50 (Alexander et al., 1987). Alexander et al. (1988) screened proteolytic fragments of bovine B-50 and found a single cyanogen bromide-chymotryptic peptide (designated M1-C1, $B-50_{43-51}$) that

bound to CaM-Sepharose in the presence or absence of calcium. From studies performed with two synthetic peptides ($B-50_{39-55}$ and Trp_{42}-$B-50_{39-55}$) they concluded that residues 42–51 form most of the CaM-binding domain of B-50. Therefore, it is not surprising that phosphorylation of Ser_{41} affects B-50 binding to CaM.

Specific proteolysis of B-50

A wealth of information is available about the synthesis of B-50, in particular under conditions of growth and regrowth of nervous tissue. Numerous chapters of this book will undoubtedly deal with all the details. We have initiated studies aimed to address the question of what happens to B-50 after its synthesis, especially since its precise function is still unknown. It is too early to accept that the intact translation product $B-50_{1-226}$ is the only physiologically active form of the protein. It is conceivable that one or more proteolytic products might have certain functions as well. Here we will briefly describe our initial efforts to characterize certain proteolysis products of B-50. The detergent/KCl extracted synaptic membrane fraction (T-ASP) described earlier (Zwiers et al., 1980b, 1982) also contained a protease that cleaved B-50 to the smaller and more acidic protein B-60 (the major form of B-60 was characterized as $B-50_{41-226}$; McMaster et al., 1988) (see Fig. 4). While the major cleavage site has been identified at the N-terminal side of the PKC-phosphorylation site, the protease itself has not been properly characterized so far. Preliminary data suggest that it may be a member of the thiol-dependent protease subtype (activity inhibited by p-hydroxymercuriphenylsulphonic acid and stimulated by dithiothreitol). Further characterization (MW, IEP) is hampered by the fact that the B-50/protease complex is apparently quite strong, resisting further separation into homogeneous components.

The N-terminal metabolite of the B-50 to B-60 cleavage (B-50$_{1-40}$) was originally (Zwiers et al., 1980b) discovered as a dialyzable factor and characterized as a B-50 phosphorylation-inhibiting peptide (PIP). In retrospect, that original finding was probably due to the fact that PIP was cleaved at or around the B-50 phosphorylation site, and addition to a B-50 phosphorylation system could have caused inhibition by competition with intact B-50 for the active site of the protein kinase. Specific proteolysis of B-50 to B-60 and PIP takes place in the absence of calcium (Zwiers et al., 1982) and, perhaps not surprisingly, can be inhibited by the addition of apo-calmodulin (Zwiers and Coggins, 1990). In light of the evidence that the extreme N-terminus of B-50 is largely responsible for membrane association (Zuber et al., 1989a), it is very likely that B-60 should be found in the soluble fraction, if specific proteolysis occurs in vivo, whereas PIP would remain in the membrane particulate fraction. We have looked for B-60 in high speed supernatant fractions of rat brain synaptosomes. As shown in Fig. 7, SDS-PAGE analysis definitely shows that B-60 can be detected in the soluble protein fraction, and more surprisingly B-50 is also present. At this moment we have not yet assessed if this represents l-B-50 (missing Cys-3, Cys-4, supposedly essential for thioacyl fatty acid linkage to the membrane bilayer) or if it represents a small proportion of soluble, but intact, B-50. This result strongly suggests that the specific proteolysis of B-50 is not solely an in vitro purification artefact,

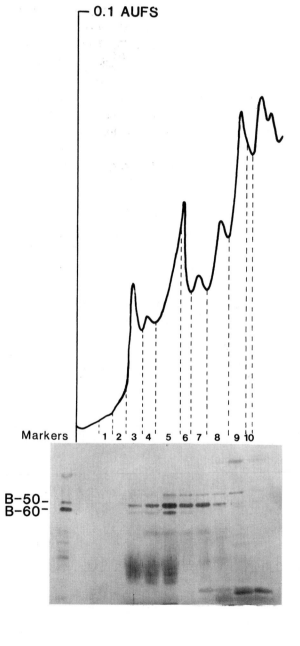

Fig. 7. Analysis of high speed supernatant protein fraction of rat synaptosomes by HPLC (top) and SDS-PAGE (bottom). Freshly prepared synaptosomes (50 mg total protein) were osmotically shocked in 4 ml of H$_2$O, spun for 30 min at 95 krpm (370 000 $\times g_{av}$) in a Beckman TL 100 microultracentrifuge. The supernatant fraction was carefully removed, heated for 10 min at 70°C, acidified with the HPLC starting buffer, and the precipitate was removed by high speed centrifugation. Under the conditions applied here, B-50 and B-60 are chemically stable and do not precipitate. HPLC fractions were collected manually based on continuous monitoring at 206 nm. SDS-PAGE analysis of fractions, using rat B-50 and B-60 as markers, strongly suggests that both B-50 and B-60 are present in the synaptosomal cytosolic protein fraction.

but also occurs in intact tissue and in purified synaptosomes. Having established this, it might be useful to briefly discuss the implications of the processing of B-50 to B-60, and to l-B-50. Both cleavages most likely should result in detachment of the major part of the protein from the synaptic membrane leaving behind, $B-50_{1-4}$ and $B-50_{1-40}$, respectively. Functionally, we discovered that B-60 does not bind to CaM (Coggins and Zwiers, 1990a), whereas it is not known if the membrane anchored PIP has retained the CaM binding properties of B-50. In addition to other documented functions of Ser_{41} phosphorylation (i.e. regulation of calmodulin binding and DPI-kinase activity), the possibility of modulation of the extent and/or specificity of proteolytic processing of B-50 should certainly not be overlooked.

Relationship with ACTH revisited

Three different target proteins apparently have similar affinities for the same region of B-50. The protein kinase phosphorylating Ser_{41} and the protease producing B-60, with apparent affinity for the region $B-50_{40-50}$, and calmodulin with a binding domain reported to span residues 43–51 (Storm and coworkers) but probably extending N-terminally to include Ser_{41} at least (Coggins and Zwiers, 1990a). We have noticed certain structural similarities between this N-terminal region of B-50 and $ACTH_{1-24}$. They include similar groups of basic residues ($B-50_{48-52}$ and $ACTH_{15-18}$, Fig. 8) and a region adjacent to the B-50 phosphorylation site which has almost identical composition, though only limited homology, to $ACTH_{6-10}$ (Fig. 8). The actual homology is limited to Phe_7-Arg_8 in ACTH and $Phe_{42}-Arg_{43}$ in B-50. If this limited structural homology between ACTH and B-50 interferes with the ability of PKC to phosphorylate B-50, then it is equally plausible that ACTH would interfere with the other two enzymes which have affinity for this same region. Indeed, we found that $ACTH_{1-24}$, in a dose related manner inhibited both dp-B-50 binding to calmodulin-Sepharose, and the conversion of B-50 to B-60 by the B-50 protease (Table I and Fig. 9). The ACTH structure/activity relationships for the three distinct interactions are quite similar. $ACTH_{1-10}$ is not an inhibitor of phosphorylation, calmodulin binding, or proteolysis; $ACTH_{11-24}$ does not inhibit phosphorylation or proteolysis but is slightly active on calmodulin

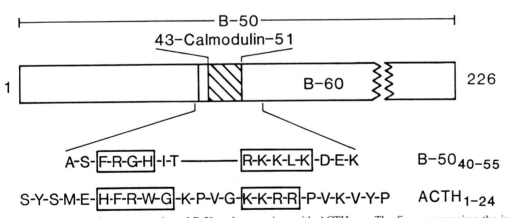

Fig. 8. A schematic representation of B-50 and comparison with $ACTH_{1-24}$. The figure summarizes the important biochemical features of B-50: the phosphorylation site Ser_{41} (vertical line), the proposed calmodulin binding domain 43–51 (hatched box) and the relative position of B-60 within the B-50 sequence. The expanded sequence 40–55 is compared to a similar region of $ACTH_{1-24}$ containing basic residues and -F-R-homology.

14

TABLE I

Inhibitory interactions between ACTH and B-50

ACTH	Phospho-rylation	Proteol-ysis	CaM binding
1–24	+ +	+ +	+ +
1–10	–	–	–
11–24	–	–	±
1–10 + 11–24	–	–	±
IC$_{50}$, ACTH$_{1-24}$	1–3 μM	1–2 μM	5 μM

+ + = effective at 10^{-6} M; ± = effective 10^{-5} M; – = not effective 10^{-4} M.

binding. Equimolar concentrations of ACTH$_{1-10}$ and ACTH$_{11-24}$ together, were not inhibitors of phosphorylation or proteolysis, implying that full inhibitory activity depends on the primary and secondary structure of the peptide (Table I). Interestingly, apo-calmodulin, binding to dp-B-50

can also protect B-50 from proteolysis (Coggins and Zwiers, 1990b).

It is difficult to get a comprehensive idea about the relevance of these new findings. Strictly speaking the biochemical consequences of a B-50/ACTH interaction, however they are mediated, would be to increase, or in the absence of synthesis to at least maintain, a relatively high concentration of free uncomplexed dephosphorylated B-50 by inhibiting phosphorylation, proteolysis and binding of calmodulin. However, with the risk of becoming too speculative, one could envision that the structural similarities between a certain region of B-50 and ACTH are meaningful and in fact are an explanation of a certain amount of overlap between the physiological actions of ACTH and B-50. The induction of filopodia in non-neuronal cells following transfection with a B-50/GAP-43 cDNA vector (Zuber, 1989b) could be taken to indicate that B-50 has trophic activity, reminiscent of the trophic activity of ACTH in

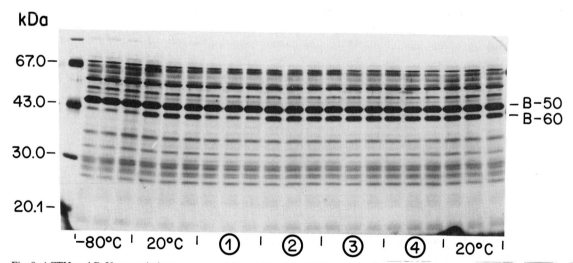

Fig. 9. ACTH and B-50 proteolysis: structure-activity relationship. Triplicate samples of T-ASP prepared from rat brain were either stored at −80°C or incubated at 20°C for 16 h. B-50 to B-60 conversion was detected by SDS-PAGE followed by silver staining. At 10 μM the following peptides were coincubated: ACTH$_{1-24}$ (1); ACTH$_{1-10}$ (2); ACTH$_{11-24}$ (3); ACTH$_{1-10}$ plus ACTH$_{11-24}$ (4). Only ACTH$_{1-24}$ inhibits the formation of B-60.

neurite growth of primary explants of fetal brain tissue (Van der Neut et al., 1988). One could envision that proteolysis of B-50, under conditions of growth or regrowth, results in the local supply of certain trophic factors. In this line of thought, ACTH then, may just be a substitute for one of these natural B-50 derived factors. Consequently, the peptide might well have a marked effect on neuronal plasticity in general.

Concluding remarks

The exact function of the protein is still unknown, the reason being that while other proteins are usually purified on the basis of certain enzymatic activities, this protein was selected because of certain peculiar features: specifically, the appearance of an interesting band on an autoradiogram, and an extremely bright shining spot on a two-dimensional fluorogram. In this chapter we have highlighted some aspects of the protein's structure, and have shown that it is a substrate for a number of different enzymes (lipid and protein kinases, a protease, a phosphatase, and a fatty acyltransferase) and calmodulin, all of which might serve to control some aspect of B-50's function(s). Since evolution has had ample time to remove unnecessary molecules, such a complicated regulation of a relatively simple 24 kDa protein might be an indication of the physiological importance of this protein. Our working hypothesis is that B-50 might serve different functions, each depending on the physiological state of the neuron (growth, regrowth, steady state; i.e., a signal transmitting synaptosome). In this view, regulation of transmitter release or intracellular signal transduction, is placed in a different category from, for instance, its function in nerve regeneration and sprouting, which may be related to B-50's putative role as a precursor of neurotropic factors.

Acknowledgements

We would like to thank Dr. V.J. Aloyo for carefully reading the manuscript, Drs. Vogel and McIntyre for their assistance in the NMR studies, and Andy Nagy and Kashif Qureshi for excellent technical assistance. We thank Pam Johnson for secretarial assistance. This research was supported by the Alberta Heritage Foundation for Medical Research and the Medical Research Council of Canada.

References

Alexander, K.A., Cimler, B.M., Meier, K.E. and Storm, D.R. (1987) Regulation of calmodulin binding to P-57. *J. Biol. Chem.*, 262: 6108–6113.

Alexander, K.A., Wakim, B.T., Doyle, G.S., Walsh, K.A. and Storm, D.R. (1988) Identification and characterization of the calmodulin-binding domain of neuromodulin, a neurospecific calmodulin binding protein. *J. Biol. Chem.*, 263: 7544–7549.

Andreasen, T.J., Luetje, C.W., Heideman, W. and Storm, D.R. (1983) Purification of a novel calmodulin binding protein from bovine cerebral cortex membranes. *Biochemistry*, 22: 4615–4618.

Basi, G.S., Jacobson, R.D., Virag, I., Schilling, J. and Skene, J.H.P. (1987) Primary structure and transcriptional regulation of GAP-43, a protein associated with nerve growth. *Cell*, 49: 785–791.

Benowitz, L.I., Perrone-Bizzozero, N.I. and Finklestein, S.P. (1987) Molecular properties of the growth-associated protein GAP-43 (B-50). *J. Neurochem.*, 48: 1640–1648.

Coggins, P.J. and Zwiers, H. (1989) Evidence for a single protein kinase C-mediated phosphorylation site in rat brain protein B-50. *J. Neurochem.*, 53: 1895–1901.

Coggins, P.J. and Zwiers, H. (1990a) Binding of the neuronal protein B-50, but not the metabolite B-60 to calmodulin. *J. Neurochem.*, 54: 274–277.

Coggins, P.J. and Zwiers, H. (1990b) Corticotropin (ACTH) inhibits the binding of B-50/GAP-43 to calmodulin. *Neurosci. Res. Commun.*, 6: 105–110.

Coggins, P.J., McIntyre, D.D., Vogel, H. and Zwiers, H. (1989) Neuronal protein B-50: a proton NMR study. *Biochem. Soc. Trans.*, 17: 785–786.

House, C., Wettenhall, R.E.H. and Kemp, B.E. (1987) The influence of basic residues on the substrate specificity of protein kinase C. *J. Biol. Chem.*, 262: 772–777.

Jolles, J., Zwiers, H., van Dongen, C.J., Schotman, P., Wirtz, K.W.A. and Gispen, W.H. (1980) Modulation of brain

16

polyphosphoinositide metabolism by ACTH-sensitive protein phosphorylation. *Nature,* 286: 623–625.

Karns, L.R., Ng, S.C., Freeman, J.A. and Fishman, M.C. (1987) Cloning of complementary DNA for GAP-43, a neuronal growth-related protein. *Science,* 236: 597–600.

Kuenzal, E.A., Mulligan, J.A., Sommercorn, J. and Krebs, E.G. (1987) Substrate specificity determinants for casein kinase II as deduced from studies with synthetic peptides. *J. Biol. Chem.,* 262: 9136–9140.

La Bate, M.E. and Skene, J.P.H. (1989) Selective conservation of GAP-43 structure in vertebrate evolution. *Neuron,* 3: 299–310.

Masure, H.R., Alexander, K.A., Wakim, B.T. and Storm, D.R. (1986) Physiochemical and hydrodynamic characterization of P-57, a neurospecific calmodulin binding protein. *Biochemistry,* 25: 7553–7560.

McMaster, D., Zwiers, H. and Lederis, K. (1988) The growth associated neuronal phosphoprotein B-50: improved purification/partial primary structure and characterization and localization of proteolysis products. *Brain Res. Bull.,* 21: 265–276.

Nielander, H.B., Schrama, L.H., Van Rozen, A.J., Kasperaitis, M., Oestreicher, A.B., De Graan, P.N.E., Gispen, W.H. and Schotman, P. (1988) *Neurosci. Res. Commun.,* 1: 163–172.

Oestreicher, A.B., Van Duin, M., Zwiers, H. and Gispen, W.H. (1984) Cross-reaction of anti-rat B-50: characterization and isolation of a "B-50" phosphoprotein from bovine brain. *J. Neurochem.,* 43: 935–943.

Pisano, M.R., Hegazy, M.G., Reimann, E.M. and Dokas, L.A. (1988) Phosphorylation of protein B-50 (GAP-43) from adult rat brain cortex by casein kinase II. *Biochem. Biophys. Res. Commun.,* 155: 1207–1212.

Skene, J.H.P. and Virag, I. (1989) Posttranslational membrane attachment and dynamic fatty acylation of a neuronal growth cone protein GAP-43. *J. Cell Biol.,* 108: 613–624.

Steinbach, J.H. and Zempel, J. (1987) What does phosphorylation do for the nicotinic acetylcholine receptor. *Trends Neurosci.,* 10: 61–65.

Van der Neut, R., Bär, P.R., Sodaar, P. and Gispen, W.H. (1988) Trophic influences of alpha-MSH and $ACTH_{4-10}$ on neuronal outgrowth in vitro. *Peptides,* 9: 1015–1020.

Van Dongen, C.J., Zwiers, H., De Graan, P.N.E., and Gispen, W.H. (1985) Modulation of the activity of purified phosphatidylinositol 4-phosphate kinase by phosphorylated and dephosphorylated B-50 protein. *Biochem. Biophys. Res. Commun.,* 128: 1219–1227.

Wakim, B.T., Alexander, K.A., Masure, H.R., Cimler, B.M., Storm, D.R. and Walsh, K.A. (1987) Amino acid sequence of P-57, a neurospecific calmodulin-binding protein. *Biochemistry,* 26: 7466–7470.

Zuber, M.X., Strittmatter, S.M. and Fishman, M.C. (1989a) A membrane-targeting signal in the amino terminus of the neuronal protein GAP-43. *Nature,* 341: 345–348.

Zuber, M.X., Goodman, D.W., Karns, L.R. and Fishman, M.C. (1989b) The neuronal growth-associated protein GAP-43 induces filopodia in non-neuronal cells. *Science,* 244: 1193–1195.

Zwiers, H. and Coggins, P.J. (1990) Corticotropin (ACTH) inhibits the specific proteolysis of the neuronal phosphoprotein B-50/GAP-43. *Peptides,* in press.

Zwiers, H., Schotman, P. and Gispen, W.H. (1980a) Purification and some characteristics of an ACTH-sensitive protein kinase and its substrate protein in rat brain membranes. *J. Neurochem.,* 34: 1689–1699.

Zwiers, H., Verhoef, J., Schotman, P. and Gispen, W.H. (1980b) A new phosphorylation-inhibiting peptide (PIP) with behavioral activity from rat brain membranes. *FEBS Lett.,* 112: 168–172.

Zwiers, H., Gispen, W.H., Kleine, L. and Mahler, H.R. (1982) Specific proteolysis of a brain membrane phosphoprotein (B-50): effects of calcium and calmodulin. *Neurochem. Res.,* 7: 127–137.

Zwiers, H., Verhaagen, J., Van Dongen, C.J., De Graan, P.N.E. and Gispen, W.H. (1985) Resolution of rat brain synaptic phosphoprotein B-50 into multiple forms by two-dimensional electrophoresis: evidence for multisite phosphorylation. *J. Neurochem.,* 44: 1083–1090.

W.H. Gispen and A. Routtenberg (Eds.)
Progress in Brain Research, Vol. 89
© 1991 Elsevier Science Publishers B.V.

CHAPTER 2

GAP-43: purification from a prokaryotic expression system, phosphorylation in cultured neurons, and regulation of synthesis in the central nervous system

S. Kathleen Doster [1,*], Andres M. Lozano [2], Susan M. Schuh [1], Susan Spencer [1] and Mark B. Willard [1]

[1] *Department of Anatomy and Neurobiology, Washington University School of Medicine, 660 South Euclid Avenue, St. Louis, MO 63110, U.S.A., and* [2] *Toronto Western Hospital, Toronto, Ont. M5T 258, Canada*

Our laboratory first encountered the protein GAP-43 in the course of experiments investigating the role of altered gene expression in regulating axon growth. After radiolabeling newly synthesized proteins in the cell bodies of neurons, we analyzed the composition of radioactive proteins that were axonally transported into normal adult axons and compared them to proteins transported by neurons that were elaborating new axons, either in response to injury in the adult, or during development of the nervous system. The relative labeling of a small number of rapidly axonally transported proteins was much greater in neurons that were extending axons than in those that were not. We designated these proteins growth-associated proteins, or GAPs (Skene

* *Present address:* Department of Neurosurgery, University of Minnesota School of Medicine, Minneapolis, MN, U.S.A.
Correspondence: Mark Willard, Department of Anatomy and Neurobiology, Box 8108, Washington University School of Medicine, 660 S. Euclid Avenue, St. Louis, MO 63110, U.S.A. Tel.: (314) 362-3462.

and Willard, 1981a,b). Among these, the relative labeling of a small (24 kDa) acidic ($pI = 4.3$) protein was consistently elevated (10–30-fold relative to the average) in neurons that were growing axons. We designated this protein GAP-43, to reflect its electrophoretic mobility on 10% SDS polyacrylamide gels, which corresponds to an apparent molecular weight of 43 kDa. (Presumably because of its high negative charge, its electrophoretic mobility does not predict the true molecular weight of GAP-43, which is 24 kDa (Jacobson et al., 1986; Benowitz et al., 1987).) The correlation between the labeling of newly synthesized GAP-43 and axon growth suggested that the synthesis of GAP-43 may respond to the same signals that regulate axon growth, and, in addition, raised the possibility that GAP-43 itself may perform a function that facilitates axon growth. Consistent with this possibility, GAP-43 is a major protein of growth cone membranes (De Graan et al., 1985; Meiri et al., 1986; Skene et al., 1986) suggesting that it may be important for some aspect of growth cone function.

GAP-43 is identical to the protein designated F1 (Nelson and Routtenberg, 1984) B-50 (Zwiers et al., 1976) pp46 (Katz et al., 1985) and P-57 or neuromodulin (Andreasen et al., 1983). The independent characterization of these synonymous proteins described elsewhere in this volume show that, in addition to growth cones, GAP-43 is a significant component of certain mature synapses in adult animals (Gispen et al., 1985; Nelson et al., 1987b; Neve et al., 1987; Benowitz and Routtenberg, 1988; McGuire et al., 1988). The distribution of these synapses, as well as a reported correlation between the phosphorylation of GAP-43 and both transmitter release in vitro (Dekker et al., 1989) and the duration of long term potentiation in the hippocampus (Nelson and Routtenberg, 1985; Lovinger et al., 1986; Nelson et al., 1987a) has generated the hypothesis that GAP-43 also plays a role in synaptic function and plasticity.

Several characteristics of GAP-43 may provide clues to its function in growth cones and presynaptic terminals. It is an elongated protein that is associated with the inner aspect of the plasma membrane, suggesting that it is a component of the membrane skeleton (Masure et al., 1986; Skene and Willard, 1981c; Meiri and Gordon-Weeks, 1990); its association with the plasma membrane may be mediated by fatty acids covalently linked to two cysteine residues near the protein's amino terminus (Skene and Virag, 1989). The following properties suggest that GAP-43 may influence the second messengers of several signaling systems: (1) it binds to apocalmodulin, and the interaction is antagonized by calcium (Andreasen et al., 1983; Cimler et al., 1987); (2) phosphorylation of GAP-43 by protein kinase C inhibits the binding of calmodulin, and vice versa (Andreasen et al., 1983; Cimler et al., 1987); (3) GAP-43 may influence the metabolism of phosphoinositides (Gispen et al., 1985); (4) it can stimulate the hydrolysis of GTP by the G protein

G_0 (Strittmatter et al., 1990). These properties are consistent with a role for GAP-43 in mediating interactions between the membrane skeleton and certain second messenger systems that may be responsive to the extracellular environment. In turn, the influence of GAP-43 upon these second messenger systems may influence other elements of the cell that are responsive to changes in calcium, calmodulin, phosphoinositides, and G proteins.

Among the outstanding questions concerning GAP-43 are the details of its structure, the role of its phosphorylation, and the source and nature of the signals that regulate its synthesis. Here we summarize the results of recent investigations relevant to each of these questions. These studies concern (1) the purification of GAP-43; (2) the phosphorylation of GAP-43 in living neurons in culture; (3) the regulation of GAP-43 synthesis in the central nervous system.

A prokaryotic expression system for purifying GAP-43

Initially, our laboratory purified GAP-43 from neonatal rat brain by a technique involving preparative two-dimensional gel electrophoresis (Meiri et al., 1986). Purification by this technique was cumbersome; it was expedited by the replacement of gel electrophoresis with HPLC reversed phase chromatography on a C4 column (Changellian et al., 1990). The yield of GAP-43 was approximately 10 μg per rat brain. To provide a more abundant source of purified GAP-43 we have utilized the expression system of Studier (Studier and Moffat, 1986). This vector features a "fusion cassette", comprising the nucleic acid sequence that codes for 11 amino acids of a capsid protein of the phage T7 located down stream from a promotor for the phage T7 RNA polymerase. DNA that is inserted at the downstream side of the fusion cassette is fused to the T7 capsid protein sequence; alternatively, the cassette can be removed by digestion with appropriate restriction endonucleases and replaced entirely with the DNA to be expressed. When bacteria that have been engineered to express the T7 RNA polymerase are transformed with the plasmid, the foreign DNA is transcribed from the T7 promoter, and the resulting message is translated.

We first inserted a large fragment of the GAP-43 cDNA described by Changellian et al. (1989) downstream of the fusion cassette. When bacteria were transformed with the plasmid, they synthesized a fusion protein in which the first 16 amino acids of GAP-43 were replaced by 13 amino acids coded by the fusion cassette, and which composed 10–20% of their total protein. Approximately 50 mg of the fusion protein could be purified from 1 liter of bacterial culture by reversed-phase chromatography (Changellian et al., 1990). When injected into rabbits, the purified fusion protein was effective in eliciting polyclonal antibodies that were specific for GAP-43.

To express authentic full length GAP-43, we replaced the fusion cassette of the vector with the entire coding sequence of GAP-43. This required first generating a full length GAP-43 coding sequence with an Nde1 site at the initiator codon by means of the polymerase chain reaction technique, using the cloned GAP-43 cDNA (Changellian et al., 1989) as substrate. Bacteria that were transfected with this vector produced authentic GAP-43, as was confirmed by amino acid sequence analysis of the amino-terminal end. This authentic GAP-43 was expressed with only one-fifth of the efficiency of the fusion protein described above; 1 liter of culture yielded approximately 10 mg of GAP-43 after purification by reversed-phase chromatography. The product provides a source of unmodified GAP-43 that should facilitate studies of its structure, posttranslational modification, and function. The effi-

TABLE I

GAP-43 purification procedures

Procedure	Material	Time	Yield
1. gel electrophoresis	100 rat brains	5 days	$\sim 100\ \mu g$
2. reversed phase chromatography	100 rat brains	2 days	$\sim 300\ \mu g$
3. bacteria-complete	10 ml culture	2 days	$1\,000\ \mu g$
4. bacteria-fusion	10 ml culture	2 days	$5\,000\ \mu g$

ciency of producing GAP-43 by various methods is compared in Table I.

Phosphorylation of GAP-43 by living neurons in culture

Several observations indicate that modification by phosphorylation is an important aspect of the function of GAP-43. It is a major substrate for protein kinase C (Aloyo et al., 1983; Akers and Routtenberg, 1989) and its phosphorylation is reported to be correlated with changes in synaptic efficacy (Nelson and Routtenberg, 1985) and with transmitter release (Dekker et al., 1989). Furthermore, phosphorylation of GAP-43 by protein kinase C affects its ability to bind calmodulin in vitro (Andreasen et al., 1983; Cimler et al., 1987). In addition, GAP-43 is reported to be a substrate in vitro for casein kinase II (Pisano et al., 1988). To analyze the phosphorylation of GAP-43 in living cells, we have undertaken an investigation of the incorporation of phosphate into GAP-43 in primary cultures of neurons, as well as several neuronal cell lines.

Preliminary to these experiments in living neurons, we identified the site that is phosphorylated in vitro by protein kinase C. We incubated purified GAP-43 with protein kinase C and γ-^{32}P-ATP, digested the products with trypsin, and separated the resulting peptides by reversed phase chromatography. Only one peptide was phosphorylated; its sequence, IQASFR, indicated that serine 41 is the only major phosphorylation site for protein kinase C in GAP-43 in vitro under these conditions. This site is located next to the domain that binds to calmodulin; this proximity is likely related to the mutual inhibitory effects of calmodulin binding and phosphorylation by protein kinase C (Coggins and Zwiers, 1989; Schu et al., 1989; Apel et al., 1990).

To investigate whether serine 41 is also phosphorylated in living neurons, we incubated primary cultures of rat superior cervical ganglion cells with ortho[^{32}P]phosphate, and, after isolating GAP-43 by means of precipitation with antibodies followed by gel electrophoresis of the immune complexes, analyzed the tryptic peptides by reversed-phase chromatography. One peak of ^{32}P was eluted at the same time as the GAP-43 peptide IQASFR, the protein kinase C substrate. However, as many as five additional peaks of ^{32}P were eluted from the column. When the cultures were incubated with phorbol ester to stimulate the activity of protein kinase C, only the radioactive peak corresponding to the IQASFR peptide increased. A similar spectrum of phosphopeptides was produced in similar experiments investigating the phosphorylation of GAP-43 in O1A1, an embryonal carcinoma cell line that undergoes neuronal differentiation in response to retinoic acid (Edwards and McBurney, 1983), and N2A, a neuroblastoma cell line. These results show that serine 41 is a major phosphorylation site for protein kinase C in several types of neurons in culture. Furthermore, they suggest that at least one (and probably several) additional sites are phosphorylated by protein kinases other than protein kinase C in these living neurons.

Regulation of GAP-43 synthesis in neurons of the central nervous system

The expression of GAP-43 tends to be elevated in neurons that are extending axons, either during development or during regeneration of injured axons, compared to intact adult neurons (Skene and Willard, 1981a,b; Benowitz and Lewis, 1983; Kalil and Skene, 1986; Grafstein, 1987; Perry et al., 1987, 1989; Moya et al., 1988; Biffo et al., 1990; Jones and Aguayo, 1990). This correlation has suggested the possibility that GAP-43 synthesis may respond to the same signals that regulate axon growth, and furthermore may itself facilitate axon growth (e.g. Willard et al., 1985; Skene, 1989). The central nervous system (CNS) of mammals provides an interesting case for considering the parameters that can influence the expression of GAP-43. Neurons do not normally regenerate injured axons within the CNS. However, under certain circumstances, i.e. when they are injured close to the cell body, they can regenerate axons for a distance of several cm into an experimentally grafted segment of peripheral nerve (David and Aguayo, 1981; Benfey and Aguayo, 1982; Richardson et al., 1982, 1984; Aguayo, 1985; So and Aguayo, 1985; Vidal-Sanz et al., 1987; Aguayo et al., 1990). On the other hand, regeneration into peripheral nerve grafts does not occur when the axons are injured far from the cell bodies. These requirements for successful regeneration raise the question of whether the potential for axon growth that is initiated by axon injury close to the cell body of CNS neurons is accompanied by an increase in GAP-43. To address this question, we collaborated with Dr. Albert Aguayo to assay GAP-43 in adult rat retinal ganglion cells after injury to their axons in the optic nerve at various distances from the cell bodies (Lozano et al., 1987; Doster et al., 1988, 1991). GAP-43 was assayed both by staining of retinal whole mounts with anti-GAP-43 antibody, as well as by measuring the amount of newly synthesized GAP-43 (labeled with [^{35}S]methionine) that was axonally transported into the optic nerve. When the optic nerve was cut or crushed intracranially (more than 6 mm from the eye) there was no change in expression of GAP-43 that could be detected by these measures. On the other hand, when the optic nerve was cut or crushed within 3 mm of the eye, the staining of the retina with anti-GAP-43 increased according to the following time course: after a delay of 6 days, the intensity of staining increased, reaching a maximum at approximately 12 days, and then declined to background levels by approximately 24 days. The amount of newly synthesized GAP-43 that was transported into the optic nerve also increased, following a similar time course; by 12 days after axotomy its labeling had increased approximately 20-fold relative to another rapidly axonally transported protein (protein 20 in Willard et al., (1974), also designated superprotein). Similar changes occurred when retinal ganglion cell axons injured close to the eye were permitted to regenerate into grafts of peripheral nerve (Lozano et al., 1987; Doster et al., 1988; Lozano, Doster, Willard and Aguayo, in preparation).

These results show that GAP-43 expression can be induced in CNS neurons even in the absence of axon regeneration. Moreover, axotomy provides a sufficient signal to initiate this induction; it does not depend upon inducing factors provided by the peripheral nerve graft. Furthermore, the induction of GAP-43 may be accompanied by the increased potential for axon growth, considering that both depend similarly upon the position of axotomy. This potential is realized only when an appropriate substrate, such as a peripheral nerve graft, is provided. These observations are consistent with the hypothesis that GAP-43 may reflect metabolic changes that enhance the potential for axon growth (Skene and Willard, 1981a,b; Willard et al., 1985).

The dependence of increased expression of GAP-43 upon the position of axotomy suggests that the mechanism that regulates GAP-43 expression in the visual system of mammals is responsive to the length of intact axon that remains attached to the cell body after axotomy. A simple hypothetical mechanism that might explain this relationship is that non-neuronal cells of the optic nerve normally supply agents that repress the synthesis of GAP-43, and that these agents require an intact axon to serve as a conduit to the cell bodies of the retinal ganglion cell. If so, the effective level of repressor reaching the cell body after axotomy might well depend upon the length of remaining axon that is accessible to the non-neuronal cells of the optic nerve. Following axotomy sufficiently close to the eye to reduce the supply of repressor to ineffectively low levels, GAP-43 (and perhaps other elements involved in the potential for axon growth) would be derepressed. Other mecanisms (e.g., the production by axotomy of positive influences that are only effective in inducing GAP-43 over small distances) could also explain the dependence of expression of GAP-43 on axon length. However, in the remainder of this chapter, we consider certain potential implications of axon-dependent repression of GAP-43 by non-neuronal cells, keeping in mind that it is one of several hypothetical and ad hoc models that could explain the dependence of GAP-43 induction upon the position of axotomy in retinal ganglion cells.

Non-neuronal cells that would be candidates for providing the repressive agents required by this model would include the three major glial cells of the optic nerve: type I and type II astrocytes and oligodendrocytes; none of these are components of peripheral nerve, where GAP-43 is apparently not effectively repressed by non-neuronal cells, as considered later. One criterion for evaluating a possible role of these glial cells in repressing GAP-43 is their time of appearance in the optic nerve, because it would be important that repression be initiated only after the developmental period of high GAP-43 expression, which may last until the time of birth in rats. By this criterion, the appearance of type I astrocytes in the optic nerve at embryonic day 16 (Raff et al., 1987) may be premature. Although the time of appearance of type II astrocytes (postnatal day 8–10) is too late for their participation in the decline prior to postnatal day 10, it is quite possible that this developmental decline is mediated by other mechanisms such as repression by elements obtained from target cells, as considered previously (Willard and Skene, 1982). Thus, the type II astrocyte would be a candidate for a source of repression in the adult optic nerve. However, of these three types of cells, the time of appearance of oligodendrocytes best fits the time course of repression of GAP-43. Oligodendrocytes appear in the optic nerve on the first postnatal day (Raff et al., 1987) and would be available to participate in the subsequent developmental decline of GAP-43 as well as in the repression of GAP-43 in the adult. Moreover, certain membrane components of oligodendrocytes have been observed to influence negatively axon growth and adhesivity (Caroni and Schwab, 1987; Schwab, 1990). Although the response of neurons to these components is rapid, suggesting that they directly affect the function of growth cones, perhaps they (or other components of oligodendrocytes) also lead to the repression of synthesis of proteins such as GAP-43.

A consideration of the expression of GAP-43 in several different neuronal systems suggests that this hypothetical mechanism involving axon-dependent repression of GAP-43 by non-neuronal cells could provide an explanation for several previous observations, but that other mechanisms are also important in regulating GAP-43 expression. For example, the dependence upon the position of axotomy relative to the cell bodies may explain why induction of GAP-43 was not observed previously after axotomy of the rabbit optic nerve near the optic chiasm (Skene and Willard, 1981b), or axotomy of the corticospinal tracts of rats (Kalil and Skene, 1986) and hamsters (Reh et al., 1987) in the brainstem at the level of the pyramids, but was observed after axotomy of the rat optic nerve (Freeman et al., 1986). It also provides a potential explanation for why CNS neurons close to infarcts have been observed to react strongly with anti-GAP-43 antibodies (Ng et al., 1988); this reactivity may represent increased expression in neurons whose axons were compromised close to the cell body. On the other hand, the expression of GAP-43 is apparently not repressed to the same degree in all CNS neurons; for example, neurons of the association areas and hippocampus appear to express higher levels than do primary sensory neurons (Benowitz et al., 1988). Thus, such repression, if it occurs, appears to be differentially effective in different CNS systems.

The dorsal root ganglion cells provide a particularly suitable situation to compare regulation of GAP-43 in the central and peripheral nervous system, because one branch of their axons enters the spinal cord (CNS) whereas the other branch enters the peripheral nerve. Injury to the peripheral branch is followed by its regeneration, and the enhanced expression of GAP-43 (Scheyer and Skene, 1988; Woolf et al., 1990). However, there is no indication that this response depends upon the position of the axotomy with relation to the cell body. Thus, axon-dependent repression of GAP-43 by non-neuronal cells does not appear to be an important means for regulating GAP-43 expression in the peripheral nervous system. One simple explanation for this apparent difference would be that the repressive elements responsible for this type of regulation in the CNS are generated only by non-neuronal cells that are absent from the peripheral nerve – for example, oligodendrocytes or astrocytes.

The induction of GAP-43 in dorsal root ganglion cells by injury to the peripheral nerve raises an additional complication: it occurs in spite of the continued access of the central axon branch to the postulated non-neuronal repressive elements. This suggests that such repression, if it occurs, can be overridden by changes that follow axotomy of the peripheral branch of the axon. In fact, these changes influence the regeneration potential of the central branch as well as the peripheral branch; the transport of GAP-43 to the central branch is augmented after axotomy of the peripheral branch (Schreyer and Skene, 1988; Wolf et al., 1990), and the propensity of the central branch to regenerate into a peripheral nerve graft is greatly enhanced (Richardson and Issa, 1984).

These observations suggest that repression of GAP-43 expression by non-neuronal cells, if it occurs, is one of several mechanisms for regulating the expression of GAP-43, and that other mechanisms are more important in the peripheral nervous system. For example, an attractive possibility is that GAP-43 is also regulated by repressive factors supplied by target cells (e.g., Willard and Skene, 1982); GAP-43 expression regulated by this mechanism might be induced by interruption of the axon (and the supply of repressor) by axotomy, independent of the distance of the injury from the cell body. This mechanism is consistent with the report that GAP-43 is induced in dorsal root ganglion cells when axonal transport, including the retrograde transport of such putative target derived inhibitory factors, is blocked by application of cholchicine to the peripheral nerve (Woolf et al., 1990).

It is possible that the repression of GAP-43 in the CNS during development might also be mediated by such target-derived repressive elements; this would provide a potential means for coupling growth processes to target selection (Willard and Skene, 1982; Willard et al., 1985). This consideration raises the question of why an additional source of inhibition would be deployed along the length of the adult optic nerve if other mechanisms, such as target-generated inhibition, were

available during development and were sufficient in the peripheral nervous system of the adult. One hypothetical rationale would be that axon-dependent inhibition by non-neuronal cells provides a fail safe mechanism for insuring that metabolic processes related to axon growth are not spuriously expressed in systems with low tolerance for reorganization, and where maintaining stability is more important than providing for recovery when the system is damaged.

References

Aguayo, A.J. (1985) Axonal regeneration from injured neurons in the adult mammalian central nervous system. In W. Cotman (Ed.), *Synaptic Plasticity*, pp. 457–484, Guilford, NY.

Aguayo, A.J., Carter, D.A., Zwinpfer, T.J., Vidal-Sanz, M. and Bray, G.M. (1989) Axonal regeneration and synapse formation in the injured CNS of adult mammals. In A. Björklund, A. Aguayo and D. Ottoson (Eds.), *Brain Repair*, p. 251, Stockton Press, Stockholm.

Akers, R.F. and Routtenberg, A. (1985) Protein kinase C phosphorylates a 47 Mr protein (F1) directly related to synaptic plasticity. *Brain Res.*, 334: 147–151.

Aloyo, V.J., Zwiers, H. and Gispen, W.H. (1983) Phosphorylation of B-50 protein by calcium-activated, phospholipid-dependent protein kinase and B-50 protein kinase. *J. Neurochem.*, 41: 649–653.

Andreasen, T.J., Leutie, C.W., Heideman, W. and Storm, D.R. (1983) Purification of a novel calmodulin binding protein from bovine cerebral cortex membranes. *Biochem.*, 22: 4615–4618.

Apel, E.D., Byford, M.F., Au, D., Walsh, K.A. and Storm, D.R. (1990) Identification of the protein kinase C phosphorylation site in neuromodulin. *Biochemistry*, 29: 2330–2335.

Benfey, M. and Aguayo, A.J. (1982) Extensive elongation of axons from rat brain into peripheral nerve grafts. *Nature*, 296: 150–152.

Benowitz, L.I. and Lewis, E.R. (1983) Increased transport of 44,000 to 49,000-dalton acidic proteins during regeneration of the goldfish optic nerve: a two-dimensional gel analysis. *J. Neurosci.*, 3: 2153–2163.

Benowitz, L.I. and Routtenberg, A. (1988) A membrane phosphoprotein associated with neural development, axonal regeneration, phospholipid metabolism and synaptic plasiticity. *Trends Neurosci.*, 10: 527–532.

Benowitz, L.I., Perrone-Bizzozero, N. and Finklestein, S. (1987) Molecular properties of the growth associated protein GAP-43 (B-50). *J. Neurochem.*, 48: 1640–1647.

24

Benowitz, L.I., Apostolides, P.J., Perrone-Biozzozero, N., Finklestein, S. and Zwiers, H. (1988) Anatomical distribution of the growth-associated protein GAP-43/B-50 in the adult rat brain. *J. Neurosci.*, 8: 339–352.

Biffo, S., Verhaagen, J., Schrama, L.H., Schotman, P., Danho, W. and Margolis, F.L. (1990) B-50/GAP-43 expression correlates with process outgrowth in the embryonic mouse nervous-system. *Eur. J. Neurosci.*, 2(6): 487–499.

Carbonetto, S., Evans, D. and Cochard, P. (1987) Nerve fiber growth in culture on tissue substrata from central and peripheral nervous systems. *J. Neurosci.*, 7: 610–620.

Caroni, P. and Schwab, M.E. (1987) Two membrane proteins of myelin and oligodendrocytes inhibit neurite outgrowth and fibroblast migration. *Experientia*, 43: 654.

Caroni, P. and Schwab, M.E. (1988) Two membrane protein fractions from rat central myelin with inhibitory properties for neurite growth and fibroblast spreading. *J. Cell Biol.*, 106: 1281–1288.

Changellian, P.S., Meiri, K., Soppet, D., Valenza, H., Loewy, A. and Willard, M. (1990) Purification of the growth-associated protein GAP-43 by reversed phase chromatography: amino acid sequence analysis and cDNA identification. *Brain Res.*, 510: 259–268.

Cimler, B.M., Giebelhaus, D.H., Wakim, B.T., Storm, D.R. and Moon, R.T. (1987) Characterization fo murine cDNAs encoding P-57, a neural-specific calmodulin-binding protein. *J. Biol. Chem.*, 262: 12158–12163.

Coggins, P.J. and Zwiers, H. (1989) Evidence for a single protein kinase C-mediated phosphorylation site in rat brain protein B-50. *J. Neurochem.*, 53: 1895–1901.

David, S. and Aguayo, A.J. (1981) Axonal elongation into peripheral nervous system bridges after central nervous system injury in adult rats. *Science*, 214: 931–933.

David, S. and Aguayo, A.J. (1985) Axonal regeneration after crush injury of rat central nervous system fibres innervating peripheral nerve grafts. *J. Neurocytol.*, 14: 1–12.

De Graan, P.N.E., Van Hoof, C.O.M., Tilly, B.C., Oestricher, A.B., Schotman, P. and Gispen, W.H. (1985) Phosphoprotein B-50 in nerve growth cones from fetal rat brain. *Neuroci. Lett.*, 61: 235–241.

Dekker, L.V., DeGraan, P.N.E., Versteeg, D.H.G., Oestreicher, A.B. and Gispen, W.H. (1989) Phosphorylation of B-50 (GAP-43) is correlated with neurotransmitter release in rat hippocampal slices. *J. Neurochem.*, 52: 24–30.

Doster, S.K., Lozano, A.M., Willard, M. and Aguayo, A.J. (1988) Axonal transport of GAP-43 in injured and regenerating retinal ganglion cell axons of adult rats. *Soc. Neurosci. Abstr.*, 14: 802.

Doster, S.K., Lozano, A.M., Aguayo, A.J. and Willard, M.B. (1991) Expression of the growth-associated protein GAP-43 in adult rat retinal ganglion cells following axon injury. *Neuron*, 6: 635–647.

Edwards, M.K. and McBurney, M.W. (1983) The concentration of retinoic acid determines the differentiated cell types formed by a teratocarcinoma cell line. *Dev. Biol.*, 98: 187–191.

Freeman, J.A., Bock, S., Deaton, M., McGuire, B., Norden, J.J. and Snipes, G.J. (1986) Axonal and glial proteins associated with development and response to injury in the rat and goldfish optic nerve. *Exp. Brain Res., Suppl.*, 13: 34–47.

Gispen, W.H., Van Dongen, C.J., De Graan, P.N.E., Oestreicher, A.B. and Zwiers, H. (1985a) The role of phosphoprotein B-50 in phosphoinositide metabolism in synaptic plasma membranes. In J.E. Blaesdale, G. Hauser and Eichberg (Eds.), *Inositol and Phosphoinositides*, pp. 399–413, Humana Press, Clifton, NJ.

Gispen, W.H., Leunissen, J.L.M., Oestreicher, A.B., Verkleij, A.J. and Zwiers, H. (1985b) Presynaptic localization of B-50 phosphoprotein: the (ACTH)-sensitive protein kinase substrate involved in rat brain polyphosphoinositide metabolism. *Brain Res.*, 328: 381–385.

Grafstein, B., Burmeister, D.W., McGuinness, C.M., Perry, G.W. and Sparrow, J.R. (1987) Role of fast axonal transport in regeneration of goldfish optic axons. *Progr. Brain Res.*, 71: 113–120.

Jacobson, R.D., Virag, I. and Skene, J.H.P. (1986) A protein associated with axon growth, GAP-43, is widely distributed and developmentally regulated in rat CNS. *J. Neurosci.*, 6: 1843–1855.

Jones, P.S. and Aguayo, A.J. (1990) Localization of specific mRNAs in differentiating retinal cell types. *Soc. Neurosci. Abstr.*, in press.

Kalil, K. and Skene, J.H.P. (1986) Elevated synthesis of an axonally transported protein correlates with axon outgrowth in nomal and injured pyramidal tracts. *J. Neurosci.*, 6: 2563–2570.

Katz, F., Ellis, L. and Pfenninger, K.H. (1985) Nerve growth cones isolated from fetal rat brarin. III. Calcium-dependent protein phosphorylation. *J. Neurosci.*, 5: 1402–1411.

Lozano, A.M., Doster, S.K., Aguayo, A.J. and Willard, M.B. (1987) Immunoreactivity to GAP-43 in axotomized and regenerating retinal ganglion cells of adult rats. *Soc. Neurosci. Abstr.*, 13, 1389.

Masure, H.R., Alexander, K.A., Wakim, B.T. and Storm, D.R. (1986) Physicochemical and hydrodynamic characterization of P-57, a neurospecific calmodulin binding protein. *Biochemistry*, 25: 7553–7560.

McGuire, C.B., Snipes, G.J. and Norden, J.J. (1988) Light-microscopic immunolocalization of the growth- and plasticity-associated protein GAP-43 in the developing rat brain. *Dev. Brain Res.*, 41: 277–291.

Meiri, K.F. and Gordon-Weeks, P.R. (1990) GAP-43 in growth cones is associated with areas of membrane that are tightly bound to substrate and is a component of a membrane skeleton subcellular fraction. *J. Neurosci.*, 10(1): 256–266.

Meiri, K.F., Pfenninger, K.H. and Willard M.B. (1986) Growth-associated protein, GAP-43, a polypeptide that is induced when neurons extend axons, is a component of growth cones and corresponds to pp46, a major polypeptide of a subcellular fraction enriched in growth cones. *Proc. Natl. Acad. Sci. USA*, 83: 3537–3541.

Moya, K.L., Benowitz, L.I., Jhaveri, S. and Schneider, G.E. (1988) Changes in rapidly transported proteins in developing hamster retinofugal axons. *J. Neurosci.*, 8: 4445–4454.

Nelson, R. and Routtenberg, A. (1985) Characterization of protein F1 (47 kDa, 4.5 pI): a kinase C substrate directly related to neural plasticity. *Exp. Neurol.*, 89: 213–244.

Nelson, R.B., Friedman, D.P., O'Neill, J.B., Mishkin, M. and Routtenberg, A. (1987b) Gradients of protein kinase C substrate phosphorylation in primate visual system peak in visual memory storage areas. *Brain Res.*, 416: 387–392.

Neve, R.L., Perrone-Bizzozero, N.I., Finklestein, S., Zwiers, H., Bird, E., Kurnit, D.M. and Benowitz, I. (1987) The neuronal growth-associated protein GAP-43 (B-50,F1): neuronal specificity, developmental regulation and regional distribution of the human and rat mRNAs. *Mol. Brain Res.*, 2: 177–183.

Ng, S., de la Monte, S.M., Conboy, G.L., Karns, L.R. and Fishman, M.C. (1988) Cloning of human GAP-43: growth association and ischemic resurgence. *Neuron*, 1: 133–139.

Perry, G.W., Burmeister, D.W. and Grafstein, B. (1987) Fast axonally transported protein in regenerating goldfish optic axons. *J. Neurosci.*, 7: 792–806.

Pisano, M.R., Hegazy, M.G., Reimann, E.M. and Dokas, L.A. (1988) Phosphorylation of protein B-50 (GAP-43) from adult rat brain cortex by casein kinase II. *Biochem. Biophys. Res. Commun.*, 155(3): 1207–1212.

Raff, M.C., Temple, S. and French-Constant, C. (1987) Glial cell development and function in the rat optic nerve. *Prog. Brain Res.*, 71: 435–438.

Reh, T.A., Redshaw, J.D. and Bisby, M.A. (1987) Axons of the pyramidal tract do not increase their transport of growth-associated proteins after axotomy. *Mol. Brain Res.*, 2: 1–6.

Richardson, P.M. and Issa, V.K.M. (1984) Peripheral injury enhances regeneration of spinal axons. *Nature*, 284: 264–265.

Richardson, P.M., Issa, V.M.K. and Schemle, S. (1982) Regeneration and retrograde degeneration of axons in the rat optic nerve. *J. Neurocytol.*, 11: 949–966.

Richardson, P.M., Issa, V.M.K. and Aguayo, A.J. (1984) Regeneration of long spinal axons in the rat. *J. Neurocytol.*, 13: 165–182.

Schreyer, D.J. and Skene, J.H.P. (1988) GAP-43 induction in regenerating dorsal root ganglion cells: an analysis of sorting in axonal transport. *Soc. Neurosci. Abstr.*, 14: 803.

Schwab, M.E. (1990) Myelin-associated inhibitors of neurite growth. *Mental Neurol.*, 109: 2–5.

Skene, J.H.P. (1989) Axonal growth-associated proteins. *Annu. Rev. Neurosci.*, 12: 127–156.

Skene, J.H. and Virag, I. (1989) Posttranslational membrane attachment and dynamic fatty acylation of a neuronal growth cone protein, GAP-43. *J. Cell Biol.*, 108(2): 613–624.

Skene, J.H.P. and Willard, M.B. (1981a) Changes in axonally transported protein during axon regeneration in toad retinal ganglion cells. *J. Cell Biol.*, 89: 86–95.

Skene, J.H.P. and Willard, M.B. (1981b) Axonally transported proteins associated with axon growth in rabbit central and peripheral nervous systems. *J. Cell Biol.*, 89: 96–103.

Skene, J.H.P. and Willard, M. (1981c) Characteristics of growth-associated polypeptides in regenerating toad retinal ganglion cell axons. *J. Neurosci.*, 1: 419–426.

Skene, J.H.P., Jacobson, R.D., Snipes, G.J., McGuire, C.B., Norden, J. and Freeman, J.A. (1986) A protein induced during nerve growth GAP-43 is a major component of growth-cone membranes. *Science*, 233: 783–785.

So, K.-F. and Aguayo, A.J. (1985) Lengthy regrowth of cut axons from ganglion cells after peripheral nerve transplantation into the retina of adult rats. *Brain Res.*, 328: 349–354.

Strittmatter, S.M., Valenzuela, D., Kennedy, T.E., Neer, E.J. and Fishman, M.C. (1990) G_0 is a major growth cone protein subject to regulation by GAP-43. *Nature*, 344: 836–841.

Studier, F.W. and Moffat, B.A. (1986) Use of bacteriophage T7 RNA polymerase to direct selective high-level expression of cloned genes. *J. Mol. Biol.*, 189: 113–130.

Vidal-Sanz, M., Bray, G.M., Villegas-Perez, M.P., Thanos, S. and Aguayo, A.J. (1987) Axonal regeneration and synapse formation in the superior colliculus by retinal ganglion cells in the adult rat. *J. Neurosci.*, 7: 2894–2909.

Willard, M. and Skene, J.H.P. (1982) Molecular events in axonal regeneration. In J.G. Nicholls (Ed.), *Repair and Regeneration of the Nervous System*, pp. 71–89, Dahlem Konferenzen, Springer-Verlag, Berlin, Heidelberg, New York.

Willard, M., Cowan, W.M. and Vagelos, P.R. (1974) The polypeptide composition of intra-axonally transported proteins: evidence for four transport velocities. *Proc. Natl. Acad. Sci. USA*, 71: 2183–2187.

Willard, M., Skene, J.H.P., Simon, C., Meiri, K., Hirokawa, N. and Clicksman, M. (1984) Regulation of axon growth and cytoskeletal development. In J.S. Elam and P. Cancalon (Eds.), *Axonal Transport in Neuronal Growth and Regeneration*, Plenum Press, New York.

Willard, M., Meiri, K. and Glicksman, M. (1985) Changes of state during neuronal development: regulation of axon elongation. In Edelman et al. (Eds.), *Molecular Bases of*

Neuronal Development, pp. 341–361, Neurosciences Research Foundation Inc.

Woolf, C.J., Reynolds, M.L., Molander, C., O'Brien, C., Lindsay, R.M. and Benowitz, L.I. (1990) The growth-associated protein GAP-43 appears in dorsal root ganglion cells and in the dorsal horn of the rat spinal cord following peripheral nerve injury. *J. Neurosci.*, 34: 465–478.

Zwiers, H., Veldhuis, H.D., Schotman, P. and Gispen, W.H. (1976) ACTH, cyclic nucleotides and brain protein phosphorylation in vitro. *Neurochem. Res.*, 1: 669–677.

W.H. Gispen and A. Routtenberg (Eds.)
Progress in Brain Research, Vol. 89
© 1991 Elsevier Science Publishers B.V.

CHAPTER 3

Selective phosphorylation and dephosphorylation of the protein B-50

Linda A. Dokas [1,2], Michael R. Pisano [2] and Yi-fan Han [2]

Departments of [1] Neurology, and [2] Biochemistry and Molecular Biology, Medical College of Ohio, Toledo, OH 43699-0008, U.S.A.

The phosphorylated presynaptic membrane-bound protein B-50 (GAP-43, F1, pp46, P57) has been correlated with aspects of neuronal second messenger systems (VanDongen et al., 1985; Alexander et al., 1987; Strittmatter et al. 1990), synaptic functions (Dekker et al., 1989) and plasticity (Lovinger et al., 1987; Neve et al., 1988). Since protein kinase C-mediated phosphorylation of B-50 has been clearly documented (Aloyo et al., 1983), it is modification by this enzyme that is best defined with regard to the functional characteristics of the protein within these contexts. The evidence derived from peptide mapping studies (Coggins and Zwiers, 1989; Oestreicher et al. 1989) and direct sequence analysis (Apel et al. 1990) does establish serine-41 as the unique site in B-50 which can be phosphorylated by protein kinase C. Moreover, in those instances where alteration of protein kinase C activity with selective activators or inhibitors can be shown to cause parallel changes in the phosphorylation of B-50

and a synaptic function such as neurotransmitter release (Dekker et al. 1989), the case for a causative role for the region of B-50 surrounding serine-41 is strengthened.

However, there are indications of more than one phosphorylatable site in B-50. Zwiers et al. (1985) demonstrated microheterogeneity of B-50 with narrow range isoelectric focusing. Although protein subtypes might be accounted for by N-terminal variability (McMaster et al., 1988), reversal of the heterogeneity by treatment with alkaline phosphatase suggests at least some of it results from the phosphorylated state of B-50. The combination of phosphorylation at more than one site by more than one kinase and the differential action of protein phosphatases upon these sites would yield dephospho-, mono- and multiply-phosphorylated forms of B-50. In at least one intact system (cultured neurons), tryptic peptide mapping of endogenously phosphorylated B-50 demonstrates multiple phosphopeptides (Schuh et al., 1989). Since the labeling of only one of these is stimulated by phorbol esters which activate protein kinase C, phosphorylation of B-50 by other protein kinases is implied.

Correspondence: Linda A. Dokas, Department of Biochemistry and Molecular Biology, Medical College of Ohio, Toledo, OH 43699-0008, U.S.A.

The absence of any tyrosine residues in the amino acid sequence of B-50 (Basi et al., 1987; Karns et al., 1987; Nielander et al., 1987) eliminates it as a potential substrate for tyrosine kinases. Of a number of serine/threonine protein kinases tested for their abilities to phosphorylate B-50, only casein kinase II has been reported to phosphorylate B-50 with stoichiometry similar to that seen with protein kinase C (Aloyo et al., 1983; Pisano et al., 1988). This chapter summarizes this work, compares the properties of casein kinase II and protein kinase C-mediated phosphorylation of B-50 and discusses the potential relationship between casein kinase II-mediated phosphorylation of B-50 and growth-associated processes. In addition, work from this and other laboratories on dephosphorylation of B-50 by a number of identified protein phosphatases is summarized.

Casein kinase II-mediated phosphorylation of B-50

Realization that B-50 might be a substrate for casein kinase II began with examination of the full amino acid sequence of this protein derived from the published cDNA nucleotide sequence (Basi et al., 1987; Karns et al., 1987; Nielander et al., 1987). Of the 14 serine and 14 threonine residues in the 226 amino acids of rat B-50, several are surrounded by sequences with required properties for phosphorylation by casein kinase II. These characteristics, based on kinetic studies with model peptide substrates, include a cluster of acidic amino acids positioned on the C-terminal side of the phosphorylated residue, with the most critical determinant being an acidic residue in the $N + 3$ position (Marin et al., 1986; Kuenzel et al., 1987). On the N-terminal side of the phosphorylation site, the presence of an acidic amino acid increases substrate efficiency, markedly decreasing the Km of casein kinase II

TABLE I

Comparison of B-50 190–197 to peptide substrates of casein kinase II

Peptide	K_m (mM) [a]
Ser-Glu-Glu-Glu-Glu-Glu	0.30
Ser-Glu-Glu-Ala-Glu-Glu	1.70
Glu-Ser-Glu-Glu-Glu-Glu-Glu	0.13
Arg-Ser-Glu-Glu-Glu-Glu-Glu	2.45
Ser-Glu-Glu-Glu-Lys-Glu	6.67
Thr-Glu-Glu-Glu-Glu-Glu	6.25
Arg-Arg-Arg-Glu-Glu-Glu-Ser-Glu-Glu-Glu	0.18
Arg-Arg-Arg-Glu-Glu-Glu-Ser-Ala-Ala-Glu	0.51
Arg-Arg-Arg-Glu-Glu-Glu-Ser-Glu-Glu-Ala	negligible
-Glu-Ser-Ser-Gln-Ala-Glu-Glu-Glu- (B-50 190–197)	0.004 [b]

[a] The K_m values of casein kinase II for the peptide substrates are from the following references: Marin et al., 1986; Kuenzel et al., 1987; Marchiori et al., 1988; Litchfield et al., 1990.
[b] This K_m value was determined using full length B-50, not B-50 190–197, as a substrate for casein kinase II (Pisano et al., 1988).

for the peptide, while basic residues must not be present within the acidic cluster or on the C-terminal side of the phosphorylation site (Marchiori et al., 1988). A serine residue in association with these features is preferred to a threonine residue (Marin et al., 1986; Kuenzel et al., 1987; Litchfield et al., 1990). A summary of selected peptides that demonstrate the relationship between these structural features and the K_m values of casein kinase II for the peptides is provided in Table I.

Within the sequence of B-50 (Fig. 1), a number of serines possess at least some of the required characteristics. A repeating pattern is the presence of serine residues in proximity to two or more glutamate residues. But of these, only the sequence around serine 192 possesses all of the necessary structural features to determine optimal phosphorylation by casein kinase II (Fig. 1

and Table I). Serine 192 is positioned three amino acids in the N-terminal direction from a group of glutamate residues. There is a glutamate in the critical N + 3 position from serine 192 and no basic amino acid on either side of serine 192 or within the sequence of glutamate residues. On the N-terminal side of serine 192 there is an acidic residue (glutamate 190). In contrast, the other potential casein kinase II sites have less favorable configurations. Most notably, serine 191, which is immediately adjacent to serine 192, lacks almost all of these characteristics because of its position relative to the determinant residues and would appear to be a much less optimal substrate site for casein kinase II. Serine 145, although positioned correctly with regard to acidic residues at 148 and 149, is less likely to be a substrate for casein kinase II since a lysine residue (146) is positioned between them. The possibility that B-50 is a substrate for casein kinase II is further strengthened by the similarities between the sequence 190–197 and those contained within other naturally-occurring proteins demonstrated to be casein kinase II substrates (Marin et al., 1986; Kuenzel et al., 1987). In particular, the sequence B-50 190–197 (Glu-Ser-Ser-Gln-Ala-Glu-Glu-Glu-) closely resembles a casein kinase II substrate site in protein phosphatase inhibitor-2 (Glu-Ser-Ser-Gly-Glu-Glu-Asp), as characterized by Holmes et al. (1986). Consistent with the idea that phosphorylation by casein kinase II may regulate functions of B-50, the sequence around serine 192 is conserved through the second glutamate residue of the cluster across species, including chicken, rat, mouse, cow and human forms (Skene, 1989; Baizer et al., 1990).

With these theoretical considerations in mind, phosphorylation of B-50 by purified casein kinase II was examined *in vitro* (Pisano et al., 1988). Purified B-50 was prepared from rat brain by the

AMINO ACID SEQUENCE OF RAT B-50/GAP-43

M L C C M R R T K Q V E K N D E D Q K I E Q D G V K P E D K A H K

A A T K I Q A S F R G H I T R K K L K D E K K G D A P A A E A E A

K E K D D A P V A D G V E K K E G D G S A T T D A A P A T S P K A

E E P S K A G D A P S E E K K G E G D A A P S E E K A G S A E T E

S A A K A T T D N S P S S K A E D G P A K E E P K Q A D V P A A V

T D A A A T T P A A E D A A K A A Q P P T E T A E S S Q A E E E K

E A V D E A K P K E S A R Q D E G K E D P E A D Q E H A

Fig. 1. The amino acid sequence of rat B-50/GAP-43. The serine residues presumed to be phosphorylated by protein kinase C (serine-41) and casein kinase II (serine-192) are indicated by asterisks. Also indicated by lines are the residues involved in membrane binding and the calmodulin-binding site.

30

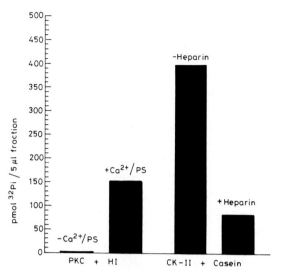

Fig. 2. Selectivity of protein kinase preparations. Purified protein kinase C (PKC) was incubated with histone (0.25 mg/ml) as substrate in the presence (+) or absence (−) of 200 μM CaCl$_2$ and 20 μg/ml of phosphatidylserine (PS). Purified casein kinase II (CKII) was assayed with casein (1 mg/ml) as substrate in the presence (+) or absence (−) of heparin at 5 μg/ml. The data are expressed as pmoles of phosphate incorporated per 5 μl of the reaction volume and were corrected for autophosphorylation.

protocol of Zwiers et al. (1985) which involves alkaline extraction, precipitation and preparative flat-bed isoelectric focusing. The final protein fraction is homogeneous and devoid of protein kinase or phosphatase activity. A number of purified protein kinases were assayed for their ability to phosphorylate B-50. These included protein kinase C, casein kinases I and II, the catalytic subunit of cyclic AMP-dependent protein kinase, phosphorylase kinase and glycogen synthase kinase 3. Only protein kinase C and casein kinase II were able to phosphorylate B-50 at a stoichiometry close to 1 mole of phosphate per mole of protein substrate. Because these enzyme preparations were partial, rather than homogeneous, purifications, it was important to establish at the outset that the ability of casein kinase II to phosphorylate B-50 did not result from contami-

nation of this preparation with protein kinase C. The protocols chosen for purification of protein kinase C and casein kinase II included a number of column chromatography steps that by themselves should have minimized cross-contamination of either enzyme preparation. In addition, the optimal assay conditions adopted for each enzyme confirmed its identity and suggested little activity of the alternate enzyme in each buffer system (Fig. 2). Activity of protein kinase C, with histone as substrate was stimulated by the presence of calcium and phosphatidylserine. Given the dependence of protein kinase C on calcium, its activity was minimized in assays of casein kinase II activity which lacked calcium and contained divalent cation chelators. Conversely, with casein as substrate, the preparation of casein kinase II was inhibited by low concentrations of heparin. Since the small amount of casein kinase II activity observed with histone as substrate was also inhibited by heparin, no contamination of the preparation with the proteolyzed form of protein kinase C that is calcium-independent was indicated.

Phosphorylation of B-50 by casein kinase II occurs solely on serine residues and, in the best preparations, reaches 1 mol of phosphate incorporated per mol of B-50. Although this implies a single phosphorylation site, the addition of phosphate onto more than one serine cannot be excluded since purified B-50 is not completely in the dephospho- form (Zwiers et al., 1985). Phosphorylation of more than one partially-filled site to completion would yield an apparent stoichiometry of 1 mole of phosphate per mole of protein. The K_m of casein kinase II for B-50 was found to be 4 μM, a value considerably lower than that for the best of the model peptide substrates and similar to those of several protein substrates for casein kinase II (Carmicheal et al., 1982; Serrano et al., 1987), including casein (40 μM), tubulin (2 μM) and the regulatory subunit of cyclic

AMP-dependent protein kinase (36 μM). Moreover, phosphorylation of B-50 by casein kinase II was inhibited completely by low concentrations (5 μg per ml) of heparin, stimulated by the polyamine spermine and inhibited by the casein kinase II substrate peptide RRREEETEEE. All of these characteristics suggest specificity in the manner by which casein kinase II phosphorylates this protein.

To compare the sites of phosphorylation in B-50 following phosphorylation by protein kinase C or casein kinase II, the ^{32}P-labeled protein was digested with *Staphylococcus aureus* protease or α-chymotrypsin and the digests were analyzed with SDS-polyacrylamide gradient gel electrophoresis and autoradiography. Molar ratios of *S. aureus* protease (1:30) and α-chymotrypsin (1:100) were chosen that resulted in a time-dependent processing of fragments over a 180-min incubation period at 30°C. Similar to the results of Oestreicher et al. (1989), *S. aureus* protease cleaves protein kinase C-phosphorylated B-50 to two ^{32}P-labeled products of approximately 28 and 15 kDa in size (Fig. 3). Under these assay conditions, production of these fragments is somewhat sequential, with more of the larger form produced at earlier time points and increasing amounts of the smaller one generated over the entire incubation period. Digestion of casein kinase II-phosphorylated B-50 with *S. aureus* protease produced two fragments of about the same sizes as those from protein kinase C-phosphorylated B-50, but with different time courses. Both peptide products are seen at the earliest time of digestion, but the larger form is completely processed by 30 min of digestion. The smaller fragment appears to be extensively digested by 180 min of incubation, a situation not seen with the similarly-sized fragment from protein kinase C-phosphorylated B-50.

Further differences between protein kinase C and casein kinase II-phosphorylated B-50 were

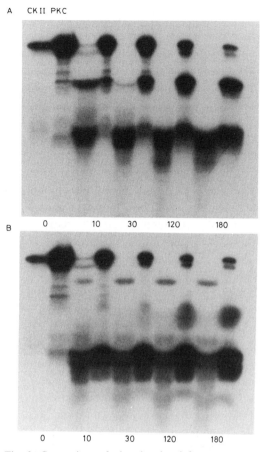

Fig. 3. Comparison of phosphorylated fragments generated from B-50 by *S. aureus* protease (A) or α-chymotrypsin (B) following phosphorylation by protein kinase C or casein kinase II. Proteolytic digestions were performed at 30°C with *S. aureus* protease at a 1:30 ratio and α-chymotrypsin at a 1:100 molar ratio to B-50 for 0, 10, 30, 120 and 180 min as indicated. Each enzyme incubation contained equal amounts of ^{32}P-labeled B-50. At each time point, the sample for casein kinase II is shown to the left of that for protein kinase C.

apparent following α-chymotryptic digestion of the labeled protein (Fig. 3B). α-Chymotrypsin was chosen for these analyses because it preferentially cleaves proteins at aromatic residues. Because B-50 contains a single phenylalanine at position 42 and no tryptophan or tyrosine residues, it was assumed that α-chymotrypsin would produce only two fragments from it (B-50

1–42 and B-50 42–226). The protein kinase C phosphorylation site (serine-41) is contained in the small N-terminal fragment and the presumed casein kinase II site on the larger C-terminal fragment. The digestion pattern following incubation of protein kinase C- or casein kinase II-phosphorylated B-50 does produce such peptides. Most notably, a 39 kDa fragment is seen only after casein kinase II phosphorylation of B-50. This fragment is not observed at any time during the digestion of protein kinase C-phosphorylated B-50. Instead, small molecular weight phosphorylated peptides predominate in the latter case. Multiple small forms varying slightly in length may result from cleavage at the same site of several related forms of B-50 differing in their N-termini (McMaster et al., 1988). The α-chymotryptic digest from casein kinase II-phosphorylated B-50 also contains small molecular weight fragments. These may result from digestion of B-50 at sites other than phenylalanine-42 by α-chymotrypsin.

To determine if casein kinase II could phosphorylate B-50 in a more intact system, synaptic plasma membranes containing B-50 were incubated with purified casein kinase II. This experimental protocol was chosen to mimic in vivo conditions, since B-50 is membrane-bound and casein kinase II is predominantly a cytosolic enzyme (Hathaway and Traugh, 1982). Because B-50 is attached to the inner face of synaptic plasma membranes by N-terminal attachment (Skene and Virag, 1989) and extends into the cytoplasm, it is possible for it to serve as an endogenous substrate for soluble casein kinase II. Synaptic plasma membranes were prepared from rat cortex and mildly heat-treated (10 min at 50°C) to inactivate endogenous protein kinase C while not significantly disrupting membrane structure. Subsequent incubation of membrane proteins with γ-[^{32}P]ATP at 30°C demonstrated no endogenous kinase activity. However, addition of purified casein kinase II to these membranes resulted in significant labeling of B-50. Moreover, like the case for the in vitro phosphorylation of B-50 by casein kinase II, ^{32}P-labeling of the protein was inhibited by heparin and stimulated by spermine.

Phosphorylation by casein kinase II would be consistent with the growth-associated function of B-50, since in other systems, casein kinase II activity has been shown to be stimulated by growth-related stimuli. For example, in several cell types, insulin and epidermal growth factor rapidly activate casein kinase II, an effect resulting from phosphorylation of the β-subunit (Sommercorn et al., 1987; Klarlund and Czech, 1988; Ackerman et al., 1990). Even more dramatically, addition of serum to quiescent fibroblasts produces a rapid and transient elevation (6-fold) of casein kinase II activity (Carroll and Marshak, 1989). In tumors and in rapidly-proliferating nonmalignant tissues, casein kinase II activity is elevated as much as 8-fold compared to less proliferative tissues (Munstermann et al., 1990). Conversely, in brain tissue from Alzheimer's disease, characterized by its degenerative state, casein kinase II possesses aberrant characteristics, being reduced in amount and distributed abnormally in neurons (Iimoto et al., 1989).

If casein kinase II-mediated phosphorylation were to occur in vivo, the possibility exists for selective regulation of B-50 phosphorylation. For example, polyamines such as spermine, which stimulate casein kinase II, have been shown to inhibit protein kinase C activity (House and Kemp, 1987). Selective phosphorylation of B-50 by casein kinase II might occur under conditions where ornithine decarboxylase, the rate-limiting enzyme of polyamine synthesis, is induced. Because this induction frequently correlates with growth responses (Russell, 1980), the phosphorylation of B-50 by casein kinase II might correlate with its growth-associated functions. In addition, there could be interactions between casein kinase

II and protein kinase C in phosphorylating B-50. It is notable in this regard that casein kinase II-mediated phosphorylation of several substrates has been shown to potentiate phosphorylation by other kinases, and in all cases, the sites phosphorylated by casein kinase II are C-terminal to those labeled by the other kinase. The relative positions of the protein kinase C phosphorylation site (serine-41) and the presumed casein kinase II site (serine-192) in B-50 match this pattern.

Dephosphorylation of B-50

The functions of B-50 that have been correlated with phosphorylation of the protein include synaptic processes that are either readily reversible or under flexible regulation. Implied in these characteristics is the requirement for protein phosphatase activity to rapidly remove phosphate groups from B-50, so that its phosphorylation state reflects the regulatory state of the synapse.

Since synaptic membranes possess high protein phosphatase activity (Maeno and Greengard, 1972), the enzymes within this fraction that possess the ability to dephosphorylate B-50 were characterized (Dokas et al., 1990; Han and Dokas, 1991). As an experimental framework, the characteristics of these enzymes were compared to those of the major classes of cellular protein phosphatases in the classification of Cohen (1989). These fall generally into two classes with the type 1 enzyme being sensitive to heat-stable protein phosphatase inhibitors, while the type 2 enzymes are not. Type 2 protein phosphatases are further distinguished by several characteristics, but most notably by cation sensitivity. Type 2A is insensitive to divalent cations, while type 2B (calcineurin) is regulated by calcium and type 2C by magnesium.

Dephosphorylation of B-50 in synaptic plasma membranes can be demonstrated following incubation with a low concentration of γ-[^{32}P]ATP.

Net phosphorylation of proteins occurs only for a few seconds under these conditions due to the high membrane-bound ATPase activity and subsequently ^{32}P-labeling of B-50 is rapidly reduced. The loss of radioactivity does not reflect protease activity as it is not sensitive to proteolytic inhibitors and because no loss of immunoreactive B-50 is seen on Western blots following incubation of membranes with this protocol. The major portion of synaptic membrane-bound B-50 phosphatase activity is insensitive to divalent cations. The heat-stable protein phosphatase inhibitors block enzyme activity, with the degree of effect determined by the ratio of inhibitor to membrane protein. Adenine nucleotides (ATP, ADP and a nonhydrolyzable form of ATP), stimulate dephosphorylation of B-50, while guanine nucleotides are ineffective. These characteristics suggest type 1 and/or 2A protein phosphatases as the synaptic membrane-bound enzymes that dephosphorylate B-50, with the former being the predominant form.

Further characterization using the more specific protein phosphatase inhibitor, okadaic acid, supports this assignment of activity. Okadaic acid is a polyether fatty acid which has tumor promotor activity (Cohen et al., 1990). Its tumor-promoting activity correlates with its ability to increase the phosphorylation level within cells, not by stimulating protein kinase activity as do the phorbol esters, but by inhibiting protein phosphatase activity. Its actions in this regard are quite selective and have been of enormous practical use in the characterization of cellular protein phosphatase activities. At 1–2 nM, okadaic acid is a selective inhibitor of protein phosphatase 2A, while at the higher concentration of 1 μM, it inhibits both types 1 and 2A. It has virtually no effect on the other types of protein phosphatases nor on protein kinase activities.

Okadaic acid was used to compare the dephosphorylation of B-50 in synaptic plasma mem-

34

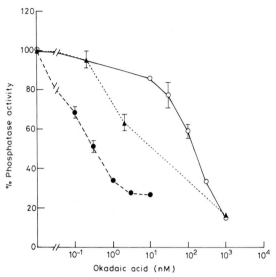

Fig. 4. Inhibition of protein dephosphorylation by okadaic acid. The ability of okadaic acid to inhibit the dephosphorylation of B-50 in synaptic plasma membranes (▲ ——— ▲) was compared to its effects on dephosphorylation of ^{32}P-labeled casein (● ——— ●) and phosphorylase A (○ ——— ○) by synaptic plasma membrane-bound protein phosphatase(s). Synaptic plasma membrane proteins were labeled with 7.5 μM γ-[^{32}P]ATP for 30 sec. Okadaic acid, at the indicated concentrations, was then added for 30 min. Labeled B-50 was quantitated by counting the radioactivity of B-50 bands excised from SDS-polyacrylamide gels. With purified ^{32}P-labeled substrates, the percentage phosphatase activity was determined as the release of soluble radioactivity following incubation for 10 min at 30°C with 12.5 μg of synaptic plasma membrane proteins as a source of enzyme activity. Each value is expressed as a percentage of the control group incubated in the absence of okadaic acid.

branes to that of purified ^{32}P-labeled proteins known to be substrates for type 1 and 2A phosphatases (Fig. 4). With ^{32}P-casein as substrate, which can be dephosphorylated only by protein phosphatase 2A in the absence of magnesium, the membrane-bound activity was inhibited 75% at 2–3 nM okadaic acid. In contrast, with [^{32}P]phosphorylase A as substrate, the synaptic membrane protein phosphatase activity was dose-dependently inhibited by okadaic acid at concentrations from 10 nM to 1 μM with a maxi-

mal inhibition of 85%. The dephosphorylation of endogenous B-50, prelabeled from γ-[^{32}P]ATP, showed an intermediate sensitivity curve to okadaic acid. Approximately 30% of B-50 dephosphorylation was inhibited at concentrations of okadaic acid that would only be expected to affect type 2A protein phosphatase. At a concentration of 1 μM, the dephosphorylation of B-50 by endogenous synaptic membrane-bound phosphatase activity was almost completely inhibited. These results suggest that approximately 70% of the synaptic membrane-bound phosphatase activity which dephosphorylates B-50 is the type 1 form, while most, if not all, of the rest is type 2A. Such an assignment of activity is consistent with the measurement of the activities of the four classes of protein phosphatases in synaptic fractions (Shields et al., 1985).

Two independent reports have shown that protein phosphatase 2B (calcineurin) can also dephosphorylate B-50 (Liu and Storm, 1989; Schrama et al., 1989). Our work does not contradict these since it has characterized only synaptic membrane-bound enzyme activities. In analogy with the argument for selective phosphorylation of B-50 by membrane-bound and cytosolic protein kinases, both types 1 and 2A phosphatases, which are both membrane-bound and cytosolic, and 2B, which is predominantly cytosolic, could have access to phosphorylated sites on B-50. It should be stressed that all of these studies have used B-50 phosphorylated by protein kinase C as a substrate. The specificity of these enzymes for casein kinase II-phosphorylated B-50 remains unknown.

References

Ackerman, P., Glover, C.V.C. and Osheroff, N. (1990) Stimulation of casein kinase II by epidermal growth factor: Relationship between the physiological activity of the kinase and the phosphorylation state of its β-subunit. *Proc. Natl. Acad. Sci. USA*, 87: 821–825.

Alexander, K.A., Cimler, B.M., Meier, K.E. and Storm, D.R. (1987) Regulation of calmodulin binding to P-57. *J. Biol. Chem.*, 262: 6108–6013.

Aloyo, V.J., Zwiers, H. and Gispen, W.H. (1983) Phosphorylation of B-50 protein by calcium-activated, phospholipid-dependent protein kinase and B-50 protein kinase. *J. Neurochem.*, 41: 649–653.

Apel, E.D., Byford, M.F., Au, D., Walsh, K.A. and Storm. D.R. (1990) Identification of the protein kinase C phosphorylation site in neuromodulin. *Biochemistry,* 29: 2330–2335.

Baizer, L. Alkan, S., Stocker, K. and Ciment, G. (1990) Chicken growth-associated protein(GAP)-43: primary structure and regulated expression of mRNA during embryogenesis. *Mol. Brain Res.*, 7: 61–68.

Basi, G.S., Jacobson, R.D., Virag, I., Schilling, J. and Skene, J.H.P. (1987) Primary structure and transcriptional regulation of GAP-43, a protein associated with nerve growth. *Cell,* 49: 785–791.

Carmicheal. D.F., Geahlen, R.L., Allen, S.M. and Krebs, E.G. (1982) Type II regulatory subunit of cAMP-dependent protein kinase. Phosphorylation by casein kinase II at a site that is also phosphorylated *in vivo. J. Biol. Chem.*, 257: 10440–10445.

Carroll, D. and Marshak, D.R. (1989) Serum-stimulated cell growth causes oscillations in casein kinase II activity. *J. Biol. Chem.,* 264: 7345–7348.

Coggins, P.J. and Zwiers, H. (1989) Evidence for a single protein kinase C-mediated phosphorylation site in rat brain protein B-50. *J. Neurochem.*, 53: 1895–1901.

Cohen, P. (1989) The structure and regulation of protein phosphatases. *Annu. Rev. Biochem.*, 58: 453–508.

Cohen, P., Holmes, C.F.B. and Tsukitani, Y. (1990) Okadaic acid: a new probe for the study of cellular regulation. *Trends Biochem. Sci.,* 15: 98–102.

Dekker, L.V., DeGraan, P.N.E., Versteeg, D.H., Oestreicher, A.B. and Gispen, W.H. (1989) Phosphorylation of B-50 (GAP-43) is correlated with neurotransmitter release in rat hippocampal slices. *J. Neurochem.*, 52: 24–30.

Dokas, L.A., Pisano, M.R., Schrama, L.H., Zwiers, H. and Gispen, W.H. (1990) Dephosphorylation of B-50 in synaptic plasma membranes. *Brain Res. Bull.*, 24: 321–329.

Han, Y.F. and Dokas, L.A. (1991) Okadaic acid-induced inhibition of B-50 dephosphorylation by presynaptic membrane-associated protein phosphatases. *J. Neurochem.*, 57: in press.

Hathaway, G.M. and Traugh, J.A. (1982) Casein kinases-multipotential protein kinases. *Curr. Topics Cell. Regul.*, 21: 101–127.

Holmes, C.F.B., Kuret, J., Chisholm, A.A.K. and Cohen, P. (1986) Identification of the sites on rabbit skeletal muscle protein phosphatase inhibitor-2 phosphorylated by casein kinase II. *Biochim. Biophys. Acta*, 870: 408–416.

House, C. and Kemp, B.E. (1987) Protein kinase C contains a pseudosubstrate prototype in its regulatory domain. *Science*, 238: 1726–1728.

Iimoto, D.S., Masliah, E., DeTeresa, R., Terry, R.D. and Saitoh, T. (1989) Aberrant casein kinase II in Alzheimer's Disease. *Brain Res.,* 507: 273–280.

Karns, L.R., Ng, S.-C., Freeman, J.A. and Fishman, M.C. (1987) Cloning of complementary DNA for GAP-43, a neuronal growth-related protein. *Science,* 236: 597–600.

Klarlund, J.K. and Czech, M.P. (1988) Insulin-like growth factor I and insulin rapidly increase casein kinase II activity in BALB/c3T3 fibroblasts. *J. Biol. Chem.*, 263: 15872–15875.

Kuenzel, E.A., Mulligan, J.A., Sommercorn, J. and Krebs, E.G. (1987) Substrate specificity determinants for casein kinase II as deduced from studies with synthetic peptides. *J. Biol. Chem.,* 262: 9136–9140.

Litchfield, D.W., Arendt, A., Lozeman, F.J., Krebs, E.G., Hargrave, P.A. and Palczewski, K. (1990) Synthetic phosphopeptides are substrates for casein kinase II. *FEBS Lett.*, 261: 117–120.

Liu, Y.C. and Storm. D.R. (1989) Dephosphorylation of neuromodulin by calcineurin. *J. Biol. Chem.*, 264: 12800–12804.

Lovinger, D.M., Colley, P.A., Akers, R.F., Nelson, R.B. and Routtenberg, A. (1987) Direct relation of long-term synaptic potentiation to phosphorylation of membrane protein F1, a substrate for membrane protein kinase C. *Brain Res.*, 399: 205–211.

Maeno, H. and Greengard, P. (1972) Phosphoprotein phosphatases from rat cerebral cortex. Subcellular distribution and characterization. *J. Biol. Chem.*, 247: 3269–3277.

Marchiori, F., Meggio, F., Marin, F., Borin, G., Calderan, A., Ruzza, P. and Pinna, L.A. (1988) Synthetic peptide substrates for casein kinase 2. Assessment of minimum structural requirements for phosphorylation. *Biochim. Biophys. Acta*, 971: 332–338.

Marin, O., Meggio, F., Marchiori, F., Borin, G. and Pinna, L.A. (1986) Site specificity of casein kinase-2(TS) from rat liver cytosol. *Eur. J. Biochem.*, 160: 239–244.

McMaster, D., Zwiers, H. and Lederis, K. (1988) The growth-associated neuronal phosphoprotein B-50: improved purification, partial primary structure and characterization and localization of proteolysis products. *Brain Res. Bull.*, 21: 265–276.

Munstermann, U., Fritz, G., Seitz, G., Yiping, L., Schneider, H.G. and Issinger, O.-G. (1990) Casein kinase II is elevated in solid human tumors and rapidly proliferating non-neoplastic tissue. *Eur. J. Biochem.*, 189: 251–257.

Neve, R.L., Finch, E.A., Bird, E.D. and Benowitz, L.I. (1988) Growth-associated protein GAP-43 is expressed selectively in associative regions of the adult human brain. *Proc. Natl. Acad. Sci. USA*, 85: 3638–3642.

Nielander, H.B., Schrama, L.H., VanRozen, A.J., Kasperaitis, W., Oestreicher, A.B., DeGraan, P.N.E., Gispen, W.H.

36

and Schotman, P. (1987) Primary structure of the neuron-specific phosphoprotein B-50 is identical to growth-associated protein GAP-43. *Neurochem. Res. Commun.*, 1: 163–171.

Oestreicher, A.B., DeGraan, P.N.E., Schrama, L.H., Lamme, V.A.F., Bloemen, R.J., Schotman, P. and Gispen, W.H. (1989) The protein kinase C phosphosite(s) in B-50 (GAP-43) are confined to 15K phosphofragments produced by *Staphylococcus aureus* V8 protease. *Neurochem. Res.*, 14: 361–372.

Russell, D. (1980) Ornithine decarboxylase as a biological and pharmacological tool. *Pharmacology*, 20: 117–129.

Schrama, L.H., Heemskerk, F.M.J. and DeGraan, P.N.E. (1989) Dephosphorylation of protein kinase C phosphorylated B-50/GAP-43 by the calmodulin-dependent phosphatase calcineurin. *Neurochem. Res. Commun.*, 5: 141–147.

Schuh, S.M., Spencer, S.A. and Willard, M. (1989) Phosphorylation sites on the growth-associated protein GAP-43. *Soc. Neurosci. Abstr.*, 15: 573.

Serrano, L., Diaz-Nido, J., Wandosell, F. and Avila, J. (1987) Tubulin phosphorylation by casein kinase II is similar to that found *in vivo*. *J. Cell Biol.*, 105: 1731–1739.

Shields, S.M., Ingebritsen, T.S. and Kelly, P.T. (1985) Identification of protein phosphatase 1 in synaptic junctions: Dephosphorylation of endogenous calmodulin-dependent kinase II and synapse-enriched phosphoproteins. *J. Neurosci.*, 5: 3414–3422.

Skene, J.H.P. (1989) Axonal growth-associated proteins. *Annu. Rev. Neurosci.*, 12: 127–156.

Skene. J.H.P. and Virag, I. (1989) Posttranslational membrane attachment and dynamic fatty acylation of a neuronal growth cone protein GAP-43. *J. Cell Biol.*, 108: 613–624.

Sommercorn, J. Mulligan, J.A., Lozeman, F.J. and Krebs, E.G. (1987) Activation of casein kinase II in response to insulin and to epidermal growth factor. *Proc. Natl. Acad. Sci. USA*, 84: 8834–8838.

Strittmatter, S.M., Valenzuela, D., Kennedy, T.E., Neer, E.J. and Fishman, M.C. (1990) G_o is a major growth cone protein subject to regulation by GAP-43. *Nature,* 344: 836–841.

Van Dongen, C.J., Zwiers, H., DeGraan, P.N.E. and Gispen, W.H. (1985) Modulation of the activity of purified phosphatidylinositol 4-phosphate by phosphorylated and dephosphorylated B-50 protein. *Biochem. Biophys. Res. Commun.*, 128: 1219–1227.

Zwiers, H., Verhaagen, J., VanDongen, C.J., DeGraan, P.N.E. and Gispen, W.H. (1985) Resolution of rat brain synaptic phosphoprotein B-50 into multiple forms by two-dimensional electrophoresis: evidence for multisite phosphorylation. *J. Neurochem.*, 44: 1083–1090.

W.H. Gispen and A. Routtenberg (Eds.)
Progress in Brain Research, Vol. 89
© 1991 Elsevier Science Publishers B.V.

CHAPTER 4

Mutagenesis of the calmodulin binding domain of neuromodulin

Edwin R. Chapman, Douglas Au, Teresa A. Nicolson and Daniel R. Storm

Department of Pharmacology, University of Washington, Seattle, WA 98195, U.S.A.

Introduction

Neuromodulin (also designated P-57, GAP-43, F1 and B50) is a membrane associated, neural specific protein which is transported to axonal growth cones by rapid axoplasmic transport (Skene, 1989). While neuromodulin has been implicated in the regulation of neurite outgrowth and in synaptic plasticity (for reviews see Willard et al., 1987; Benowitz and Routtenberg, 1987; Skene, 1989; Liu and Storm, 1990), the actual role of the protein in these processes has yet to be determined. In the present study, we have focused on biochemical properties of neuromodulin which may be important for its function in neurons.

Neuromodulin binds calmodulin (CaM) in the absence of free calcium and is a major presynaptic substrate for protein kinase C (Benowitz and Routtenberg, 1987; Willard et al., 1987; Nelson et al., 1989a). The phosphorylation of neuromodulin by protein kinase C significantly reduces its affinity for CaM (Alexander et al., 1987; Apel et al., 1990) suggesting that one function of neuromodulin may be to bind CaM at specific sites in neurons, and to release CaM locally when protein kinase C is activated (Alexander et al., 1987). Furthermore, phosphoneuromodulin is a substrate for calcineurin and phosphatase 2A, indicating that the concentrations of free CaM in neurons may be controlled by a phosphorylation/dephosphorylation cycle (Liu and Storm, 1989).

The phosphorylation of neuromodulin by protein kinase C has also been implicated in the regulation of phosphatidylinositol metabolism (Van Dongen et al., 1985; Van Hooff et al., 1988), neurotransmitter release from rat cortical synaptosomes (Dekker et al., 1989) and long term potentiation in the rat hippocampus (Routtenberg and Lovinger 1985; Akers and Routtenberg, 1985; Lovinger et al., 1986; Nelson et al., 1989b).

Abbreviations: CaM, calmodulin; SDS, sodium dodecyl sulfate; PAGE, polyacrylamide gel electrophoresis; EGTA, ethylene glycol bis(β-aminoethyl ether)-*N,N,N',N'*-tetraacetic acid; DTT, dithiothreitol EDTA, ethylenediaminetetraacetic acid; Tris-HCl, tris(hydroxymethyl)aminomethane hydrochloride; FP57-Trp, CaM binding peptide containing tryptophan in place of phenylalanine.
Correspondence: Edwin R. Chapman, Department of Pharmacology, University of Washington, Seattle, WA 98195, U.S.A.

The importance of neuromodulin/calmodulin interactions in these processes is unclear, but may involve the availability of CaM for the activation of CaM-dependent kinases which have been implicated in long-term potentiation (Nelson et al., 1989b) and neurotransmitter release (Dekker et al., 1989).

The putative CaM binding domain of neuromodulin has been localized to the N-terminal end of the protein and may include amino acid residues 43–51 (Alexander et al., 1988). A synthetic heptadecapeptide corresponding to residues 39–55 was prepared in which the native phenylalanine-42 was replaced with a tryptophan residue, resulting in a fluorescent peptide. This peptide, FP57-Trp (Fig. 1b), bound to CaM with an affinity comparable to native neuromodulin and inhibited binding of CaM to neuromodulin (Alexander et al., 1988). However, it has recently been reported that a peptide corresponding to residues 41–226 did not adsorb to CaM-Sepharose (Coggins and Zwiers, 1990). In order to determine the physiological role for the calmodulin binding properties of neuromodulin, it will prove important to prepare mutants which lack these biochemical properties. In the present study, we have used site directed mutagenesis to address this issue.

a. NH$_2$---**Q A S F R G H I T R K K L K G E K K**---COOH
      39   56

b.   **Q A S W R G H I T R K K L K G E K**   FP57-Trp

c.

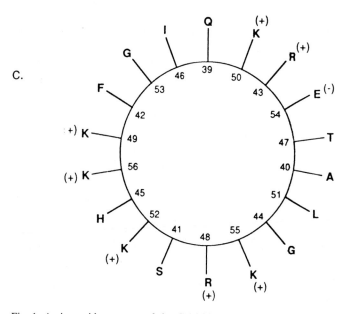

Fig. 1. Amino acid sequence of the CaM binding domain of neuromodulin. (a) CaM binding domain of neuromodulin showing sequence of amino acids deleted in NM(ΔCaM); (b) amino acid sequence of FP57-Trp (Alexander et al., 1988); (c) amino acid sequence of the CaM binding domain of neuromodulin arranged in a helical wheel.

Methods

Protein preparations

Bovine calmodulin was purified according to the method of Masure et al. (1986). CaM-Sepharose was prepared by the method of Westcott et al. (1979). Protein concentrations were determined by the method of Lowry et al. (1951).

Site directed mutagenesis and construction of in vitro transcription vectors

The cDNA sequence encoding wild type neuromodulin was excised from the bacterial expression vector pKK-NEU1 (Au et al., 1989) with *Sph*I and *Bam*HI. Because the *Sph*I restriction site flanks the start codon, the initial methionine residue was lost. An intact 5' end was regenerated by subcloning the *Sph*I-*Bam*HI fragment into the *Sph*I-*Bam*HI sites of pUC18 (Pharmacia). A 750 bp fragment containing the entire wild type neuromodulin coding sequence was subsequently excised with *Hin*dIII and *Eco*RI and inserted into the *Hin*dIII and *Eco*RI sites of M13mp18. Oligonucleotide directed site-specific mutagenesis (Kunkel et al., 1987) was carried out using the Bio-Rad Muta-Gene in vitro mutagenesis kit according to the supplier's instructions. The CaM binding domain, corresponding to amino acid residues 39–56 (Fig. 1a; Alexander et al., 1988), was deleted by priming with the oligonucleotide 5'-GCTGCGACCAAAATTGGTGATGCACC-AGCT-3'. This mutation was referred to as NM(ΔCaM). Serine-41 was substituted with an aspartate or asparagine residue by priming with the oligonucleotides 5'-TTCAGGCTGACTTC-CGTG-3' or 5'-TCAGGCTAACTTCCGTG-3' respectively. Mutants were identified by direct nucleotide sequencing. The mutant neuromodulin cDNAs were excised from the replicative form of phage M13mp18 with *Hin*dIII and *Eco*RI. These fragments, as well as a 1.2 kb *Eco*RI fragment encoding wild type neuromodulin (Cimler et al., 1985), were blunt-ended with Klenow fragment and subcloned into the *Bgl*II site of pSP64T, an SP6-dependent in vitro transcription vector (Kreig and Melton, 1984).

In vitro transcription and translation

The transcription vector, pSP64T containing cDNA inserts for wild-type, CaM binding domain deletion (NM(ΔCaM)), aspartate-41, and asparagine-41 mutant neuromodulins, was linearized 3' of the coding sequence with *Eco*RI and transcribed with SP6 RNA polymerase. Protein was translated using rabbit reticulocyte lysate in the presence of [^{35}S]methionine.

CaM-Sepharose affinity chromatography

The most convenient assay to monitor binding of neuromodulin or phosphoneuromodulins to CaM is to monitor adsorption of the proteins to CaM-Sepharose in the absence of free Ca^{2+} (Alexander et al., 1987). CaM-Sepharose columns (0.5 ml) were equilibrated with 50 mM Tris, pH 7.5, 1 mM DTT and 2 mM EDTA (buffer A). Samples (50 μl) containing [^{35}S]methionine-labeled neuromodulins from the in vitro translation reactions were applied to the CaM-Sepharose column. The resin was then rinsed sequentially with 20 ml buffer A and 20 ml buffer A containing 8 mM CaCl$_2$. One ml column fractions were collected and vacuum evaporated down to 50 μl. To each sample, an equal volume of SDS-PAGE sample buffer (0.2 M Tris-HCl, pH 6.9, 10 mM EGTA, 10% glycerol, 5% DTT, 5% SDS, and 0.2% bromophenol blue) was added. The samples were then boiled and subjected to SDS-PAGE on 10% polyacrylamide gels. As a standard, 5 μg of purified recombinant neuromodulin was also loaded onto the gels. Gels were stained with Coomassie blue to visualize the neuromodulin protein standard, and autoradiographed to visualize the ^{35}S-labeled translation reaction products.

40

Results

Definition of the CaM binding domain of neuromodulin

The amino acid sequences of at least 14 CaM binding proteins have been reported and their CaM binding domains identified (reviewed in O'Neil and DeGrado, 1990). In most cases examined thus far, CaM binding domains of these proteins are determined by linear sequences of amino acids of approximately 20 residues in length and synthetic peptides corresponding to these sequences bind to CaM with high affinity. Many of these sequences, but not all, conform to the Baa-helical (basic amphiphilic α-helix) model with hydrophobic residues clustered on one side of the helix and positively charged amino acids on the opposing face (O'Neil and DeGrado, 1990). Arrangement of the sequence of the putative CaM binding domain of neuromodulin in a classical amphiphilic helical wheel shows some clustering of positive charges but the overall motif does not conform strongly to the Baa-helical model (Fig. 1c). However, this region does contain the greatest concentration of positively charged amino acids in the sequence of a protein that is very acidic, with a high level of negative charge. In addition, Chou-Fasman analysis (1978) of the neuromodulin sequence predicts no α-helical structure in the first 172 amino acid residues.

The interaction of neuromodulin with CaM is different from most other CaM binding proteins in several respects. Under conditions of low ionic strength, neuromodulin has higher affinity for CaM in the absence of Ca^{2+} than in its presence (Andreasen et al., 1981, 1983; Alexander et al., 1987). Furthermore, the affinity of neuromodulin for CaM both in the presence and absence of Ca^{2+} is considerably lower than other CaM binding proteins (Alexander et al., 1987). We have reported that the amino acid sequence 43–51 is critical for CaM binding and that a synthetic peptide comprised of the sequence 39–55, with phenylalanine-42 replaced by a tryptophan residue (Fig. 1b), bound to CaM and inhibited CaM/neuromodulin interactions (Alexander et al., 1988).

It has been reported that a polypeptide derived from neuromodulin corresponding to the amino acid sequence 41–226 does not bind to CaM (Coggins and Zwiers, 1990). Because this sequence contains most of the putative CaM binding domain, it was of interest to determine if the sequence from 39–55 was requisite for CaM binding. We addressed this issue by preparing a deletion mutant of neuromodulin, designated NM(ΔCaM), which lacked the amino acid residues from 39–56 (Fig. 1a). Wild-type neuromodulin and NM(ΔCaM) were synthesized in an in vitro translation system and the [35S]methionine-labeled proteins were examined for binding to CaM-Sepharose (Fig. 2). The wild-type neuromodulin adsorbed to CaM-Sepharose in the presence of excess EDTA and was eluted from CaM-Sepharose with calcium (Fig. 2A), consistent with previous observations (Andreasen et al., 1983; Alexander et al., 1988; Au et al., 1989). In contrast, NM(ΔCaM) did not adsorb to CaM-Sepharose in the presence or absence of Ca^{2+} (Fig. 2B). These data establish that the amino acid sequence 39–56 contains residues required for binding to CaM and are consistent with the assignment of this sequence as the CaM binding domain of neuromodulin.

Aspartate-41 neuromodulin does not bind to CaM-Sepharose

The location of the protein kinase C phosphorylation site, serine-41 (Coggins and Zwiers, 1989; Apel et al., 1990; Chapman et al., 1991), within the CaM binding domain of neuromodulin very likely explains why protein kinase C phosphorylation of neuromodulin lowers its affinity for CaM. Presumably, introduction of one or more negative charges at this site is sufficient to perturb interactions with CaM, particularly hydrophobic interactions between phenylalanine-42 and CaM (Chap-man et al., 1991). Recently, it was demonstrated that the functional effects associated with autophosphorylation of CaM-dependent protein ki-

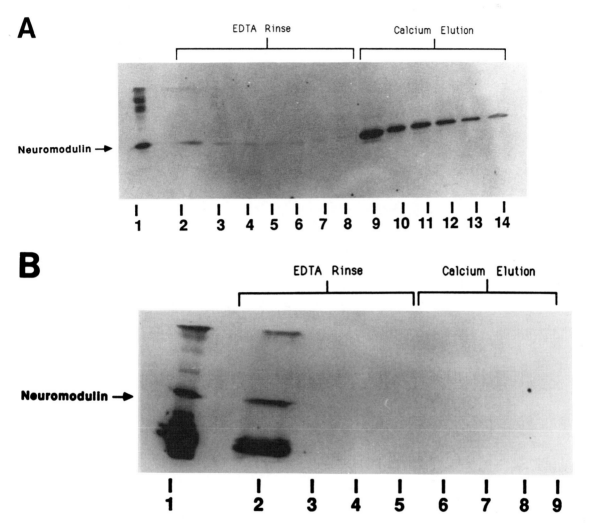

Fig. 2. Interactions of wild type recombinant neuromodulin and NM(ΔCaM) with CaM-Sepharose. Wild-type neuromodulin and the CaM binding domain deletion mutant, NM(ΔCaM), were synthesized in an in vitro translation system as described in Methods. CaM-Sepharose columns (0.5 ml) were equilibrated with a buffer containing 50 mM Tris, pH 7.5, 1 mM DTT, and 2 mM EDTA (buffer A). Fifty μl samples containing [^{35}S]methionine-labeled neuromodulins from the in vitro translation reaction mixture were applied to the CaM-Sepharose column. The resin was then rinsed sequentially with 20 ml buffer A and 20 ml buffer A containing 8 mM CaCl$_2$. One ml fractions were collected and vacuum evaporated down to 50 μl. The samples were subjected to SDS-PAGE on 10% polyacrylamide gels. Five μg of purified recombinant neuromodulin, as a standard, was also loaded onto the gel and visualized by staining with Coomassie blue. The position of the neuromodulin standard is denoted by the arrow. The translation products were visualized by autoradiography. (A) Wild-type neuromodulin: lane 1, 1 μl reaction mixture pre-load; lanes 2–8, every other fraction of EDTA rinse; lanes 9–14, every other fraction from calcium elution. (B) NM(ΔCaM): lane 1, 1 μl reaction mixture pre-load; lanes 2–5, fractions 2–5 of the EDTA rinse; lanes 6–9, fractions 1–4 from the calcium elution.

nase II can be mimicked by mutation of the threonine autophosphorylation site to aspartate (Fong et al., 1989). These data suggested that the major consequence of the autophosphorylation of this enzyme is to introduce a negative charge at a specific site in the protein. In order to determine if the introduction of a single negative charge is sufficient to block neuromodulin/CaM interactions, mutant neuromodulins containing an aspartate or asparagine residue at position 41 were prepared by site directed mutagenesis. The interaction of the ^{35}S-labeled mutant proteins with

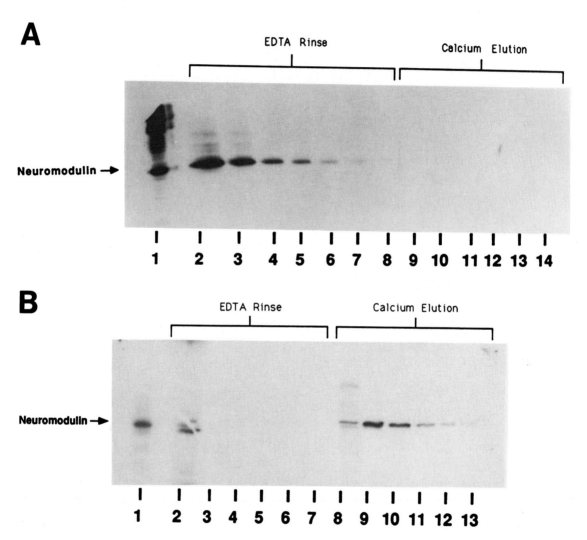

Fig. 3. Interactions of aspartate and asparagine-41 mutant neuromodulins with CaM-Sepharose. Mutant neuromodulins, in which serine-41 was substituted with an aspartate or asparagine residue, were synthesized in an in vitro translation system as described in Methods and assayed for binding to CaM-Sepharose as described in Fig. 2. (A) Aspartate-41 mutant neuromodulin: lane 1, 1 μl reaction mixture pre-load; lanes 2–8, every other fraction of the EDTA rinse; lanes 9–14, every other fraction from the calcium elution. (B), asparagine-41 mutant neuromodulin: lane 1, 1 μl reaction mixture pre-load; lanes 2–7, every other fraction of the EDTA rinse; lanes 8–13, every other fraction from the calcium elution.

calmodulin was examined using CaM-Sepharose affinity chromatography. In contrast to the asparagine mutant (Fig. 3B), the aspartate-41 neuromodulin did not adsorb to CaM-Sepharose (Fig. 3A). These data indicate that the loss in CaM binding accompanying phosphorylation of neuromodulin can be explained by the introduction of one or more negative charges near or within the CaM binding domain of neuromodulin.

Discussion

Two interesting biochemical properties which may be important for the physiological function of neuromodulin are its affinity for CaM and the effect of protein kinase C phosphorylation on CaM interactions. The concentration of neuromodulin in brain and the affinity of the protein for CaM in the absence of free calcium, are sufficient to complex the majority of CaM present (Cimler et al., 1985). Consequently, interactions between neuromodulin and CaM as well as the regulation of CaM binding by protein kinase C phosphorylation are of considerable interest.

In a previous study, we screened a series of proteolytic fragments of neuromodulin for CaM-Sepharose binding activity and the amino acid sequence from residue 43 to 51 was implicated for CaM binding (Alexander et al., 1988). A heptadecapeptide, FP57-Trp (Fig. 1b), containing this sequence with a tryptophan residue replacing the native phenylalanine, bound to CaM with high affinity. Antibodies specific to this peptide blocked binding of neuromodulin to CaM. It was recently reported, however, that a larger peptide containing amino acid residues from 41–226 did not bind to CaM-Sepharose (Coggins and Zwiers, 1990). In this study, we prepared a mutant neuromodulin lacking the putative CaM binding domain by deleting amino acid residues 39–56 (Fig. 1a). This deletion mutant, designated NM(ΔCaM), did not adsorb to CaM-Sepharose, con-

sistent with the assignment of residues 39–56 as the CaM binding domain of neuromodulin. Alternatively, the decreased affinity of NM(ΔCaM) for CaM may be secondary to higher ordered structural effects on the protein. This may not be a major concern because neuromodulin has little periodic secondary or tertiary structure in solution (Masure et al., 1986). Furthermore we have previously demonstrated that phenylalanine-42 of neuromodulin interacts directly with calmodulin (Chapman et al., 1991). These data, in conjunction with the original peptide screening data described above (Alexander et. al., 1988), strongly indicate that the amino acid sequence 39–56 contains residues critical for CaM binding.

The protein kinase C phosphorylation site of neuromodulin is serine-41 (Coggins and Zwiers, 1989; Apel et al., 1990; Chapman et al., 1991). This phosphorylation significantly reduces the affinity of neuromodulin for CaM (Alexander et al., 1987; Apel et al. 1990), suggesting that the introduction of negative charge may abolish CaM binding. We addressed this issue by substituting serine-41 with an aspartate or an asparagine residue. The aspartate-41 mutant neuromodulin did not bind to CaM-Sepharose. In contrast, the asparagine-41 mutant bound to CaM-Sepharose in a manner indistinguishable from the wild type protein. These data demonstrate that the introduction of negative charge into the CaM binding domain of neuromodulin at amino acid position 41, either by phosphorylation of the serine or substitution with an aspartate residue, is sufficient to block CaM binding.

The potential functional significance of the amino terminus of neuromodulin is underscored by its precise conservation in all vertebrate neuromodulins sequenced to date (LaBate et al., 1989). We have shown that the deletion of a portion of this conserved region, residues 39–56 results in the loss of calmodulin binding and thereby assign a functional role to this domain.

Furthermore, a more refined mutation, the substitution of serine-41 with an aspartate, may prove to be an especially useful tool for evaluating the importance of CaM binding for the physiological functions of neuromodulin because it completely abolishes CaM binding with a minimal structural change.

Acknowledgements

This work was supported, in part, by National Institutes of Health grants GM-33708 and HL-23606. E.R.C. was supported by National Institutes of Health Predoctoral Training grant GM-07270.

References

Akers, R.F. and Routtenberg, A. (1985) *Brain Res.*, 344, 147–151.

Alexander, K.A., Cimler, B.M., Meier, K.E. and Storm, D.R. (1987) *J. Biol. Chem.*, 262: 6108–6113.

Alexander, K.A., Wakim, B.T., Doyle, G.S., Walsh, K.A. and Storm, D.R. (1988) *J. Biol. Chem.*, 263: 7544–7549.

Andreasen, T.J., Keller, C.H., LaPorte, D.C., Edelman, A.M. and Storm, D.R. (1981) *Proc. Natl. Acad. Sci. USA*, 78: 2782–2785.

Andreasen, T.J., Leutje, C.W., Heideman, W. and Storm, D.R. (1983) *Biochemistry*, 22: 4615–4618.

Au, D., Ape, E.D., Chapman, E.R., Estep, R.P., Nicolson, T.A., and Storm, D.R. (1989) *Biochemistry*, 28: 8142–8148.

Apel, E.D., Byford, M.F., Au, D., Walsh, K.A. and Storm, D.R. (1990) *Biochemistry*, 29: 2330–2335.

Benowitz, L.I., and Routtenberg, A. (1987) *Trends Neurosci.*, 12: 527–532.

Chapman, E.R., Au, D., Alexander, K., Nicolson, T.A. and Storm, D.R. (1991) *J. Biol. Chem.*, 266: 207–213.

Chou, P.Y. and Fasman, G.D. (1978) *Adv. Enzymol.*, 47: 45–147.

Cimler, B.M., Andreasen, T.J., Andreasen, K.I. and Storm, D.R. (1985) *J. Biol. Chem.*, 262: 10784–10788.

Coggins, P.J. and Zwiers, H. (1989) *J. Neurochem.*, 53: 1895–1901.

Coggins, P.J. and Zwiers, H. (1990) *J. Neurochem.*, 54: 274–277.

Dekker, L.V., De Graan, P.N.E., Oestreicher, A.B., Versteeg, D.H.G. and Gispen, W.H. (1989) *Nature*, 342: 74–76.

Enyedi, A., Vorherr, T., James, P., McCormick, D.J., Filoteo, A.G., Carafoli, E. and Penniston, J.T. (1989) *J. Biol. Chem.*, 264: 12313–12321.

Fong, Y.L., Taylor, W.L., Means, A.R., and Soderling, T.R. (1989) *J. Biol. Chem.*, 264: 16759–16763.

Kreig, P.A. and Melton, D.A. (1984) *Nucleic Acids Res.*, 12: 7057–7070.

Kunkel, T.A., Roberts, J.D. and Zakour, R.A. (1987) *Methods Enzymol.*, 154: 367–382.

LaBate, M.E. and Skene, J.H.P. (1989) *Neuron*, 3: 299–310.

Liu, Y. and Storm, D.R. (1989) *J. Biol Chem.*, 264: 12800–12804.

Liu, Y.L. and Storm, D.R. (1990) *Trends Pharmacol. Sci.*, 11: 107–111.

Lovinger, D.M., Colley, P.A., Akers, R.F., Nelson R.B. and Routtenberg, A. (1986) *Brain Res.*, 399: 205–211.

Lowry, O.H., Rosebrough, N.J., Farr, A.L. and Randall, R.J. (1951) *J. Biol. Chem.*, 251: 265–275.

Masure, M.R., Alexander, K.A., Wakim, B.T. and Storm, D.R. (1986) *Biochemistry*, 25: 7553–7560.

Nelson, R.B., Linden, D.J., Hyman, C., Pfenninger, K.M. and Routtenberg, A. (1989a) *J. Neurosci.*, 9: 381–389.

Nelson, R.B., Linden, D.J. and Routtenberg, A. (1989b) *Brain Res.*, 497: 30–42.

Olwin, B.B., Titani, K., Martins, T.J. and Storm, D.R. (1983) *J. Biol. Chem.*, 22: 5390–5395.

O'Neil, K.T., Wolfe, Jr., M.R., Erickson-Viitanen, S.E. and DeGrado, W.F. (1987) *Science*, 236: 1454–1456.

O'Neil, K.T. and DeGrado, W.F. (1990) *Trends Biol. Sci.*, 15: 59–64.

Routtenberg, A. and Lovinger, D.M. (1985) *Behav. Neural Biol.*, 43: 3–11.

Skene, J.M.P. (1989). Axonal growth-associated proteins. *Annu. Rev. Neurosci.*, 12: 127–156.

Van Dongen, C.J., Zwiers, M., De Graan, P.N.E. and Gispen, W.M. (1985) *Biochem. Biophys. Res. Commun.*, 128: 1219–1227.

Van Hooff, C.O.M., DeGraan, P.N.E., Oestreicher, A.B. and Gispen, W.M. (1988) *J. Neurosci.*, 8: 1789–1795.

Westcott, K.R., LaPorte, D.C. and Storm, D.R. (1979) *Proc. Natl. Acad. Sci. USA*, 76: 204–208.

Willard, M., Meiri, K.F. and Johnson, M.I. (1987) In R.S. Smith and M.A. Bishop (Eds.), Axonal Transport, pp. 407–420, Alan R. Liss, New York.

Regulation of Expression of
B-50/F1/GAP-43/Neuromodulin

W.H. Gispen and A. Routtenberg (Eds.)
Progress in Brain Research, Vol. 89
© 1991 Elsevier Science Publishers B.V.

CHAPTER 5

Expression of the growth- and plasticity-associated neuronal protein, GAP-43, in PC12 pheochromocytoma cells

Brian Costello, Li-Hsien Lin, Afshin Meymandi, Susan Bock, Jeanette J. Norden and John A. Freeman

Department of Cell Biology, Vanderbilt Medical School, Nashville, TN 37232, U.S.A.

Introduction

GAP-43 is a neuronal protein which exhibits elevated synthesis and axonal fast-transport during nerve growth (reviewed in Freeman et al., 1986; Snipes et al., 1987b; Benowitz and Routtenberg, 1987; Skene, 1989; Norden et al., 1991a). This correlation of increased GAP-43 expression with neuronal growth has been observed in the CNS of fish, amphibians and neonatal mammals (Benowitz et al., 1981; Skene and Willard, 1981a; Skene and Kalil, 1984; Freeman et al., 1986). Increases relative to normal mature nerves also are seen in mammals in the developing CNS and in the regenerating PNS (Skene and Willard, 1981b; Freeman et al., 1986). Likewise, culture conditions which promote neurite outgrowth cause increased GAP-43 expression by neurons in vitro (Perrone-Bizzozero et al., 1986). Injury to

the CNS of higher vertebrates, however, where successful regeneration does not occur, fails to increase GAP-43 expression (Skene and Willard, 1981b).

Temporal and spatial changes in GAP-43 content and localization in brain also are positively correlated with neuronal growth and synaptic plasticity. The levels and rates of synthesis of GAP-43 in rat cerebellum and cerebral cortex are about 10-fold greater in neonates compared with adults (Jacobson et al., 1986). GAP-43 is localized to growing neuronal processes during development, and in the mature brain is enriched in neuropil areas, especially those exhibiting synaptic plasticity (Neve et al., 1987; McGuire et al., 1988; Benowitz et al., 1988, 1989). Furthermore, the cell bodies of neurons synapsing in such areas contain elevated levels of GAP-43 messenger RNA (Rosenthal et al., 1987; Neve et al., 1987; De la Monte et al., 1989). This observation, together with the greater cell free translation of GAP-43 from mRNA isolated from neonatal compared to adult cortex (Jacobson et al., 1986)

Correspondence: Jeanette J. Norden, Department of Cell Biology, Vanderbilt Medical School, Nashville, TN 37232, U.S.A.

suggests that growth- and plasticity-associated changes in GAP-43 synthesis rate and steady state levels are mediated via changes in its message levels.

Identification of GAP-43 with the independently studied neuronal phosphoproteins F1, B-50 and pp46 (Snipes et al., 1987a; Karns et al., 1987; Rosenthal et al., 1987; Nielander et al., 1987; Basi et al., 1987) suggests that its function in synaptic plasticity may be dependent on posttranslational modification as well as its level of expression. For example, induction or maintenance of long-term potentiation in the hippocampus is correlated with increased phosphorylation of GAP-43 (protein F1; Routtenberg et al., 1986; Malenka et al., 1986). This change in GAP-43 phosphorylation suggests that it may play a direct role in the processes underlying synaptic plasticity. Hence, high levels of GAP-43 expression in certain areas of the adult brain may be a prerequisite for synaptic plasticity which is modulated in a rapid, activity-dependent fashion by altering GAP-43 phosphorylation. The observation that phosphorylation of GAP-43 (protein B-50) by protein kinase C also is involved in plasma membrane polyphosphoinositide metabolism (Jolles et al., 1980; Van Hooff et al., 1988) is intriguing, but the functional significance of this effect remains unclear. However, it has been demonstrated that GAP-43 antibodies which interfere with B-50 phosphorylation also inhibit Ca^{2+}-induced neurotransmitter release from synaptosomes (Dekker et al., 1989). GAP-43 also is identical to p-57 or neuromodulin (Wakim et al., 1987), a calmodulin-binding protein whose activity is manifest most strongly under non-physiological conditions of ionic strength (Alexander et al., 1987). Very recently it has been shown that GAP-43 stimulates GTP binding to Go, a member of the signal-transducing G-protein family and a major growth cone component (Strittmatter et al., 1990). It appears that this effect may be dependent on

the degree of GAP-43 palmitylation at its two N-terminal cysteines. These observations suggest that, by modulating Go activity, GAP-43 might be influencing voltage-dependent Ca^{2+} channels (Hescheler et al., 1987; Ewald et al., 1988) and thereby affecting neurotransmitter release (reviewed in Norden et al., 1991b). Finally, recent transfection experiments in non-neuronal cells and in nerve growth factor (NGF) treated PC12 cells have demonstrated an effect of GAP-43 on the formation of filopodia and on neurite outgrowth (Zuber et al., 1989a; Yankner et al., 1990). Involvement of GAP-43 in both process formation and outgrowth, and in neurotransmitter release, suggests a regulatory role in vesicle fusion and membrane addition at presynaptic membranes and in growth cones.

Although the correlation of elevated GAP-43 synthesis and posttranslational modification with neuronal growth states and synaptic plasticity is well established, the specific factors which control its expression are poorly understood. In recent work from our laboratory and others, the PC12 line of pheochromocytoma cells has been used to study GAP-43 regulation. These cells are very useful for such studies since NGF stimulation causes them to cease dividing and undergo a program of neuronal differentiation which includes neurite outgrowth (Greene and Tischler, 1976). NGF also causes a concomitant increase in GAP-43 expression (Costello et al., 1986, 1987; Van Hooff et al., 1986). Furthermore, the localization of GAP-43 to growth cones in NGF-treated PC12 cells (Van Hooff et al., 1986; Costello et al., 1987) is very similar to that seen in cultured neurons (Meiri et al., 1986, 1988) and suggests that GAP-43 is serving the same growth-associated function in both cases. These findings, together with the obvious advantages of using a cell line, make PC12 cells an excellent model for studying the molecular mechanisms underlying GAP-43 expression.

In this report we have summarized our observations on NGF induction of GAP-43 in PC12 cells (Costello et al., 1986; Costello and Freeman, 1988; Costello et al., 1990). We have characterized this NGF response in terms of its time course, dose–response relationship, sensitivity to variations in culture conditions which affect neurite outgrowth, dependence on protein synthesis, and sensitivity to methyltransferase inhibitors and glucocorticoids. Many of these parameters also were examined for a number of other agents that were found to increase GAP-43 expression. Finally, our results allow us to draw some conclusions regarding the mechanisms underlying induction of GAP-43.

Results

PC12 cells exhibit a wide range of pleiomorphic responses to NGF stimulation (Greene and Tischler, 1982). These include alterations in gene expression, biochemical makeup and morphology, which effect a conversion from a contiuously dividing cell resembling precursors of the adrenal medullary lineage, to a non-dividing, relatively differentiated cell type resembling sympathetic neurons. This differentiated phenotype includes extensive neurite outgrowth (Fig. 1).

Since GAP-43 expression had been shown to be correlated with neurite outgrowth in the nervous system (Skene, 1989), we looked to see

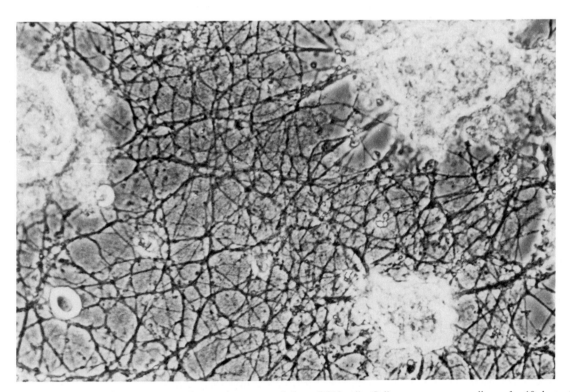

Fig. 1. NGF stimulation causes extensive neurite outgrowth from PC12 cells. Cells were grown on collagen for 10 days with 100 ng/ml 7S NGF. At this time all cells have stopped dividing and an extensive meshwork of neurites has been established. Three large clumps of cell bodies, all extending neurites, are visible in this field. Magnification = 710 × .

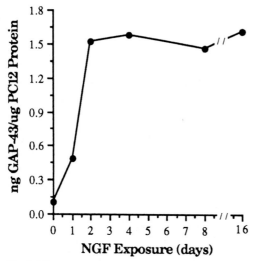

Fig. 2. Time course of GAP-43 induction in PC12 cells. Cells were grown for varying periods in the presence of 100 ng/ml 7S NGF. Whole cell samples were prepared (Laemmli, 1970) and analyzed by Western blot using GAP-43 antiserum and ^{125}I-labeled second antibody. Autoradiographs were analyzed by quantitative densitometry. A calibration curve produced in similar fashion using varying amounts of purified GAP-43 on Western blots (Snipes et al., 1987a) was used to normalize the data.

whether GAP-43 expression also coincided with neurite growth in PC12 cells (Costello et al., 1986). Cells were stimulated with NGF for varying periods and GAP-43 levels were determined by quantitative densitometry of autoradiographs. A typical analysis is shown in Fig. 2. The GAP-43 level increases about 16-fold after 2 days exposure to NGF and remains elevated in the continued presence of NGF. This increase, which is qualitatively similar to that observed by others (Van Hooff et al., 1986), coincides with the initia-

tion of neurite outgrowth (Greene and Tischler, 1976).

GAP-43 has been characterized as a fast axonally-transported protein in neurons (Skene and Willard, 1981a). In order to determine whether GAP-43 in PC12 cells is also primarily localized to neurites, NGF-treated cells were stained immunohistochemically for GAP-43 (Fig. 3). Specific immunoreactivity is present throughout the neurites but is not evident in cell bodies. Recent immunogold labeling experiments have shown that the very low levels of GAP-43 (B50) in unstimulated PC12 cells is mostly associated with lysosomal structures and Golgi apparatus (Van Hooff et al., 1989). In stimulated cells, the much higher GAP-43 content is localized mostly to the plasma membrane, especially around processes, and to membranous structures in neurites. To determine whether the GAP-43 in NGF-stimulated cells is in fact membrane-bound as originally reported for toad nerves (Skene and Willard, 1981a), we isolated both particulate fractions (low- and high-speed pellets; LSP, HSP) and a soluble fraction (high-speed supernatant; HSS), and analyzed them for GAP-43 by Western blot (Fig. 4). Although GAP-43 is present in the membranous fractions (LSP, HSP) it also is present at high levels in the supernatant fraction (HSS). This distribution is similar to that seen in rat sciatic nerve (unpublished results, this laboratory), goldfish optic tectum (Benowitz and Lewis, 1983) and bovine brain (Cimler et al., 1985), but is in marked contrast to the partitioning observed in toad nerves (Skene and Willard, 1981a) or in neonatal rat brain (Fig. 4) where very little GAP-43 is

→

Fig. 3. GAP-43 is localized to neurites in NGF stimulated PC12 cells. Cells were grown for 7 days in the presence of 100 ng/ml 7S NGF. After fixation with methanol, they were treated with either (A) GAP-43 antiserum which had been preblocked with purified GAP-43 or (B) GAP-43 antiserum, followed by peroxidase-coupled anti-IgG. The peroxidase reaction product is specifically localized to neurites. Magnification = 723 × .

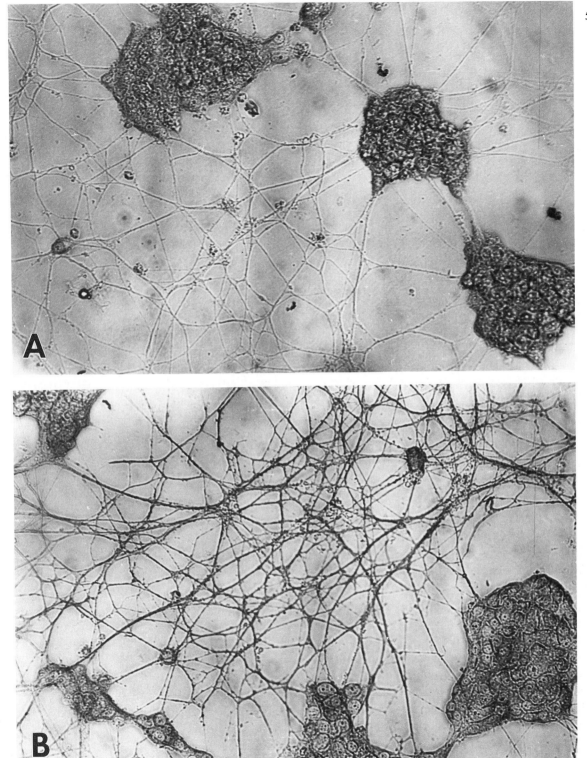

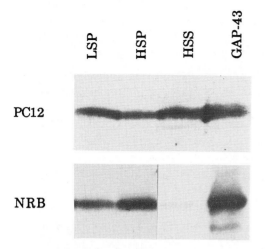

Fig. 4. GAP-43 is only partially associated with membranes in PC12 cells. Cells were stimulated with 100 ng/ml 7S NGF for 4 days, homogenized and fractionated using differential centrifugation. Three fractions were produced: a low speed pellet (10 000 $g \times$ 10 min; LSP), a high speed pellet (100 000 $g \times 60$ min; HSP) and a high speed supernatant (100 000 $g \times 60$ min; HSS). These three fractions, along with purified GAP-43, were analyzed by Western blot using GAP-43 antiserum. GAP-43 immunoreactivity is present in both the particulate (LSP and HSP) and soluble (HSS) fractions. Also shown is an analysis of the same fractions derived from neonatal rat brain (NRB). In this case, almost all GAP-43 is associated with the particulate fractions.

present in soluble form. Membrane association of GAP-43 is at least partly mediated via fatty acylation at the two cysteine residues within its short, amino-terminus hydrophobic region (Skene and Virag, 1989; Zuber et al., 1989b). Perhaps this post-translational modification is less complete in NGF-stimulated PC12 cells than in some types of nerve cells.

The basis for the increased GAP-43 levels in PC12 cells was investigated by measuring synthesis rates. Control and NGF-stimulated cells were pulse-labeled with [35S]-methionine and incorporation of label was determined by two-dimensional gel electrophoresis followed by fluorography and quantitative densitometry (Fig. 5). NGF stimulated cells showed a 19-fold increase (± 1.8,

$n = 3$) in incorporation of methionine into GAP-43. This result is consistent with the observed increases in synthesis and fast axonal transport of GAP-43 in developing optic nerves and in regenerating peripheral nerves in rabbits (Skene and Willard, 1981b). Furthermore, the magnitude of this NGF-induced labeling increase suggests that the increased steady-state GAP-43 levels in NGF-treated PC12 cells result from an increased synthesis rate.

In addition to NGF, N^6-2'-O-dibutyryladenosine 3',5'-cyclic monophosphate (dBcAMP) and fibroblast growth factor (FGF) also have effects on process outgrowth in PC12 cells. dBcAMP does not induce true neurite outgrowth, but rather transcription-independent spiking which is believed to result from a rapid and unstable reorganization of cytoskeletal elements (Gunning et al., 1981). In contrast, both acidic and basic FGF mimic all effects of NGF on PC12 cells (Togari et al., 1983; Rydell and Greene, 1987; Neufeld et al., 1987), although these effects have been reported to be transient (Togari et al., 1985). We tested the effect on GAP-43 levels of these agents, and of two others, epidermal growth factor (EGF) and the phorbol ester, 12-O-tetradecanoylphorbol-13-acetate (TPA), which have been shown to influence gene expression in PC12 cells (Greenberg et al., 1985) (Fig. 6). The very small induction by EGF is consistent with previous observations that EGF mimics some, but not all of the effects of NGF on PC12 cells (Seeley et al., 1984). In particular, EGF does not elicit neurite outgrowth. TPA also has very little if any effect on neurite outgrowth in PC12 cells (Burstein et al., 1982; Reinhold and Neet, 1989; but see Hall et al., 1988) and displayed no induction of GAP-43. However, this apparent lack of TPA effect may be due to down-regulation of protein kinase C (PKC) with chronic phorbol ester exposure (Matthies et al., 1987). The inductions by dBcAMP, FGF and NGF are generally proportional to their effects on process outgrowth, with NGF being the most efficacious.

In order to determine the basis for the increased GAP-43 synthesis rate in NGF-treated PC12 cells and to study the mechanisms controlling GAP-43 expression, we measured GAP-43 message levels in PC12 cells treated with NGF and various other effectors (Fig. 7A). Of the agents tested, only insulin and retinoic acid were entirely without effect, and only dexamethasone inhibited GAP-43 mRNA expression. EGF, dB-cAMP, FGF and NGF had effects on message levels that closely paralleled those on protein levels. In contrast, the effect of TPA on message was greater than its effect on protein levels, perhaps because of the different treatment periods (1 day for message, 2 days for protein) and the

rapid down-regulation of PKC by TPA. We also found that elevated KCl was somewhat effective in inducing GAP-43 message, presumably by raising $[Ca^{2+}]_i$ (Meldolesi et al., 1984). In these experiments, the positive effectors were all used at optimal concentrations as determined from dose–response curves. Saturable dose–response relationships were obtained for dBcAMP, FGF and NGF, with half-maximal responses at approximately 10^{-4} M, 4 ng/ml and 5 ng/ml, respectively. In contrast, the dose–response relationships for TPA and EGF showed optima at about 10^{-7} M and 1–3 ng/ml, respectively, with decreasing responses at higher doses. Again, the TPA response is no doubt attributable to more rapid PKC down regulation at higher doses. The dose–response relationship for EGF is very simi-

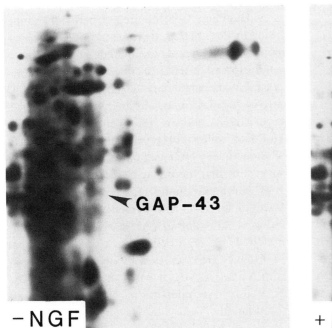
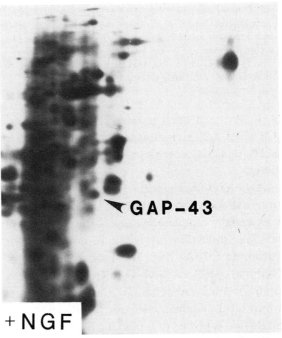

Fig. 5. NGF stimulation increases GAP-43 synthesis in PC12 cells. PC12 cells were grown for 3 days in the presence (100 ng/ml) or absence of 7S NGF. They were then pulse labeled for 1 h with [^{35}S]methionine and whole cell samples were prepared for IEF-SDS PAGE. Fluorographs (shown) were prepared from the gels and the radioactivity incorporated into GAP-43 was determined by quantitative densitometry. NGF-stimulated cells showed 19-fold (± 1.8, $n = 3$) more radioactivity in GAP-43, indicating a correspondingly increased synthesis rate.

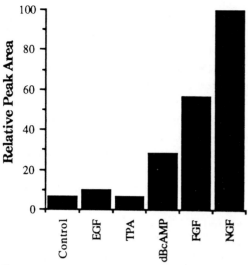

Fig. 6. GAP-43 levels in PC12 cells are sensitive to hormones and to agents that activate intracellular signaling pathways. PC12 cells were stimulated for two days with several agents that are known to influence gene expression. Whole cell samples were then prepared for Western blot analysis using GAP-43 antiserum and ^{125}I-coupled second antibody. Autoradiographs were scanned to obtain the relative peak areas which represent relative amounts of GAP-43 present in the samples. dBcAMP, FGF and NGF are especially effective in inducing GAP-43 expression, whereas EGF and TPA failed to induce GAP-43 expression above control levels.

lar to that seen when assaying substrate adhesion (Chandler and Herschman, 1980) and induction of ornithine decarboxylase activity (Guroff et al., 1981).

An interesting characteristic of the more effective GAP-43 inducers, dBcAMP, FGF and NGF, is essentially complete inhibition of their GAP-43 mRNA inductions by 5'-S-(2-methylpropyl) adenosine (SIBA). Fig. 7B shows the effect of NGF plus SIBA on GAP-43 mRNA levels. SIBA is an S-adenosylhomocysteine hydrolase inhibitor which blocks methyltransferase activity (Enouf et al., 1979). This class of compound has been shown to block all tested responses of PC12 cells to FGF and NGF, probably by interfering at some very early stage of signal transduction (Seeley et al., 1984; van Calker et al., 1989; Togari et al., 1985).

In contrast, EGF actions on PC12 cells are all either unaffected or enhanced by these compounds (Seeley et al., 1984; Landreth and Rieser, 1985; Acheson and Thoenen, 1987; van Calker et al., 1989). We found no effect of SIBA on EGF induction of GAP-43 and observed no consistent effect on the TPA response. The mechanism of SIBA inhibition of hormone action on PC12 cells remains unclear (Seeley et al., 1984), but may not involve blockage of methylation reactions (Acheson and Thoenen, 1987).

Dexamethasone was the only agent tested that inhibited GAP-43 mRNA expression. This was expected since PC12 cells resemble immature chromaffin cells and NGF and glucocorticoids have opposite and antagonistic effects on the differentiation of such cells into either sympathetic neurons or mature chromaffin cells, respectively (Doupe et al., 1985). This antagonism is evident (Fig. 8) in the greater percentage inhibition by dexamethasone of basal GAP-43 expression (75% decrease) compared to NGF-stimulated expression (30% decrease). Others have reported essentially complete inhibition of GAP-43 mRNA expression by dexamethasone (Federoff et al., 1988) but we have never observed this. However, we also observed much stronger GAP-43 induction by NGF compared to Federoff et al., suggesting that there may be clonal differences in NGF sensitivity of the PC12 cells we are using. GAP-43 levels in the presence of the weaker inducers, TPA and dBcAMP, also are powerfully inhibited by dexamethasone (-65%) whereas FGF levels are decreased by only about 30%, as with NGF. This suggests that dexamethasone and FGF also have antagonistic effects on PC12 differentiation.

Although induction of GAP-43 message by NGF coincides with initiation of neurite outgrowth on permissive substrates, its expression does not require neurite outgrowth. This was determined by stimulating cells grown on three different substrates that vary in their ability to support neurite outgrowth: glass (poor adhesion, no neurites), tissue culture plastic (good adhesion, poor neurites) and collagen-coated plastic (good adhesion, good neurites). We also grew cells at very high density (500 000/cm²), which inhibits neurite growth. None of these variations had any effect on GAP-43 message levels, with or without NGF stimulation (not shown). In contrast, the NGF-stimulated induction in PC12 cells of the neuronal gene SCG10, which displays a time course very similar to that of GAP-43, is powerfully inhibited at high cell density (Stein et al., 1988).

The only substrate which has shown an effect on GAP-43 message levels is native extracellular matrix (ECM). To show this, ECM-coated plates were produced by growing bovine corneal endothelial cells to confluence (Stocker et al., 1958; Gospodarowicz et al., 1977). These cells deposit a layer of ECM which remains attached to the culture dishes after removal of the endothelial cells by detergent lysis. When PC12 cells are grown on this surface, they extend neurites (Fujii et al., 1982; Vlodavsky et al., 1982) and their GAP-43 message levels increase without additional stimulation (Fig. 9). This induction of GAP-43 is likely the effect of FGF which is produced and secreted by the endothelial cells and remains bound to the ECM after their removal (Vlodavsky et al., 1987). In fact, this bound FGF has been shown to be responsible for neurite extension on ECM (Rogelj et al., 1989). We found that none of several artificial substrate

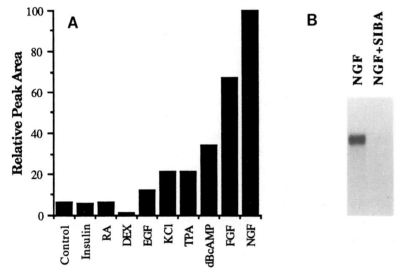

Fig. 7. GAP-43 message levels in PC12 cells are sensitive to hormones and to agents that activate intracellular signaling pathways. (A) PC12 cells were stimulated for 24 h prior to isolation of cytoplasmic RNA, which was analyzed by Northern blot using a GAP-43 cDNA probe. Message levels were measured by quantitative densitometry of autoradiographs and are expressed as relative peak areas of the GAP-43 band. Effector concentrations were as follows: insulin (5 μg/ml), retinoic acid (RA; 50 nM), dexamethasone (DEX; 1 μM), EGF (1 ng/ml), KCl (40 mM), TPA (0.1 μM), dBcAMP (1 mM), pituitary FGF (40 ng/ml) and NGF (100 ng/ml). For determination of the KCl effect, cells were grown in medium which was 75% normal growth medium + 25% 160 mM KCl, or 75% normal growth medium + 25% 160 mM NaCl as control. This control medium had no effect on basal GAP-43 expression. (B) Inhibition of NGF induction of GAP-43 message by simultaneous treatment with the methyltransferase inhibitor SIBA. One dish was pretreated for 30 min with 3 mM SIBA followed by exposure to 100 ng/ml NGF in the continued presence of SIBA. Another dish only received the NGF treatment. Cytoplasmic RNA was isolated and analyzed by Northern blot as in (A). The autoradiograph shows complete inhibition by SIBA of GAP-43 message induction by NGF.

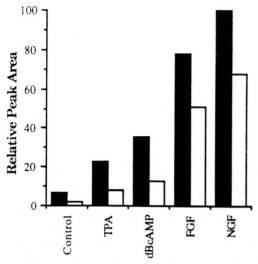

Fig. 8. Glucocorticoid inhibition of GAP-43 mRNA induction. Cells were grown for 24 h in medium with (open bars) or without (solid bars) 1 μM dexamethasone, followed by another 24 h in the continued presence or absence of dexamethasone plus 0.1 μM TPA, 1 mM dBcAMP, 37 ng/ml pituitary FGF, 100 ng/ml NGF, or nothing (control). GAP-43 message content was analyzed as in Fig. 7. Dexamethasone inhibited basal expression and induction by all of these effectors.

compounds or purified ECM components used to coat cell culture dishes (polylysine, polyornithine, types I or IV collagen, laminin, fibronectin, hyaluronic acid, chrondroitin sulfate A, heparan sulfate, heparin) had any effect on GAP-43 message levels. The apparent ability of ECM-bound FGF to induce expression of a neuronal growth associated protein is of great importance because it is now evident that FGF is very effective in promoting neurite outgrowth and survival in CNS neurons in culture (Morrison et al., 1986; Walicke et al., 1986).

Despite the growing realization that NGF and FGF are likely to be important for survival and growth of neurons in the CNS, their mechanisms of action still remain obscure after years of study. Their effects on PC12 cells may be broadly categorized as either transcriptionally-dependent or independent (Greene et al., 1980). To determine whether protein synthesis is necessary for induction of GAP-43 message by FGF and NGF, cells were stimulated in the presence of the translational inhibitor, cycloheximide (CHX). The effect of CHX on induction by TPA and dBcAMP was also investigated since the signaling pathways these agents affect (i.e., those involving protein kinases C and A, respectively) might be involved in mediating NGF and/or FGF actions. Although the CHX treatment blocked 95% of protein synthesis, none of the inductions was fully blocked (Fig. 10). In fact, the induction by dBcAMP was completely unaffected. This suggests that dBcAMP has its effects via post-translational modification, presumably PKA-mediated phosphorylation, of a preexisting protein. Hence, the CHX resistant components of the NGF and FGF inductions, the magnitudes of which are similar to the dBcAMP response, also could be mediated via PKA. In contrast, PKC alone is unlikely to be involved in mediating expression stimulated by FGF or NGF since its action is almost completely blocked by CHX.

To shed additional light on the involvement of PKA and PKC on NGF action, additivity experiments were performed (Fig. 11). The combined effect of TPA plus dBcAMP is superadditive and actually matches that of NGF, suggesting that their target signaling pathways could mediate NGF induction. However, NGF potentiates both TPA and dBcAMP actions and all three together are even more effective. These additive or superadditive effects suggest complicated, interdependent mechanisms of induction for these three effectors. Synergistic effects of NGF and cAMP also have been observed for induction of ornithine decarboxylase activity (Hatanaka et al., 1978; Guroff et al., 1981) and for elevation of cellular protein and RNA levels and extension of neurites in PC12 cells (Gunning et al., 1981). As a final test of PKC involvement in GAP-43 induction by NGF, we downregulated PKC via chronic TPA treatment (Matthies et al., 1987). This resulted in GAP-43 message levels that were indistinguishable from control and actually had a slight enhancement in NGF induction of GAP-43 (Fig. 12). These results argue strongly that neither the low basal expression nor the much greater NGF-stimulated expression are dependent in any way on PKC activity.

The elevated GAP-43 message levels in stimulated PC12 cells could arise via one or a combination of several mechanisms including increased transcription, decreased degradation of message, or increased processing and export of transcripts from the nucleus. To test for transcriptional acti-

vation, we performed nuclear runoff transcription assays (Marzluff, 1978; Greenberg and Ziff, 1984). In these experiments cells were treated with effectors for 6 h before isolation of nuclei. The 6-h treatment period was chosen because, for effective agents (i.e., see Fig. 7A), message levels are already elevated and increasing rapidly at that time and any effect on transcription rate should be correspondingly pronounced. For example, a typical time course for GAP-43 message induction by NGF is shown in Fig. 13A. As positive control in these experiments, we assayed increase in c-fos transcription after 20 min stimulation with NGF and, as reported previously (Greenberg et al., 1985), observed a very large increase

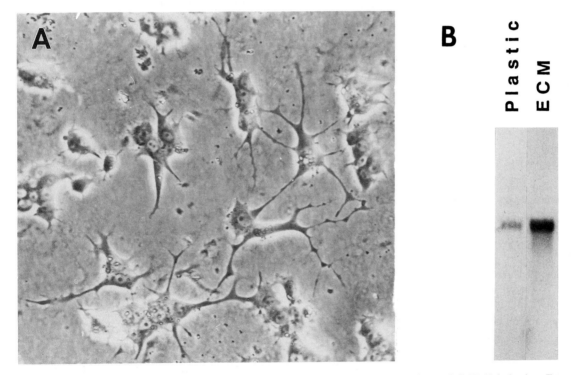

Fig. 9. Growth of PC12 cells on extracellular matrix causes morphological differentiation and GAP-43 induction. Extracellular matrix (ECM) coated dishes were produced by establishing cultures of bovine corneal endothelium (Gospodarowicz et al., 1977) which were grown to confluence in the presence of 20 ng/ml EGF. The endothelial cells were then removed by lysis with 0.5% Triton X-100 in PBS, followed by repeated PBS washes. PC12 cells were then plated on the ECM surface and were photographed 24 h later (A) and analyzed by Northern blot to determine GAP-43 message levels (B). Compared to tissue culture plastic, the ECM surface stimulates both neurite outgrowth and expression of GAP-43 message. Magnification in A = 832 ×.

58

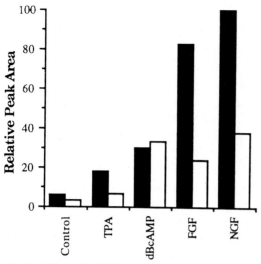

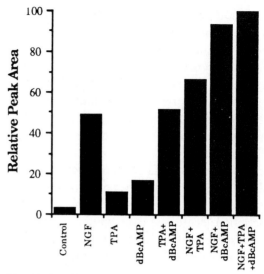

Fig. 10. Effect of inhibition of protein synthesis on GAP-43 message induction. Cells were grown for 2 h in medium with (open bars) or without (solid bars) 1 μg/ml cycloheximide, followed by 24 h in the continued presence or absence of cycloheximide plus inducer as indicated. Inducer concentrations and analysis of GAP-43 message levels were as described in Fig. 7. Inhibition of protein synthesis reduced basal GAP-43 message levels and induction by all of the effectors except dBcAMP.

Fig. 11. Combined effects of TPA, dBcAMP and NGF on GAP-43 message. Cells were treated for 24 h with various combinations of maximally effective concentrations of TPA (0.1 μM), dBcAMP (1 mM) and NGF (100 ng/ml). GAP-43 message levels then were analyzed as described for Fig. 7. Additive or superadditive effects were observed for all combinations of these effectors.

(Fig. 13B). However, none of the agents tested (NGF, basic FGF, dBcAMP, EGF, TPA or dexamethasone) had any effect on GAP-43 transcription rate (not shown). These results agree with those reported by Federoff et al. (1988) for NGF but differ with respect to dexamethasone. They observed very strong inhibition by dexamethasone of basal transcription and of NGF-induced message levels in PC12 cells. As discussed above we see only partial inhibition by dexamethasone of NGF-induced message increases. These reported differences in dexamethasone effect may reflect clonal variation in the PC12 cells used.

These results suggest that induction of GAP-43 message must result from alterations in transcript processing and/or in message stability. In preliminary studies we have measured message levels in

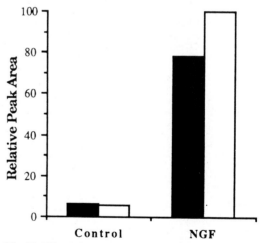

Fig. 12. Effect of down-regulation of PKC on NGF induction of GAP-43 message. Cells were treated for three days with either DMSO vehicle (solid bars) or TPA (open bars). TPA concentration was 1 μM for the first 2 days and 0.1 μM for the third day. During the third day, some cells were also treated with 100 ng/ml NGF as indicated. Down-regulation of protein kinase C had no effect on basal GAP-43 message levels and actually increased NGF stimulated expression.

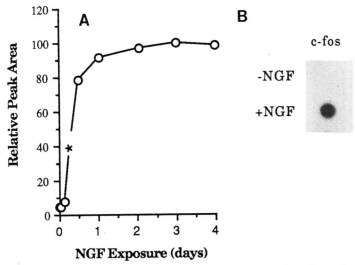

Fig. 13. Nuclear runoff transcription analysis in PC12 cells. PC12 cells were stimulated for 6 h with maximally effective concentrations (see Fig. 7) of NGF, bFGF, dBcAMP, EGF, TPA or dexamethasone. Nuclei were then isolated and used for runoff transcription analysis as described by Greenberg and Ziff (1984). None of these reagents had any discernable effect on GAP-43 transcription. (A) The 6-h stimulation period was chosen because, for effective inducers such as NGF, GAP-43 message levels are increasing rapidly at that time (*) (B) As a positive control, cells were stimulated with NGF for 20 min prior to nuclei isolation. These nuclei showed a very large induction of c-*fos* message as reported previously (Greenberg et al., 1985).

cells treated with NGF, FGF or dBcAMP for 1 day, followed by treatment with the transcriptional inhibitor actinomycin D, with or without effector, for a second day (Fig. 14). It is clear that NGF does increase message stability whereas bFGF and dBcAMP appear not to. Therefore, the basis for increased message levels in FGF- and dBcAMP-treated cells may be mostly altered processing (i.e., decreased nuclear degradation) and increased export from the nucleus. Similar changes might also accompany the increased message stability in NGF-treated cells.

Discussion

The correlation of increased synthesis and fast axonal transport of GAP-43 with neuronal growth, during both development and regeneration, has been firmly established in in vivo experiments (Freeman et al., 1986; Benowitz and Routtenberg, 1987; Snipes et al., 1987b; Skene, 1989; Norden et al., 1991a). This increased synthesis appears to result from elevated message levels. Hence, cell-free translation of GAP-43 is greater using mRNA isolated from neonatal compared to adult cortex (Jacobson et al., 1986). Also, in situ hybridization experiments show that GAP-43 message levels are greatest in cell bodies of neurons which synapse in so-called plastic areas of the adult brain, and these areas are especially rich in GAP-43 (Neve et al., 1987; Rosenthal et al., 1987; Benowitz et al., 1988; McGuire et al., 1988). Otherwise, very little is known about the molecular mechanisms controlling GAP-43 expression. PC12 cells are an excellent model system for studying these mechanisms. Treatment

60

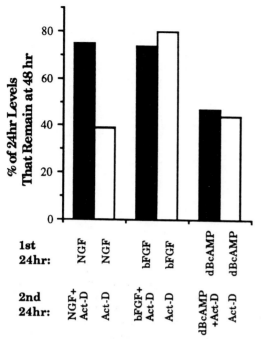

**1st
24hr:** NGF   NGF   bFGF   bFGF   dBcAMP   dBcAMP

**2nd
24hr:** NGF+
Act-D   Act-D   bFGF+
Act-D   Act-D   dBcAMP
+Act-D   Act-D

Fig. 14. Effects on GAP-43 message stability in PC12 cells. PC12 cells were treated for 24 h with either 100 ng/ml NGF, 15 ng/ml bFGF or 1 mM dBcAMP and messenger RNA was isolated. As indicated, some dishes were treated for an additional 24 h with 5 μg/ml actinomycin, either alone or in the continued presence of inducer, before isolation of messenger RNA. All RNA samples were then analyzed to determine GAP-43 message levels as described for Fig. 7. The levels remaining at 48 h are indicated as a percentage of those present after 24 h treatment with effector. Only NGF appears to have a stabilizing effect on GAP-43 message.

with NGF causes the cells to cease division and to initiate a program of neuronal differentiation which includes extensive neurite outgrowth. The availability of this unlimited, homogeneous cell population, which is easily manipulated in culture, greatly facilitates studies of neuronal gene expression at the molecular level.

We and others have shown that increased GAP-43 expression accompanies NGF-stimulated neurite outgrowth in PC12 cells (Costello et al., 1986; Van Hooff et al., 1986). Since our measurements of the maximally induced GAP-43 level

and synthesis rate were made under steady-state conditions, the following relationship holds:

$$ks = kd \, [\text{GAP-43}]$$

where ks and kd are the rate constants for GAP-43 synthesis and degradation, respectively, and [GAP-43] is the concentration of GAP-43. From this equation it is clear that the NGF-induced 16-fold increase in steady-state GAP-43 level is entirely accounted for by the 19-fold increase in synthesis rate. In fact, these numbers suggest that NGF also might slightly increase the rate constant for GAP-43 degradation.

It is clear from our immunolocalization studies that most of the elevated GAP-43 in NGF-stimulated cells resides in neurites and in the growth cones. This is consistent with the identification of GAP-43 as a fast axonally-transported protein in neurons (Skene and Willard, 1981a). Although most fast-transported proteins are membrane-bound (Willard et al., 1974), we found that a significant fraction of the GAP-43 in NGF-stimulated PC12 cells is soluble. However, its original characterization as a membrane-bound protein notwithstanding (Skene and Willard, 1981c), a fraction of the GAP-43 in some neuronal tissues also partitions into soluble fractions (Benowitz and Lewis, 1983; Cimler et al., 1985; unpublished results, this laboratory). These data suggest that N-terminal fatty acylation of GAP-43, which is believed to be responsible for its membrane attachment (Skene and Virag, 1989; Zuber et al., 1989b), is incomplete in PC12 cells as well as in some neurons.

It is interesting to consider how GAP-43 expression is correlated with neurite growth in PC12 cells. We addressed this question in two ways. First, we examined the effects of varying the artificial growth substrate and cell density, parameters which influence neurite outgrowth. We found that the NGF-stimulated increase in GAP-

43 message level was independent of cell density or adhesion, and of the artificial substrate's ability to support neurite growth. We also examined the time course of GAP-43 expression and found that it was induced quite rapidly, such that maximal steady-state message and protein levels were achieved within approximately one and two days NGF exposure, respectively. At those times, very few cells have ceased division and begun to extend neurites. In fact, the initial response of PC12 cells to NGF actually includes increased proliferation (Boonstra et al., 1983). Furthermore, under the culture conditions we employed, complete cessation of cell division and initiation of morphological differentiation does not occur until about 7 days of NGF treatment (Greene and Tischler, 1976), indicating that NGF induces full expression of GAP-43 in proliferating cells. Therefore, our data suggest that elevated GAP-43 synthesis in PC12 cells is not necessarily sufficient to induce neurite growth. However, recent transfection experiments have shown that expression of GAP-43 in some non-neuronal cell lines, which normally do not express GAP-43, causes dramatic process formation (Zuber et al., 1989a). In contrast, elevated GAP-43 expression in transfected PC12 cells has only been shown to enhance neurite growth in NGF-stimulated or primed cells (Yankner et al., 1990). Therefore, it is possible that in PC12 cells the enhancing effect of GAP-43 on process formation/neurite growth requires the simultaneous expression of other rate limiting proteins which are more slowly induced by NGF. These might include Thy-1 (Dickson et al., 1986), the neurofilament subunits (Dickson et al., 1986; Lindenbaum et al., 1988; Leonard et al., 1987, 1988) or proteins corresponding to the unidentified clones studied by Leonard et al. (1987).

The observed effects of SIBA on GAP-43 induction by the growth factors is consistent with previous work. Seeley et al. (1984) first examined the effect of methyltransferase inhibitors on NGF actions in PC12 cells because it had been discovered that treatment of SCG neurons with NGF causes a rapid burst of methyl group incorporation into phospholipids (Pfenninger and Johnson, 1981). It was found that these drugs inhibit all tested responses of PC12 cells to NGF while having either no effect or enhancing responses to EGF (Seeley et al., 1984; Van Calker et al., 1989). Our observations of essentially complete inhibition of NGF induction, but no effect on EGF induction of GAP-43 is consistent with these earlier observations. Since FGF generally mimics the effects of NGF on PC12 cells (Togari et al., 1983, 1985; Neufeld et al., 1987; Rydel and Greene, 1987), we expected, and observed that FGF vigorously induces GAP-43 and that this induction is inhibited by SIBA. FGF-stimulated neurite regeneration by primed cells also is blocked by methyltransferase inhibitors (Togari et al., 1985). As discussed in detail previously (Seeley et al., 1984), the mechanism(s) by which methyltransferase inhibitors block NGF (or FGF) actions is unclear. In fact, it has been argued that inhibition of NGF actions by S-adenosylhomocysteine hydrolase inhibitors is unrelated to blockage of methylation reactions (Acheson and Thoenen, 1987). Regardless of how they act, however, it is clear that these drugs interfere at a very early stage in the signal transduction sequence since they block even alterations in cell surface morphology which occur within seconds of NGF exposure (Seeley et al., 1984).

Dexamethasone also inhibits GAP-43 expression. It decreased GAP-43 mRNA levels in unstimulated cells and decreased induction by NGF, FGF, TPA and dBcAMP. The opposite effects of NGF and dexamethasone on GAP-43 expression are consistent with its identification as a neuronal marker since these agents have opposite and antagonistic effects on differentiation of PC12 cells into sympathetic neuron-like or mature chromaffin-like cells (Greene and Tischler, 1982). Their antagonistic effects are evident in the intermediate phenotypes achieved with regard to neurite outgrowth (Tocco et al., 1988) or specific gene transcriptions (Leonard et al., 1987), when PC12 cells are simultaneously stimulated with both NGF and dexamethasone, as was done in our experiments. The much smaller proportional decrease of GAP-43 RNA levels by dexamethasone in cells treated with NGF (approximate 30% decrease) compared to those in cells not treated with NGF (75% decrease) argues for NGF's ability to antagonize glucocorticoid action. Our data suggest that this is also true for FGF. In contrast, the increased but much lower GAP-43 mRNA levels in the presence of TPA or dBcAMP were decreased proportionately as much as were basal levels. Although the mechanism of receptor mediated gene induction by steroid hormones such as the glucocorticoids is well understood (Yamamoto, 1985), the negative effects of steroids on gene expression have been accounted for in only a few cases. It is known that some genes may contain glucocorticoid response elements which overlap with binding sites for positively-acting transcription factors and thereby inhibit transcriptional stimulation (Drouin et al., 1987; Akerblom et al., 1988). Presumably such an arrangement could account for the mutually antagonistic actions of dexamethasone and NGF with respect to GAP-43 expression.

We have attempted to gain insight into GAP-43 induction, especially by NGF, in several ways. The message induction experiments using CHX showed that inductions by TPA, FGF and NGF were all diminished by CHX treatment, indicating a requirement for protein synthesis for effective action by these agents. Only the dBcAMP induction was unaffected by CHX, which clearly indicates that dBcAMP does not effect increase of GAP-43 message levels through the agency of a newly synthesized protein(s). The residual inductions by FGF and NGF in the presence of CHX are approximately equal in magnitude to the entirely CHX-resistant dBcAMP induction, which at first glance suggests that these residual components may be mediated via the same mechanism as dBcAMP-stimulated induction. However, the effect on message levels of treating cells with maximally stimulating concentrations of dBcAMP plus NGF was superadditive. One way of interpreting this kind of data is to postulate basically independent mechanisms which, nevertheless, have higher order synergistic effects on one another (e.g., Hatanaka et al., 1978). In addition, our data allow the possibility that NGF is acting partially via activation of PKA, but that this activation is incomplete. Hence, simultaneous treatment with a maximally effective dose of dBcAMP would not only (1) fully activate the PKA-dependent pathway for GAP-43 induction, but also thereby (2) fully facilitate any synergistic interactions which might be manifest by activation of PKA and the one or more additional pathways mediating NGF induction of GAP-43.

The nature of this additional pathway(s) is unclear. It could involve an increase in $[Ca^{2+}]_i$ since elevated extracellular KCl, which causes opening of voltage-sensitive Ca^{2+} channels in PC12 cells (Meldolesi et al., 1984), also weakly induces GAP-43. Alternatively, it might involve activation of one of the several less-studied protein kinases, such as the NGF-regulated N kinase (Rowland et al., 1987). However, we can say that it does not involve activation of PKC. This was suggested by the truly additive interactions of TPA plus NGF on GAP-43 induction, which is consistent with simple independence. In addition, PKC activation as a component of NGF action was completely ruled out by our down-regulation experiment. Hence, when PKA activity was totally eliminated this also blocked any induction by TPA but, if anything, had a slightly enhancing effect on NGF induction. Protein kinase C involvement in NGF-induced neurite outgrowth, ornithine decarboxylase activity and SCG10 gene expression has also been ruled out based on down-regulation data (Reinhold and Neet, 1989; Sigmund et al., 1990).

Although the pathways mediating NGF induction of GAP-43 remain uncertain, it is clear that none of the positive effectors that we tested had any effect on GAP-43 transcription rate. These observations were somewhat surprising since data suggesting developmentally regulated control of GAP-43 transcription in rat brain have been reported (Basi et al., 1987). However, our findings are in agreement with those of Federoff et al. (1988) who also found no effect of NGF on GAP-43 transcription in PC12 cells. We are presently testing rat brain fractions from various developmental stages for transcriptional activation of GAP-43. It is possible that a factor unrelated to those we tested is present in neonatal rat brain to stimulate GAP-43 transcription during critical developmental periods. Another possibility is that the GAP-43 gene in PC12 cells has been modified (Cedar, 1988) such that a low level of constitutive transcription persists, but activation is no longer possible

Nevertheless, our present data indicate that GAP-43 message levels in PC12 cells are only regulated post-transcriptionally. For NGF this regulation involves message stabilization, whereas FGF and dBcAMP do not exhibit this effect. Since our measurements were not made under steady-state conditions, we do not know whether this stabilization is sufficient of itself to produce the large increases in NGF-induced message levels. We are currently performing pulse/chase labeling experiments to address this question. Presumably, FGF and dBcAMP; which appear not to increase message halflife, must act by increasing the nuclear processing and cytoplasmic transport of primary GAP-43 transcripts, thereby decreasing their nuclear degradation. This effect also may be occurring with NGF treatment. Mea-surement of nuclear GAP-43 transcript levels should yield additional information on this point.

Our results suggest that GAP-43 expression in vivo is likely to be influenced by an array of growth factors. We have shown that NGF and FGF are potent inducers. These growth factors are present at high levels in certain areas of both the peripheral and central nervous systems (Gospodarowicz et al., 1984; Korsching et al., 1985). Closely related neurotrophic factors such as brain-derived neurotrophic factor (Barde et al., 1982) and neurotrophin-3 (Hohn et al., 1990) also are present in the CNS and are very likely to influence GAP-43 expression as well. By studying the regulated expression of genes such as GAP-43 and a similarly-regulated gene, SCG10 (Stein et al., 1988), whose inductions correlate with neuronal development, we will gain insight into the mechanisms controlling differentiation of neuronal precursor cells and the development of the mature neuronal phenotype. Such studies also will help to shed light on the still poorly understood mechanisms of action of the ever growing array of recognized neurotrophic factors.

Acknowledgements

This work was supported by NIH grants NS18103 to J.A.F. and NS25150 to J.J.N.

References

Acheson, A. and Thoenen, H. (1987) Both short- and long-term effects of nerve growth factor on tyrosine hydroxylase in calf adrenal chromaffin cells are blocked by *S*-adenosylhomocysteine hydrolase inhibitors. *J. Neurochem.*, 48: 1416–1424.

Akerblom, I.E., Slater, E.P., Beato, M., Baxter, J.D. and Mellon, P.L. (1988) Negative regulation by glucocorticoids through interference with a cAMP responsive enhancer. *Science*, 241: 350–353.

Alexander, K.A., Cimler, B.M., Meier, K.E. and Storm, D.R. (1987) Regulation of calmodulin binding to P-57. *J. Biol. Chem.*, 262: 6108–6113.

64

Barde, Y.-A., Edgar, D. and Thoenen, H. (1982) Purification of a new neurotrophic factor from mammalian brain. *EMBO J.*, 1: 549–553.

Basi, G.S., Jacobson, R.D., Virag, I., Schilling, J. and Skene, J.H.P. (1987) Primary structure and transcriptional regulation of GAP-43, a protein associated with nerve growth. *Cell*, 49: 785–791.

Benowitz, L.I. and Lewis, E.R. (1983) Increased transport of 44,000- to 49,000-dalton acidic proteins during regeneration of the goldfish optic nerve: a two-dimensional gel analysis. *J. Neurosci.*, 3: 2153–2163.

Benowitz, L.I. and Routtenberg, A. (1987) A membrane phosphoprotein associated with neural development, axonal regeneration, phospholipid metabolism and synaptic plasticity. *Trends Neurosci.*, 10: 527–532.

Benowitz, L.I., Shashoua, V.E. and Yoon, M.G. (1981) Specific changes in rapidly transported proteins during regeneration of the goldfish optic nerve. *J. Neurosci.*, 1: 300–307.

Benowitz, L.I., Apostolides, P.J., Perrone-Bizzozero, N.I., Finklestein, S.P. and Zwiers, H. (1988) Anatomical distribution of the growth-associated protein GAP-43/B-50 in the adult rat brain. *J. Neurosci.*, 8: 339–352.

Benowitz, L.I., Perrone-Bizzozero, N.I., Finklestein, S.P. and Bird, E.D. (1989) Localization of the growth-associated phosphoprotein GAP-43 (B-50, F1) in the human cerebral cortex. *J. Neurosci.*, 9: 990–995.

Boonstra, J., Moolenaar, W.H., Harrison, P.H., Moed, P., van der Saag, P.T. and de Laat, S.W. (1983) Ionic responses and growth stimulation induced by nerve growth factor and epidermal growth factor in rat pheochromocytoma (PC12) cells. *J. Cell Biol.*, 97: 92–98.

Burstein, D.E., Blumberg, P.M. and Greene, L.A. (1982) Nerve growth factor-induced neuronal differentiation of PC12 pheochromocytoma cells: lack of inhibition by a tumor promoter. *Brain Res.*, 247: 115–119.

Cedar, H. (1988) DNA methylation and gene activity. *Cell*, 53: 3–4.

Chandler, C.E. and Herschman, H.R. (1980) Tumor promoter modulation of epidermal growth factor- and nerve growth factor-induced adhesion and growth factor binding of PC-12 pheochromocytoma cells. *J. Cell. Physiol.*, 105: 275–285.

Cimler, B.M., Andreasen, T.J., Andreasen, K.I. and Storm, D.R. (1985) P-57 is a neural specific calmodulin-binding protein. *J. Biol. Chem.*, 260: 10784–10788.

Costello, B. and Freeman, J.A. (1988) Regulation of GAP-43, a neuronal growth-associated phosphoprotein, in PC12 pheochromocytoma cells. *Soc. Neurosci. Abstr.*, 14: 1126.

Costello, B., Snipes, G.J., Norden, J.J., Bock, S.S. and Freeman, J.A. (1986) Regulation and phosphorylation of the growth-associated protein, GAP-43, in NGF-stimulated PC12 cells. *Soc. Neurosci. Abstr.*, 12: 500.

Costello, B., Norden, J.J. and Freeman, J.A. (1987) Induction and localization of the growth-associated protein, GAP-43, in NGF-stimulated PC12 cells. *Soc. Neurosci. Abstr.*, 13: 1480.

Costello, B., Meymandi, A. and Freeman, J.A. (1990) Factors influencing GAP-43 gene expression in PC12 pheochromocytoma cells. *J. Neurosci.*, 10: 1398–1406.

Dekker, L.V., De Graan, P.N.E., Oestreicher, A.B., Versteeg, D.H.G. and Gispen, W.H. (1989) Inhibition of noradrenaline release by antibodies to B-50 (GAP-43). *Nature*, 342: 74–76.

De la Monte, S.M., Federoff, H.J., Ng, S.-C., Grabczyk, E. and Fishman, M.C. (1989) GAP-43 gene expression during development: persistence in a distinctive set of neurons in the mature central nervous system. *Dev. Brain Res.*, 46: 161–168.

Dickson, G., Prentice, H., Julien, J.-P., Ferrari, G., Leon, A. and Walsh, F.S. (1986) Nerve growth factor activates Thy-1 and neurofilament gene transcription in rat PC12 cells. *EMBO J.*, 5: 3449–3453.

Doupe, A.J., Landis, S.C. and Patterson, P.H. (1985) Environmental influences on the development of neural crest derivatives: glucocorticoids, growth factors, and chromaffin cell plasticity. *J. Neurosci.*, 5: 2119–2142.

Drouin, J., Charron, J., Gagner, J.P., Jeannotte, L., Nemer, M., Plante, R.K. and Wrange, O. (1987) The pro-opiomelanocortin gene: a model for negative regulation of transcription by glucocorticoids. *J. Cell. Biochem.*, 35: 293–304.

Enouf, J., Lawrence, F., Tempete, C., Robert-Gero, M. and Lederer, E. (1979) Relationship between inhibition of protein methylase I and inhibition of Rous sarcoma virus-induced cell transformation. *Cancer Res.*, 39: 4497–4502.

Ewald, D.A., Sternweis, P.C. and Miller, R.J. (1988) Guanine nucleotide-binding protein G_o-induced coupling of neuropeptide Y receptors to Ca^{2+} channels in sensory neurons. *Proc. Natl. Acad. Sci. USA*, 85: 3633–3637.

Federoff, H.J., Grabczyk, E. and Fishman, N.C. (1988) Dual regulation of GAP-43 gene expression by nerve growth factor and glucocorticoids. *J. Biol. Chem.*, 263: 19290–19295.

Freeman, J.A., Bock, S., Deaton, M., McGuire, B., Norden, J.J. and Snipes, G.J. (1986) Axonal and glial proteins associated with development and response to injury in the rat and goldfish optic nerve. In G. Gilad and M. Schwartz (Eds.), *Recovery from Neural Trauma*, Springer-Verlag, Berlin, pp. 34–47.

Fujii, D.K., Massoglia, S.L., Savion, N. and Gospodarowicz, D. (1982) Neurite outgrowth and protein synthesis by PC12 cells as a function of substratum and nerve growth factor. *J. Neurosci.*, 2: 1157–1175.

Gospodarowicz, D., Mescher, A.L. and Birdwell, C.R. (1977) Stimulation of corneal cell proliferation in vitro by fibroblast and epidermal growth factors. *Exp. Eye Res.*, 25: 75–89.

Gospodarowicz, D., Cheng, J., Lui, G.-M., Baird, A. and

Bohlent, P. (1984) Isolation of brain fibroblast growth factor by heparin-sepharose affinity chromatography: identity with pituitary fibroblast growth factor. *Proc. Natl. Acad. Sci. USA*, 81: 6963–6967.

Greenberg, M.E. and Ziff, E.B. (1984) Stimulation of 3T3 cells induces transcription of the c-fos proto-oncogene. *Nature*, 311: 433–438.

Greenberg, M.E., Greene, L.A. and Ziff, E.B. (1985) Nerve growth factor and epidermal growth factor induce rapid transient changes in proto-oncogene transcription in PC12 cells. *J. Biol. Chem.*, 260: 14101–14110.

Greene, L.A. and Tischler, A.S. (1976) Establishment of a noradrenergic clonal line of rat adrenal pheochromocytoma cells which respond to nerve growth factor. *Proc. Natl. Acad. Sci. USA*, 73: 2424–2428.

Greene, L.A. and Tischler, A.S. (1982) PC12 pheochromocytoma cultures in neurobiological research. In S. Federoff and L. Hertz (Eds.), *Advances in Cellular Neurobiology*, vol. 3, Academic Press, New York, pp. 373–414.

Greene, L.A., Burstein, D.E. and Black, N.M. (1980) The priming model for the mechanism of action of nerve growth factor: evidence derived from clonal PC12 pheochromocytoma cells. In E. Giacobini et al. (Eds.), *Tissue Culture in Neurobiology*, Raven Press, New York, pp. 313–319.

Gunning, P.W., Landreth, G.E., Bothwell, M.A. and Shooter, E.M. (1981) Differential and synergistic actions of nerve growth factor and cyclic AMP in PC12 cells. *J. Cell Biol.*, 89: 240–245.

Guroff, G., Dickens, G. and End, D. (1981) The induction of ornithine decarboxylase by nerve growth factor and epidermal growth factor in PC12 cells. *J. Neurochem.*, 37: 342–349.

Hall, F.L., Fernyhough, P., Ishii, D.N. and Vulliet, P.R. (1988) Suppression of nerve growth factor-directed neurite outgrowth in PC12 cells by sphingosine, an inhibitor of protein kinase C. *J. Biol. Chem.*, 263: 4460–4466.

Hatanaka, H., Otten, U. and Thoenen, H. (1978) Nerve growth factor-mediated selective induction of ornithine decarboxylase in rat pheochromocytoma; a cyclic AMP-independent process. *FEBS Lett.*, 92: 313–316.

Hescheler, J., Rosenthal, W., Trautwein, W. and Schultz, G. (1987) The GTP-binding protein, G_o, regulates neuronal calcium channels. *Nature*, 325: 445–447

Hohn, A., Leibrock, J., Bailey, K. and Barde, Y.-A. (1990) Identification and characterization of a novel member of the nerve growth factor/brain-derived neurotrophic factor family. *Nature*, 344: 339–341.

Jacobson, R.D., Virag, I. and Skene, J.H.P. (1986) A protein associated with axon growth, GAP-43, is widely distributed and developmentally regulated in rat CNS. *J. Neurosci.*, 6: 1843–1855.

Jolles, J., Zwiers, H., van Dongen, C.J., Schotman, P., Wirtz, K.W.A. and Gispen, W.H. (1980) Modulation of brain polyphosphoinositide metabolism by ACTH-sensitive protein phosphorylation. *Nature*, 286: 623–625.

Karns, L.R., Ng, S.-C., Freeman, J.A. and Fishman, M.C. (1987) Cloning of complementary DNA for GAP-43, a neuronal growth-related protein. *Science*, 236: 597–600.

Korsching, S., Auburger, G., Heumann, R., Scott, J. and Thoenen, H. (1985) Levels of nerve growth factor and its mRNA in the central nervous system of the rat correlate with cholinergic innervation. *EMBO J.*, 4: 1389–1393.

Laemmli, U.K. (1970) Cleavage of structural proteins during the assembly of the head of bacteriophage T4. *Nature*, 227: 680–685.

Landreth, G.E. and Rieser, G.D. (1985) Nerve growth factor- and epidermal growth factor-stimulated phosphorylation of a PC12 cytoskeletally associated protein in situ. *J. Cell Biol.*, 100: 677–683.

Leonard, D.G.B., Ziff, E.B. and Greene, L.A. (1987) Identification and characterization of mRNAs regulated by nerve growth factor in PC12 cells. *Mol. Cell. Biol.*, 7: 3156–3167.

Leonard, D.G.B., Gorham, J.D., Cole, P., Greene, L.A. and Ziff, E.B. (1988) A nerve growth factor-regulated messenger RNA encodes a new intermediate filament protein. *J. Cell Biol.*, 106: 181–193.

Lindenbaum, M.H., Carbonetto, S., Grosveld, F., Flavell, D. and Mushynski, W.E. (1988) Transcriptional and post-transcriptional effects of nerve growth factor on expression of the three neurofilament subunits in PC-12 cells. *J. Biol. Chem.*, 263: 5662–5667.

Malenka, R.C., Madison, D.V. and Nicoll, R.A. (1986) Potentiation of synaptic transmission in the hippocampus by phorbol esters. *Nature*, 321: 175–177.

Marzluff, W.F. (1978) Transcription of RNA in isolated nuclei. *Methods Cell Biol.*, 19: 317–331.

Matthies, H.J.G., Palfrey, H.C., Hirning, L.B. and Miller, R.J. (1987) Down regulation of protein kinase C in neuronal cells: effects on neurotransmitter release. *J. Neurosci.*, 7: 1198–1206.

McGuire, C.B., Snipes, G.J. and Norden, J.J. (1988) Light-microscopic immunolocalization of the growth- and plasticity-associated protein GAP-43 in the developing rat brain. *Dev. Brain Res.*, 41: 277–291.

Meiri, K.F., Pfenninger, K.H. and Willard, M.B. (1986) Growth-associated protein, GAP-43, a polypeptide that is induced when neurons extend axons, is a component of growth cones and corresponds to pp46, a major polypeptide of a subcellular fraction enriched in growth cones. *Proc. Natl. Acad. Sci. USA*, 83: 3537–3541.

Meiri, K.F., Willard, M. and Johnson, M.I. (1988) Distribution and phosphorylation of the growth-associated protein GAP-43 in regenerating sympathetic neurons in culture. *J. Neurosci.*, 8: 2571–2581.

Meldolesi, J., Huttner, W.B., Tsien, R.Y. and Pozzan, T.

(1984) Free cytoplasmic Ca^{2+} and neurotransmitter release: studies on PC12 cells and synaptosomes exposed to a-latrotoxin. *Proc. Natl. Acad. Sci. USA*, 81: 620–624.

Morrison, R.S., Sharma, A., De Vellis, J. and Bradshaw, R.A. (1986) Basic fibroblast growth factor supports the survival of cerebral cortical neurons in primary culture. *Proc. Natl. Acad. Sci. USA*, 83: 7537–7541.

Neufeld, G., Gospodarowicz, D., Dodge, L. and Fujii, D.K. (1987) Heparin modulation of the neurotropic effects of acidic and basic fibroblast growth factors and nerve growth factor on PC12 cells. *J. Cell. Physiol.*, 131: 131–140.

Neve, R.L., Perrone-Bizzozero, N.I., Finklestein, S., Zwiers, H., Bird, E., Kurnit, D.M. and Benowitz, L.I. (1987) The neuronal growth-associated protein GAP-43 (B-50, F1): neuronal specificity, developmental regulation and regional distribution of the human and rat mRNAs. *Mol. Brain Res.*, 2: 177–183.

Nielander, H.B., Schrama, L.H., Van Rozen, A.J., Kasperaitis, M., Oestreicher, A.B., De Graan, P.N.E., Gispen, W.H. and Schotman, P. (1987) Primary structure of the neuron-specific phosphoprotein B-50 is identical to growth-associated protein GAP-43. *Neurosci. Res. Commun.*, 1: 163–172.

Norden, J.J., Wouters, B., Knapp, J., Bock, S. and Freeman, J.A. (1991a) The role of GAP-43 in axon growth and synaptic plasticity in the visual system: evolutionary implications. In J. Cronly-Dillon (Ed.), *Vision and Visual Dysfunction*, Vol. 11, Macmillan Press, London, chapter 12.

Norden, J.J., Lettes, A., Costello, B., Lin, L.H., Bock, S. and Freeman, J.A. (1991b) Possible role of GAP-43 in calcium regulation/neurotransmitter release. *Ann. N.Y. Acad. Sci.*, in press.

Perrone-Bizzozero, N.I., Finklestein, S.P. and Benowitz, L.I. (1986) Synthesis of a growth-associated protein by embryonic rat cerebrocortical neurons *in vitro*. *J. Neurosci.*, 6: 3721–3730.

Pfenninger, K.H. and Johnson, M.P. (1981) Nerve growth factor stimulates phospholipid methylation in growing neurites. *Proc. Natl. Acad. Sci. USA*, 78: 7797–7800.

Reinhold, D.S. and Neet, K.E. (1989) The lack of a role for protein kinase C in neurite extension and in the induction of ornithine decarboxylase by nerve growth factor in PC12 cells. *J. Biol. Chem.*, 264: 3538–3544

Rogelj, S., Klagsbrun, M., Atzmon, R., Kurokawa, M., Haimovitz, A., Fuks, Z. and Vlodavsky, I. (1989) Basic fibroblast growth factor is an extracellular matrix component required for supporting the proliferation of vascular endothelial cells and the differentiation of PC12 cells. *J. Cell Biol.*, 109: 823–831.

Rosenthal, A., Chan, S.Y., Henzel, W., Haskell, C., Kuang, W.-J., Chen, E., Wilcox, J.N., Ullrich, A., Goeddel, D.V. and Routtenberg, A. (1987) Primary structure and mRNA localization of protein F1, a growth-related protein kinase C substrate associated with synaptic plasticity. *EMBO J.*, 6: 3641–3646.

Routtenberg, A., Colley, P., Linden, D., Lovinger, D., Murakami, K. and Sheu, F.-S. (1986) Phorbol ester promotes growth of synaptic plasticity. *Brain Res.*, 378: 374–378.

Rowland, E.A., Muller, T.H., Goldstein, M. and Greene, L.A. (1987) Cell-free detection and characterization of a novel nerve growth factor-activated protein kinase in PC12 cells. *J. Biol. Chem.*, 262: 7504–7513.

Rydel, R.E. and Greene, L.A. (1987) Acidic and basic fibroblast growth factors promote stable neurite outgrowth and neuronal differentiation in cultures of PC12 cells. *J. Neurosci.*, 7: 3639–3653.

Seeley, P.J., Rukenstein, A., Connolly, J.L. and Greene, L.A. (1984) Differential inhibition of nerve growth factor and epidermal growth factor effects on the PC12 pheochromocytoma line. *J. Cell Biol.*, 98: 417–426.

Sigmund, O., Naor, Z., Anderson, D.J. and Stein, R. (1990) Effect of nerve growth factor and fibroblast growth factor on SCG10 and c-fos expression and neurite outgrowth in protein kinase C-depleted PC12 cells. *J. Biol. Chem.*, 265: 2257–2261.

Skene, J.H.P. (1989) Axonal growth-associated proteins. *Annu. Rev. Neurosci.*, 12: 127–156.

Skene, J.H.P. and Kalil, K. (1984) A "growth-associated protein" (GAP-43) in developing and severed axons of the hamster pyramidal tract. *Soc. Neurosci. Abstr.*, 10: 1030.

Skene, J.H.P. and Virag, I. (1989) Post-translational membrane attachment and dynamic fatty acylation of a neuronal growth cone protein, GAP-43. *J. Cell Biol.*, 108: 613–624.

Skene, J.H.P. and Willard, M. (1981a) Changes in axonally transported proteins during axon regeneration in toad retinal ganglion cells. *J. Cell Biol.*, 89: 86–95.

Skene, J.H.P. and Willard, M. (1981b) Axonally transported proteins associated with axon growth in rabbit central and peripheral nervous systems. *J. Cell Biol.*, 89: 96–103.

Skene, J.H.P. and Willard, M. (1981c) Characteristics of growth-associated polypeptides in regenerating toad retinal ganglion cell axons. *J. Neurosci.*, 1: 419–426.

Snipes, G.J., Chan, S.Y, McGuire, C.B., Costello, B.R., Norden, J.J., Freeman, J.A. and Routtenberg, A. (1987a) Evidence for the coidentification of GAP-43, a growth-associated protein and F1, a plasticity-associated protein. *J. Neurosci.*, 7: 4066–4075.

Snipes, G.J., Costello, B., McGuire, C.B., Mayes, B.N., Bock, S.S., Norden, J.J. and Freeman, J.A. (1987b) Regulation of specific neuronal and non-neuronal proteins during development and following injury in the rat CNS. *Prog. Brain Res.*, 71: 155–175

Stein, R., Orit, S. and Anderson, D.J. (1988) The induction of a neural-specific gene, SCG10, by nerve growth factor in PC12 cells is transcriptional, protein synthesis dependent and glucocorticoid inhibitable. *Dev. Biol.*, 127: 316–325.

Stocker, F.W., Eiring, A., Georgiade, M.S.R. and Georgiade, N. (1958) A tissue culture technique for growing corneal epithelial stromal and endothelial tissues separately. *Am. J. Ophthalmol.*, 46: 294–298.

Strittmatter, S.M., Valenzuela, D., Kennedy, T.E., Neer, E.J. and Fishman, M.C. (1990) G_o is a major growth cone protein subject to regulation by GAP-43. *Nature*, 344: 836–841.

Tocco, M.D., Contreras, M.L., Koizumi, S., Dickens, G. and Guroff, G. (1988) Decreased levels of nerve growth factor receptor on dexamethasone-treated PC12 cells. *J. Neurosci. Res.*, 20: 411–419.

Togari, A., Baker, D., Dickens, G. and Guroff, G. (1983) The neurite-promoting effect of fibroblast growth factor on PC12 cells. *Biochem. Biophys. Res. Commun.*, 114: 1189–1193.

Togari, A., Dickens, G., Kuzuya, H. and Guroff, G. (1985) The effect of fibroblast growth factor on PC12 cells. *J. Neurosci.*, 5: 307–316.

Van Calker, D., Takahata, K. and Heumann, R. (1989) Nerve growth factor potentiates the hormone-stimulated intracellular accumulation of inositol phosphates and Ca^{2+} in rat PC12 pheochromocytoma cells: comparison with the effect of epidermal growth factor. *J. Neurochem.*, 52: 38–45.

Van Hooff, C.O.M., De Graan, P.N.E., Boonstra, J., Oestreicher, A.B., Schmidt-Michels, M.H. and Gispen, W.H. (1986) Nerve growth factor enhances the level of the protein kinase C substrate B-50 in pheochromocytoma PC12 cells. *Biochem. Biophys. Res. Commun.*, 139: 644–651.

Van Hooff, C.O.M., De Graan, P.N.E., Oestreicher, A.B. and Gispen, W.H. (1988) B-50 phosphorylation and polyphosphoinositide metabolism in nerve growth cone membranes. *J. Neurosci.*, 8: 1789–1795.

Van Hooff, C.O.M., Holthuis, J.C.M., Oestreicher, A.B., Boonstra, J., De Graan, P.N.E. and Gispen, W.H. (1989) Nerve growth factor-induced changes in the intracellular localization of the protein kinase C substrate B-50 in pheochromocytoma PC12 cells. *J. Cell Biol.*, 108: 1115–1125.

Vlodavsky, I., Levi, A., Lax, I., Fuks, Z. and Schlessinger, J. (1982) Induction of cell attachment and morphological differentiation in a pheochromocytoma cell line and embryonal sensory cells by the extracellular matrix. *Dev. Biol.*, 93: 285–300.

Vlodavsky, I., Folkman, J., Sullivan, R., Fridman, R., Ishai-Michaeli, R., Sasse, J. and Klagsbrun, M. (1987) Endothelial cell-derived basic fibroblast growth factor: synthesis and deposition into subendothelial extracellular matrix. *Proc. Natl. Acad. Sci. USA*, 84: 2292–2296.

Wakim, B.T., Alexander, K.A., Masure, H.R., Cimler, B.M., Storm, D.R. and Walsh, K.A. (1987) Amino acid sequence of P-57, a neurospecific calmodulin-binding protein. *Biochemistry*, 26: 7466–7470.

Walicke, P., Cowan, W.M., Ueno, N., Baird, A. and Guillemin, R. (1986) Fibroblast growth factor promotes survival of dissociated hippocampal neurons and enhances neurite extension. *Proc. Natl. Acad. Sci. USA*, 83: 3012–3016.

Willard, M., Cowan, W.M. and Vagelos, P.R. (1974) The polypeptide composition of intra-axonally transported proteins: evidence for four transport-velocities. *Proc. Natl. Acad. Sci. USA*, 71: 2183–2187.

Yamamoto, K.R. (1985) Steroid receptor regulated transcription of specific genes and gene networks. *Annu. Rev. Genet.*, 19: 209–252.

Yankner, B.A., Benowitz, L.I., Villa-Homaroff, L. and Neve, R.L. (1990) Transfection of PC12 cells with the human GAP-43 gene: effects on neurite outgrowth and regeneration. *Mol. Brain Res.*, 7: 39–44.

Zuber, M.X., Goodman, D.W., Karns, L.R. and Fishman, M.C. (1989a) The neuronal growth-associated protein GAP-43 induces filopodia in non-neuronal cells. *Science*, 244: 1193–1195.

Zuber, M.X., Strittmatter, S.M. and Fishman, M.C. (1989b) A membrane-targeting signal in the amino terminus of the neuronal protein GAP-43. *Nature*, 341: 345–348.

W.H. Gispen and A. Routtenberg (Eds.)
Progress in Brain Research, Vol. 89
© 1991 Elsevier Science Publishers B.V.

CHAPTER 6

The expression of GAP-43 in relation to neuronal growth and plasticity: when, where, how, and why?

Larry I. Benowitz [1] and Nora I. Perrone-Bizzozero [2]

[1] *Department of Neurosurgery, Children's Hospital and Harvard Medical School, Boston, MA, U.S.A.,*
and [2] *Department of Biochemistry, University of New Mexico, Albuquerque, NM, U.S.A.*

Numerous studies have implicated the membrane phosphoprotein GAP-43 (B-50, F1, pp46, neuromodulin, GAP-48) in the growth, regeneration, and remodeling of neuronal connections (Skene and Willard, 1981a,b; Benowitz et al., 1981, 1983, 1990; Jacobson et al., 1986; McGuire et al., 1988; Dani et al., 1990; Zuber et al., 1989; Yankner et al., 1989). Studies on where and when GAP-43 gets expressed afford a unique opportunity to visualize neuronal differentiation and synaptic organization as they occur in vivo and in culture. Moreover, it is likely that by investigating how the expression of GAP-43 is regulated, and how the protein functions physiologically in the cell, we will gain a deeper understanding of the molecular mechanisms that underlie growth and plasticity in the nervous system.

Correspondence: Larry I. Benowitz, Ph.D., Laboratory for Neuroscience Research in Neurosurgery, Children's Hospital/Enders 312, 300 Longwood Avenue, Boston, MA 02115, U.S.A. Tel.: (617) 735-6368, Fax: (617) 735-0636.

When and where

Neuronal development in culture

Prior to their final cell division and process outgrowth, neurons express very little GAP-43 mRNA and protein (Perrone-Bizzozero et al., 1986; de la Monte et al., 1989; Dani et al., 1990). This can be seen in embryonic cortical neurons that are grown in culture in the absence of serum, in which case the cells remain viable but fail to extend neurites (Perrone-Bizzozero et al., 1986). Likewise, very little GAP-43 is found in the germinal zones of the developing brain, in which cells have not yet undergone their final cell division and migration, or even in postmitotic cells of the mantle zone prior to process outgrowth (Dani et al., 1990). PC-12 pheochromocytoma cells similarly express only low levels of GAP-43 before being stimulated to differentiate (Costello et al., 1990; van Hooff et al., 1989; Karns et al., 1987).

Levels of GAP-43 increase markedly with the beginning of neurite outgrowth. In cultured neu-

70

rons or PC-12 cells, high concentrations of the protein are seen throughout the cell body and in all nascent processes coincident with the onset of differentiation (Perrone-Bizzozero et al., 1986; Meiri et al., 1988; Goslin et al., 1989; van Hooff et al., 1989; Woolf et al., 1990; Fig. 1). Detailed studies in hippocampal pyramidal cells show that as one nascent process differentiates into an axon, GAP-43 withdraws from all of the other processes but remains densely concentrated along the growing axon (Goslin et al., 1989; Fig. 1). In ganglionic neurons and PC-12 cells, the concen-

tration of the protein diminishes proximally as development proceeds, but remains high in distal portions, particularly in growth cones (Meiri et al., 1988; van Hooff et al., 1989).

Development of the nervous system: peripheral ganglia, spinal cord, and brainstem

A similar progression can be observed in vivo, where the shifting pattern of GAP-43 immunostaining provides a unique picture of the sequential events that take place during brain development. Coincident with the early outgrowth of

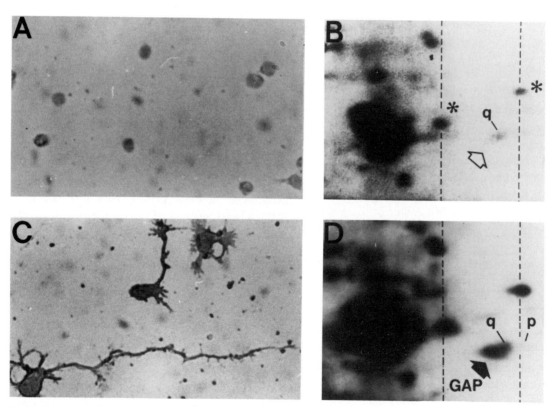

Fig. 1. The expression of GAP-43 coincides with the beginning of neurite outgrowth. Embryonic cortical neurons were cultured either in the absence of serum (but with a hormone supplement), which suppressed neurite outgrowth (top), or with serum, which promoted outgrowth (bottom). Using a monospecific antibody to GAP-43 in the indirect-peroxidase method, undifferentiated cells showed only low levels of GAP-43 immunostaining (A) and low levels of GAP-43 synthesis by metabolic labeling (position of GAP-43 indicated by open arrow on the two-dimensional gel inset, B); by contrast, cells undergoing neurite outgrowth showed intense GAP-43 immunoreactivity (C) and high levels of GAP-43 synthesis by metabolic labeling (closed arrow, D) (Perrone-Bizzozero et al., 1986; and unpublished data).

peripheral nerves and spinal pathways, levels of GAP-43 mRNA begin to rise in the peripheral ganglia and spinal cord at around embryonic day 11 (De la Monte et al., 1989). GAP-43 immunostaining is detectable in the axons that extend from the ganglia and the ventral horn of the

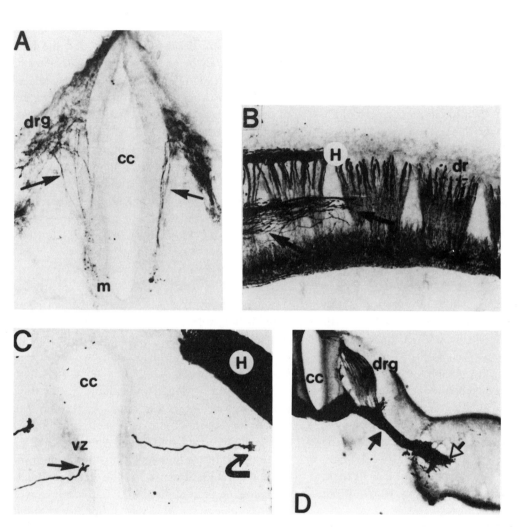

Fig. 2. GAP-43 in developing spinal cord and peripheral nerves. Using the indirect-immunoperoxidase method, GAP-43 is present in the earliest fibers appearing in the spinal cord and peripheral nerves on postnatal day 11 (A: transverse section through rostral spinal cord. Dorsal roots (arrows) grow out from the dorsal root ganglia and down the lateral sides of the cord towards the motor neuron pool.) By the following day (B), parasagittal sections through the spinal cord reveal an intensely stained plexus that includes the ingrowing dorsal roots (dr) and the bundle of His (H). Arrows point to the growth cones of axons descending in the developing lateral bundle. Other growth cones, from the dorsal lumbar cord at E14, are shown in C; these derive from stray axons that leave the bundle of His and grow towards the ventricular zone (vz), or which loop around back laterally (curved arrow). On E12, GAP-43 immunoreactivity is intense in developing peripheral nerves (D, closed arrow) as they enter a limb bud and begin to form a plexus (open arrow). (Fitzgerald et al., 1991; Reynolds et al., 1991). Other abbreviations: cc, central canal; drg, dorsal root ganglia; m, motoneuron pool.

spinal cord at this time, becoming intense one day later (i.e., E12: Reynolds et al., 1991; Fig. 2). The immunostaining of peripheral nerves follows a rostrocaudal pattern and advances proximodistally, reaching the extremities by E19. Skin innervation precedes muscle innervation throughout the neuraxis by 1–2 days, with the latter continuing to be refined through the early postnatal period. Axonal staining diminishes sharply between E21 and the beginning of the second postnatal week, though the protein remains evident in the nerve terminals throughout the period in which end arbors are being elaborated; by the end of the first postnatal month, only low levels remain.

Within the spinal cord, the incoming dorsal roots provide some of the earliest immunostaining seen, appearing first around E11 but not penetrating into the dorsal horn until El5 (Fitzgerald et al., 1991; Fig. 2). Interneurons of the spinal cord also begin to show staining as early as E11, with projections into the spinal grey matter becoming visible at around E14–15. By the late embryonic period (E18–19) the pattern of GAP-43 immunostaining in the spinal cord has become so dense and complex as to make the identification of individual fiber systems nearly impossible. Axonal staining then falls off, though the continued presence of GAP-43 in the nerve terminals gives rise to a dense staining of the neuropil through the first postnatal week. This, too, then declines, and by postnatal day 10 there is only a light background staining which all but vanishes by P29.

A developmental pattern similar to that seen in the spinal cord can be observed at higher levels of the neuraxis, though with some temporal differences (Dani et al., 1990). Densely stained axons are seen coursing through the brain stem and into the basal forebrain as early as E12–13. Many of these can be demonstrated by double-antibody labeling to be the ascending monoaminergic pathways that arise early on from well-defined cell groups in the brain stem (Adams et al., 1990). By E15 the density and complexity of the GAP-43 immunostaining increases to include most fiber systems of the brain stem. As in the spinal cord, the intensity of this staining diminishes in the brain stem and basal forebrain around the time of birth, though the neuropil now becomes densely stained, corresponding to a shift in the mode of growth from axonal elongation to the formation of end arbors and synapses. Within a week or so, most of the neuropil staining also disappears, and the adult brain stem shows only very low levels of the protein, particularly within sensory and motor nuclei (Dani et al., 1990; Benowitz et al., 1988). As might be expected from its delayed development, the pyramidal tract continues to show a higher level of GAP-43 immunostaining later than other pathways (Gorgels et al., 1988).

Optic nerve

One pathway that has been studied in some detail is the projection from the hamster retina to the superior colliculus. Studies using metabolic labeling and immunocytochemistry show that GAP-43 is expressed abundantly along the length of optic axons through the middle of the first postnatal week, as these axons grow over the surface of the midbrain and diencephalon. As they begin to branch and form collaterals, GAP-43 continues to be synthesized at high levels but now is concentrated primarily in the nerve endings. Synaptogenesis continues into the beginning of the second postnatal week, during which time alterations in the pattern of physiological activity or the competition from other inputs can induce a change in synaptic relationships. This plasticity falls off by the end of the second week, and is paralleled by a precipitous decline in GAP-43 synthesis (Moya et al., 1988, 1989). These observations suggest that the protein may play an initial role in axonal extension, then synaptogenesis, perhaps even serving as one molecular determinant of the period of plasticity. This latter possibility is supported by similar findings in other neural systems; it is also consistent with the finding that physiological manipulations at the nerve ending influence synaptic function and GAP-43 phosphorylation in parallel (DeGraan et al., 1987) and with experiments showing that interference with GAP-43 alters neurite outgrowth (Zuber et al., 1987; Yankner et al., 1990; Shea et al., 1991) (see below).

Cerebral cortex

The temporal relationship of GAP-43 expression to the development and plasticity of neuronal connections is striking in the cerebral cortex (Fig. 3). In the rat, axons in the subplate show intense immunostaining as early as E17, but do not penetrate the cortical mantle until the first postnatal week. The concentration of GAP-43 in the presumptive grey matter remains low until about P3; then, in the middle of the first postnatal week, there is an explosive increase. This reaches its maximum intensity then falls off in most regions of the cortex within a few days (Oestreicher and Gispen, 1986; McGuire et al., 1988; Dani et al., 1990). This peak in GAP-43 expression can also be demonstrated by examining the overall levels of GAP-43 mRNA (Basi et al., 1987) and protein (Jacobson et al., 1986) in the cerebral cortex. In the so-called barrel field in the primary somatosensory cortex, containing the representation of the facial vibrissae, ascending thalamic axons contain high levels of GAP-43 and show a vibrissa-related organization before they form synapses with cortical neurons on days P3–5. By 7 days, the intensity of the GAP-43 immunostaining has already begun to decrease from its peak at days 4–5, and by P8 the barrels are almost devoid of GAP-43 immunostaining (Erzurumlu et al., 1990). The time course of GAP-43 changes in the barrels coincides closely with the

critical period in which peripheral innervation can alter the pattern of cortical organization (Belford and Killackey, 1980). Likewise, in the striate cortex of the cat, peak levels of GAP-43 immunostaining appear in the cortical neuropil between the second and sixth weeks postnatally, then decline to near-background levels in the adult, coinciding with the known critical period for the development of ocular dominance columns and feature detectors (Benowitz et al., 1989b; Fig. 4).

Persistence of GAP-43 in the adult cortex

Although GAP-43 levels decline sharply in most parts of the brain after synaptic relationships have become stabilized, there are regions where high levels persist throughout life, and this may be related to an ongoing potential for structural remodeling. In the adult rat, for example, levels of GAP-43 are much higher in the forebrain than in the brainstem; within the cerebral cortex, high concentrations are found in limbic and associative regions, whereas primary sensory and motor regions show somewhat less (Benowitz et al., 1988). A greater degree of cortical differentiation is seen in the cat, where sensory and motor areas show very little GAP-43, whereas intense immunostaining persists in such limbic system areas as the hippocampus, parahippocampal gyrus, cingulate cortex, and entorhinal cortex (Benowitz et al., submitted). In the monkey visual system, a marked gradient is seen in the extent of in vitro phosphorylation of this protein (i.e., F1) in going from primary through secondary visual areas (Brodman areas 17,18,19) and up through higher integrative areas, with particularly high levels in the inferior temporal cortex (Nelson et al., 1987). And in the human cerebral cortex, the regional differentiation for GAP-43 and its mRNA is extreme: whereas only low levels of the mRNA and protein are seen throughout all primary and secondary sensory areas or in the motor

74

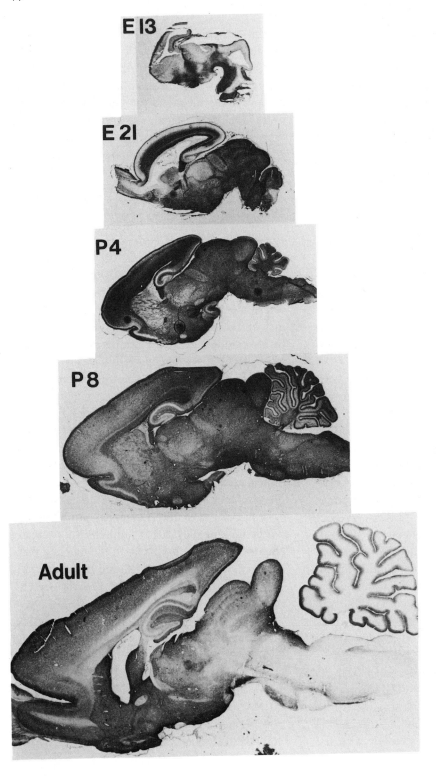

E I3

E 2I

P4

P 8

Adult

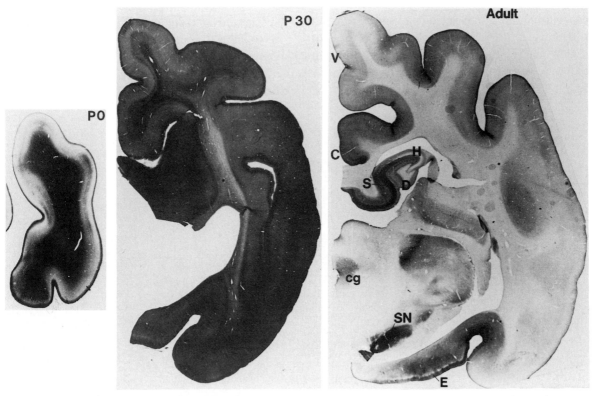

Fig. 4. Developmental changes in GAP-43 immunostaining in the cat cortex. GAP-43 immunostaining is intense in growing axons deep in the cortex at the day of birth (P0), but lighter in the mantle layer. At P30, the height of the critical period, the entire cortical neuropil is densely stained, including the primary visual cortex (V). In the adult, this region is only faintly stained, although other regions of the cortex, including the hippocampus (Hi), subiculum (S), cingulate (C) and entorhinal areas (E), as well as the substantia nigra pars reticulata (SN) of the diencephalon, remain densely stained (Benowitz et al., 1989).

cortex, high concentrations are found in associative and limbic areas. Highest levels of GAP-43 are found in such regions as the inferior temporal cortex, which is involved in higher visual associations; perisylvian structures, which, on the left side of the brain are involved in language functions; the prefrontal cortex, which has been implicated in planning and in the organization of complex behaviors; and in certain portions of the

hippocampal formation, which is important for memory storage. By in situ hybridization, the mRNA appears to be concentrated in the smaller pyramidal cells in layer 2 of the associative neocortex, and this finding accords well with the appearance of the protein in layers 1 and 6 of these same cortical areas. In the hippocampal gyrus, GAP-43 mRNA is abundant in the CA3 pyramidal cells, while the protein is most heavily

←

Fig. 3. GAP-43 is concentrated in growing axons and nerve endings during brain development. Levels peak in the neocortex at postnatal day 4, at the height of synaptogenesis (and the critical period), then decline rapidly, though moderately high levels remain in the molecular layer of the neocortex, basal forebrain, and some other regions throughout life (Dani et al., 1990).

76

concentrated in the projections arising from CA3 neurons (via the Schaeffer collaterals to the CAl region) and in the distribution of the perforant pathway. We have proposed that the presynaptic terminals of these particular neurons may be specialized in their ability to undergo structural remodeling throughout life, and that these elements may contribute to the functional plasticity of the human brain (Neve et al., 1987, 1988; Benowitz et al., 1989; Fig. 5).

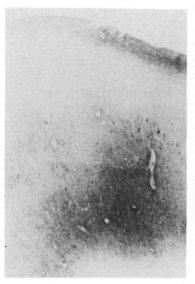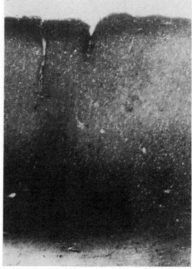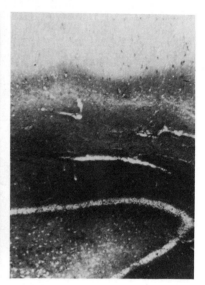

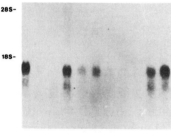

Fig. 5. Distribution of GAP-43 protein and mRNA in the human brain. In the adult human brain, very little GAP-43 remains in sensory or motor areas of the brain, as exemplified by the primary visual cortex (Brodmann area 17, top left). In contrast, higher integrative areas of the brain, such as the inferior temporal cortex (Brodmann area 20, top middle), language cortex, or frontal cortex continue to show high levels throughout life, as does the hippocampus (top right). By Northern blot analysis, the levels of the mRNA are generally found to parallel those of the protein, but with some exceptions. GAP-43 mRNA is scarce in primary and secondary sensory areas (e.g., areas 17,18) and in motor cortex (area 4), but it is extremely high in frontal cortex (area 10), inferior temporal cortex (area 20), and perisylvian language areas (40 and 44). The hippocampus is low, presumably due to the fact that the majority of cells contributing to the RNA pool are granule cells, which do not express high levels of GAP-43. By in situ hybridization, however, the CA3 pyramidal cells are seen to express high levels of the mRNA (Neve et al., 1988). These findings are consistent with a role for the protein in associative mechanisms (Neve et al., 1987, 1988; Benowitz et al., 1989), as is also suggested by its implication in long-term potentiation (Akers and Routtenberg, 1986).

Nerve regeneration

The reemergence of GAP-43 during nerve regeneration has been reviewed in detail (Skene, 1989). High levels of GAP-43 expression appear to be a characteristic feature of all instances of axonal regeneration studied to date, including the regenerating optic nerves of fish and amphibia (Benowitz et al., 1981, 1983; Skene and Willard, 1981a; Perry et al., 1987), regenerating peripheral nerves of mammals (Skene and Willard, 1981b; Bisby, 1988; Hoffman et al., 1989; van der Zee et al., 1989), and damaged CNS neurons of mammals induced to regenerate through grafts of peripheral nerves (Willard, chapter 2, this volume). In the case of peripheral nerve injury, ganglionic neurons transport high levels of GAP-43 not only down the regenerating peripheral branch but also down their intact, centrally-directed processes (Woolf et al., 1990; Erzurumlu et al., 1989). From clinical neurology, it is known that when damaged peripheral nerves regenerate, there can be an increased sensation of pain (i.e., hyperalgesia) from previously neutral stimuli. The finding that GAP-43 is transported bidirectionally from ganglionic neurons provides a possible explanation for this phenomenon, in that the increased levels of the protein conveyed down the central processes may dispose their nerve endings to undergo adventitious remodeling. If so, fibers that normally convey tactocutaneous stimuli might begin to form synapses in zones that receive noxious inputs (e.g., laminae I and II of Rexed) and hence be transmitted to higher levels of the neuraxis as being painful.

Synaptic remodeling

To examine the hypothesis that GAP-43 in the adult nervous system is associated with synapses that undergo structural remodeling, we examined a well-characterized model of synaptic reorganization: when the perforant pathway from the entorhinal cortex is lesioned, other neurons that project to the dentate gyrus sprout axon collaterals that form synapses onto denervated portions of the granule cell dendrites (Cotman et al., 1973; Lynch et al., 1976). Within the first two days after lesioning the perforant pathway, GAP-43 levels increase greatly in the terminals of the commissural-associational pathway and remain elevated for a few days, as these nerve endings begin to sprout. The immunostaining then declines somewhat, but continues to define an expanded com-

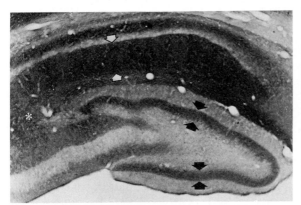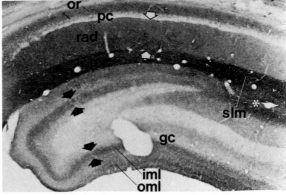

Fig. 6. GAP-43 is involved in synaptic remodeling in the adult CNS. Following lesions of the perforant pathway, levels of GAP-43 increase markedly in the inner molecular layer of the dentate gyrus ipsilateral to the lesion (black arrows, left), coincident with the collateral sprouting of commissural-associational axons originating from pyramidal cells in the hilus and CA3 region (Benowitz et al., 1990).

missural-associational projection field for at least a month. These results indicate that increases in GAP-43 synthesis occur in relation to the reorganization of synaptic relationships, even in the absence of extensive axonal elongation (Benowitz et al., 1990; Norden et al., 1989; Schrama et al., 1989; Fig. 6).

Ultrastructural localization

Electron microscopic studies point to GAP-43 having a predominantly presynaptic localization (Gispen et al., 1985; Verhaagen et al., 1989; Norden et al., 1987; van Lookeren Campagne et al., 1990), as would be expected from axonal transport studies (Benowitz et al., 1981, 1983; Skene and Willard, 1981), its distribution in cell culture (e.g., Meiri et al., 1988; Goslin et al., 1989), and from the immunolocalization pattern found at the light microscopic level (Oestreicher and Gispen, 1986; McGuire et al., 1988; Benowitz et al., 1988, 1989a,b, 1990). In an electron microscopic study of the adult rat striatum, GAP-43 was found to be localized in a subset of nerve endings, most of which were large and asymmetric. Such terminals are known to arise from the corticostriate projections, which would be consistent with the high levels of GAP-43 mRNA that are found in cortical pyramidal cells (Rosenthal et al., 1987; Neve et al., 1989). The protein was concentrated mostly in preterminal regions apposed to spines, though these often did not contain synaptic specializations. Within these regions, the reaction product was seen on the plasma membrane and on vesicles (DiFiglia et al., 1990). This suggests that the protein could be involved in vesicle translocation or fusion, and may be associated with areas in which new growth is occurring. An issue that remains unresolved, however, is whether the vesicular localization of the protein as described by several investigators (DiFiglia et al., 1990; Norden et al., 1987; Verhaagen et al., 1988) is an artifact of the DAB reaction product diffusing from sites on the plasma membrane, where studies using immunogold labeling suggest it to be (van Lookeren Campagne et al., 1990). Our understanding of the role that the protein plays in the nerve ending will clearly be enhanced by a resolution of this issue.

How is the expression of GAP-43 regulated?

During nerve development and regeneration, levels of GAP-43 mRNA increase just prior to the changes in protein levels (Basi et al., 1987; Karns et al., 1987; Federoff et al., 1988; Hoffman, 1989; Van der Zee et al., 1989; Hoffman et al., 1989; Costello et al., 1990; LaBate and Skene, 1989; Perrone-Bizzozero et al., 1990). A close parallel between levels of the protein and the mRNA is also observed regionally in the human cerebral cortex (Neve et al., 1987,1988; Benowitz et al., 1989). Thus, levels of GAP-43 synthesis appear to be determined primarily by the steady-state level of the mRNA. The question becomes, then, What mechanisms control the steady-state levels of the mRNA? Although one study reported that the changes in GAP-43 levels occurring during brain development are accompanied by increases in the rate of transcription of the gene (Basi et al., 1987), another study found that the gene was transcribed at approximately the same rate in the developing brain and in the adult (Perrone-Bizzozero et al., 1990b). Nuclear run-on experiments in PC-12 cells likewise indicate that the increases in GAP-43 mRNA levels that occur in response to NGF treatment are not paralleled by a change in the rate of gene transcription (Federoff et al., 1988; Perrone-Bizzozero et al., 1990b). Moreover, in the goldfish retina, where synthesis of the protein increases 100-fold during the first two weeks of optic nerve regeneration, the rate of transcription of the gene appears not to change appreciably (Perrone-Bizzozero et al., 1990a; Fig. 7).

Post-transcriptional regulation

Several observations support the possibility that GAP-43 levels may be regulated by a post-transcriptional mechanism, perhaps involving the rate of mRNA degradation. As mentioned above, in the intact goldfish retina or in undifferentiated PC-12 cells, although the GAP-43 gene is constitutively transcribed at a moderately high level, the mRNA cannot be detected, and this implies that it is being degraded rapidly. Sequences similar to those known to confer instability upon other mRNAs (Shaw and Kamen, 1986; Pandey and Marzluff, 1987), are found in the 3' untranslated region of rat, human, and mouse GAP-43 mRNA (Ng et al., 1989), and some of these sequences can also be traced back to the goldfish mRNA (Perrone-Bizzozero et al., 1990). Such sequences, along with the polyadenylation site, are binding sites for specific proteins that are believed to modulate ribonuclease activity (Malter et al., 1989; Bernstein, 1989). At the same time, analysis of the GAP-43 gene points to a possible absence of transcriptional regulation: no recognizable promoter or enhancer regions are seen up to 3 kb upstream from the transcription start site (Grabczyk et al., 1989; Nedivi et al., 1989). However, there are several stretches of guanine-adenosine repeats that are thought to promote the formation of H-DNA (Wells et al., 1988); this structure is characterized by a triple helix conformation that leaves a piece of DNA single-stranded, and this might always remain "open" for transcription. Data from our laboratory indicate that the changes in GAP-43 mRNA levels during NGF-induced neuronal differentiation of PC-12 cells may be due in part to an increased half-life of the mRNA. These findings, when taken together, support the hypothesis that the rate at which the GAP-43 gene is transcribed is not altered during neuronal differentiation, and that one control point for the expression of GAP-43 is likely to be post-transcriptional, involving

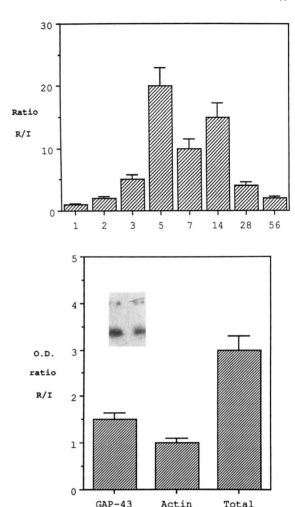

Fig. 7. GAP-43 mRNA levels change dramatically during regeneration of the goldfish optic nerve, but may not be regulated by changes in the rate of gene transcription. Top: Changes in the levels of GAP-43 mRNA in retinas at various times after nerves were crushed in vivo. Bottom: Transcription run-on assays performed in nuclei isolated from intact (I) and regenerating (R) goldfish retinas (day 5 post-crush). Inset shows slot blot hybridization to GAP-43 (top) and actin (bottom) cDNA clones. Results from quantitation of autoradiograms from four independent experiments are expressed relative to the values of the intact nerve (Perrone-Bizzozero et al., 1990).

changes in the rate of mRNA degradation. It might be noted that similar disparities between transcription and accumulation of mRNA have been reported in the regulation of other neuronal and non-neuronal proteins (Stein et al., 1988; Lindenbaum et al., 1988; Messina, 1989; Cleveland and Havercroft, 1983; Fernyhough et al., 1989).

Factors controlling GAP-43 mRNA levels

Studies in peripheral nerves indicate that the expression of the mRNA is regulated in part by an inhibitory factor that travels up the nerve in the retrograde direction. Vinblastine, a vinkyl

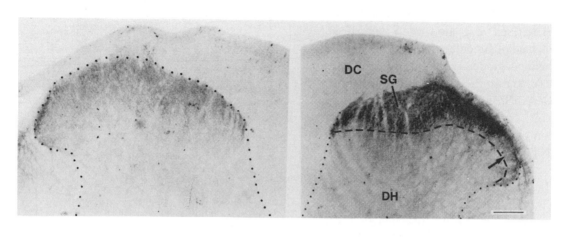

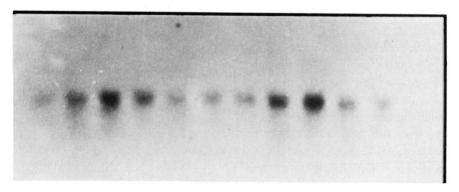

Fig. 8. The expression of GAP-43 appears to be regulated by a retrogradely transported factor. Top: levels of GAP-43 increase in the dorsal horn of the spinal cord after crushing the sciatic nerve or after applying vinblastine (left side, the control half of the spinal cord contralateral to the nerve crush; right side, ipsilateral to the experimental manipulations, showing much higher levels of the protein). Bottom: more direct evidence that GAP-43 expression is normally suppressed by a retrogradely transported factor comes from experiments in which nerves were left intact but treated with vinblastine to block axonal transport. Analysis of GAP-43 mRNA levels in ganglionic neurons was done by Northern blots at the indicated times. At 4–10 days, increases in GAP-43 mRNA levels were as great after vinblastine treatment as after nerve crush (studies done in collaboration with Drs. C.J. Woolf, C. O'Brien, N.I. Perrone-Bizzozero, and N. Irwin).

alkyloid that depolymerizes microtubules and arrests axonal transport bidirectionally, leads to elevated levels of GAP-43 mRNA in dorsal root ganglia when applied to intact peripheral nerves (and to increased levels of the protein in both the regenerating peripheral branch and in the dorsal horn of the spinal cord: Woolf et al., 1990; Fig. 8). Another finding that points to a retrogradely transported inhibitory factor is that when peripheral nerves undergoing regeneration are ligated, the elevated GAP-43 levels that are induced by axotomy fail to decrease (Bisby, 1988). The inhibitory factor could arise from the neuron itself, in which some of the proteins synthesized in the soma are transported in the anterograde direction and then return, either post-translationally modified or unchanged, depending upon local interactions along the axon or at the nerve ending (Aquino et al., 1987; Martz et al., 1989). Other possible sources of such factors are the cells that surround the axon (e.g., astrocytes, oligodendroglia, or Schwann cells), which could provide diffusible material or cell-surface contact signals; or cells with which the neuron forms synaptic contacts, e.g., muscle in the peripheral nervous system or postsynaptic neurons in the central nervous system. In any event, the regulation of GAP-43 levels must be a complex affair, and cannot be explained by any one or two factors alone. For example, although suppression of a retrogradely transported signal is sufficient to trigger a build-up of GAP-43 mRNA in peripheral nerves, it is not sufficient to trigger it in the central nervous system. This implies that additional factors must arise from the non-neuronal environment. Yet, the existence of glial factors is still insufficient to account for the fact that even within a given area of the brain, some neurons express high levels of GAP-43, while near neighbors do not (e.g., in the human cortex, where small pyramidal cells in layer 2 of associative regions express higher levels than adjacent populations). It is likely that a multiplicity of factors – some by the extracellular environment, some controlled by the establishment of synaptic connections, and some determined by the neuron's individual identity – all play upon GAP-43 expression. At this point, it is unknown whether all of these factors converge upon one common regulatory mechanism or whether they control different stages of the process (i.e., some may induce a small transcriptional change, while others might influence hnRNA processing, mRNA stability, transcriptional efficiency, or post-translational processing).

Why do neurons make GAP-43?

A detailed molecular description of how GAP-43 interacts with other constituents of the cell to affect its functional state is not yet available. Certainly, findings described in other chapters of this volume, showing that the protein interacts in vitro with calmodulin (Alexander et al., 1987), with the α-subunit of the GTP-binding, signal transduction molecule G_o (Strittmatter et al., 1990), with phospholipid metabolism (via PIP kinase: Jolles et al., 1980), and with the membrane cytoskeleton (Moss and Allsopp, 1989; Meiri and Gordon-Weeks, 1990), are all important leads, as are the findings that GAP-43 influences Ca^{2+} fluxes (Freeman et al., 1988) and transmitter release (Dekker et al., 1989). However, it may still be some time before we understand the interrelationship of these diverse observations, or before we obtain additional pieces of the puzzle that may still be essential to explain the molecular role of the protein in growth and plasticity.

Nevertheless, the case for the protein being essential for growth and plasticity is increasing. Transfection of the GAP-43 gene into a non-neuronal cell line has been reported to increase the extension of filopodia (Zuber et al., 1989), while in PC-12 pheochromocytoma cells, it seems to augment the rate at which the cells extend neurites in response to NGF (Yankner et al., 1990; Fig. 9). This enhancement is manifest both in the increased sensitivity of the cells to NGF (i.e., lower concentrations of the growth factor are required for the same level of cellular response) and by an increased rate of neurite outgrowth in response to NGF. Interestingly, although transfected PC-12 cells expressed high levels of the mRNA even without NGF, the protein was not found to be translated prior to NGF treatment. This suggests that in addition to increasing the rate of mRNA accumulation (see above), NGF could also have an effect on translational efficiency.

Effects of blocking GAP-43

Experiments complementary to these show that suppression of the protein, either by immunologic or genetic means, profoundly decreases neurite outgrowth. In NB2A/d1 neuroblastoma cells, there is a constitutive level of GAP-43 expression that increases only slightly when cells are stimulated to extend processes. When neurite outgrowth is induced by agents such as dibutyryl cyclic AMP, GAP-43 gets translocated down the growing processes. Antibodies to GAP-43, introduced into these cells by transiently permeabilizing them with lysophosphatidylcholine and glycerol, arrests process outgrowth in a dose-dependent fashion (Shea et al., 1991; Fig. 10). As a revealing control, a population of cells that was transiently permeabilized and treated with antibodies to GAP-43 before the induction of process outgrowth was subsequently reacted with an appropriately labeled secondary antibody to visualize the binding of the primary antibody. Whereas cells that had successfully taken up the primary antibody were undifferentiated, individual cells that failed to take up the primary antibody did show neurite outgrowth. The effects of the antibody treatment were only seen in the first few hours of differentiation. This implies that GAP-43

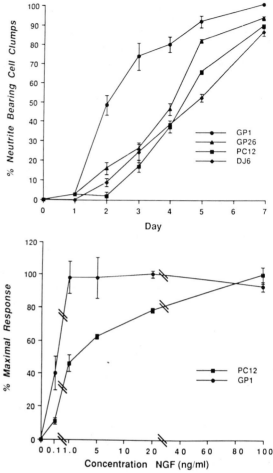

Fig. 9. Transfection of the GAP-43 gene into PC-12 pheochromocytoma cells enhances responsiveness to NGF. Top: PC-12 cells showing highest levels of GAP-43 expression (GP1) show a much more rapid extension of neurites in response to NGF than control cells (PC-12), cells transfected with the vector but without the GAP-43 insert (DJ6), or successfully transfected cells that did not express the mRNA to the same extent (GP26). Bottom: transfected GP1 cells also showed a much greater sensitivity to NGF than control PC12 cells, with maximal outgrowth occurring in response to <1 ng/ml NGF, compared with almost 100 ng/ml for control cells (Yankner et al., 1990).

is not needed for maintaining the integrity of existing neurites once formed, but is necessary for the active stage of growth (Shea et al., 1991). This in turn is consistent with the fact that although

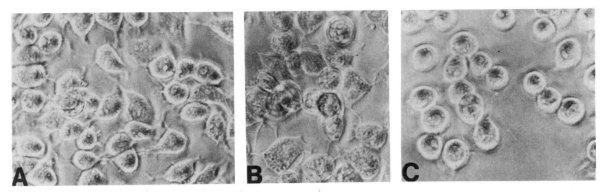

Fig. 10. Delivery of antibodies to GAP-43 into neuroblastoma cells inhibits neurite outgrowth. Cells were exposed to no antibody (A), or to 1:20 dilutions of preimmune serum (B) or a-GAP serum (C), then treated with dbcAMP for 4 h. Phase-contrast images demonstrate that antibodies to GAP-43 selectively inhibit neuritogenesis and may even cause cell bodies to revert to a nearly undifferentiated state marked by rounded somata (Shea et al., submitted).

neurons express high levels of GAP-43 during process outgrowth, most require only low levels after stable connections have formed.

Another approach to suppressing the expression of GAP-43 is through the use of oligonu-

cleotides complementary to the messenger RNA. We constructed an *antisense* 20-mer cDNA complementary to a portion of GAP-43 mRNA that includes the translation start site. In primary neuronal cultures (derived from embryonic day 17 rat

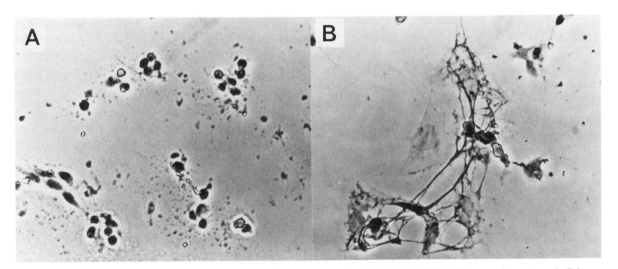

Fig. 11. A 20-kb oligonucleotide complementary to the translation start site of GAP-43 mRNA inhibits neurite outgrowth. Primary cortical embryonic cultures were induced to differentiate in the presence of serum and transfected with either an antisense oligonucleotide complementary to 20 bases of GAP-43 mRNA (left), or with a control oligonucleotide constructed in the sense orientation (right). These results point to GAP-43 as playing an essential role in the growth of neuronal processes (L.I. Benowitz, L. Dawes, C. Irwin, N.I. Perrone-Bizzozero, C. O'Brien and R.L. Neve, unpublished data).

cortex), such oligonucleotides inhibited process outgrowth, whereas oligonucleotides in the sense orientation did not have a suppressive effect (L.I. Benowitz, L. Dawes, N.I. Perrone-Bizzozero, N. Irwin, C. O'Brien and R.L. Neve, unpublished observations).

Hence, whereas introducing the GAP-43 gene into cells seems to augment neurite outgrowth, inhibition of the protein, either through immunological or genetic means, suppresses it. It might therefore be concluded that the expression of the protein is not just a correlate of outgrowth or an epiphenomenon: rather, the protein in and of itself seems to be necessary for outgrowth to occur. The data linking GAP-43 to a number of biochemical pathways and signal transduction mechanisms support some models of its molecular significance but do not yet add up to an "explanation" of how this protein, through its interactions with other cellular constituents, causes growth to occur. The finding that antibodies to GAP-43 (i.e., B-50) interfere with the release of noradrenaline from cortical synaptosomes suggests that the protein could contribute to the movement and fusion of membranous vesicles, and this might be its role in growth as well; moreover, the finding that antibodies to the protein interfere with Ca^{2+} fluxes through the nerve terminal (Freeman, 1988) suggests that the effects on transmitter release may be secondary to regulating the Ca^{2+} entry step, a process that would be expected to be critical for growth as well (Kater et al., 1988). At the very least, the relationship of GAP-43 expression to the development and reorganization of neuronal connections affords us a unique means of visualizing where and when these events take place in the nervous system; beyond this, an understanding of how the protein is regulated, and how it functions in the nerve terminal, promises to contribute significantly to a deeper understanding of the molecular bases of growth and plasticity.

References

Adams, C.E., Benowitz, L.I. and Finger, T.E. (1990) Tyrosine hydroxylase- and GAP-43-like immunoreactivities in the developing nigrostriatal system of the rat, submitted.

Akers, R.F. and Routtenberg, A. (1985) Protein kinase C phosphorylates a 47 Mr protein directly related to synaptic plasticity. *Brain Res.*, 334: 147–151.

Alexander, K.A., Cimler, B.M., Meier, K.E. and Storm, D.R. (1987) Regulation of calmodulin binding to P-57. *J. Biol. Chem.*, 263: 7544–7549.

Aloyo, V.J., Zwiers, H. and Gispen, W.H. (1983) Phosphorylation of B-50 protein by calcium-activated, phospholipid-dependent protein kinase and B-50 protein kinase. *J. Neurochem.*, 41: 649–653.

Aquino, D.A., Bisby, M.A., and Ledeen, R.W. (1987) Bidirectional transport of gangliosides, glycoproteins and neutral glycosphingolipids in the sensory neurons of the rat sciatic nerve. *Neuroscience*, 20: 1023–1029.

Basi, G.S., Jacobson, R.D., Virag, I., Schilling, J. and Skene, J.H.P. (1987) Primary structure and transcriptional regulation of GAP-43, a protein associated with nerve growth. *Cell*, 49: 785–791.

Belford, G.R. and Killackey, H.P. (1980) The sensitive period in the development of the trigeminal system of the neonatal rat. *J. Comp. Neurol.*, 193: 335–350.

Benowitz, L.I. and Lewis, E.R. (1983) Increased transport of 44,000- to 49,000-dalton acidic proteins during regeneration of the goldfish optic nerve: a two-dimensional gel analysis. *J. Neurosci.*, 3: 2153–2163.

Benowitz, L.I. and Routtenberg, A. (1987) A membrane phosphoprotein associated with neural development, axonal regeneration, phospholipid metabolism, and synaptic plasticity. *Trends Neurosci.*, 10: 527–532.

Benowitz, L.I., Shashoua, V.E. and Yoon, M.G. (1981) Specific changes in rapidly-transported proteins during regeneration of the goldfish optic nerve. *J. Neurosci.*, 1: 300–307.

Benowitz, L.I., Apostolides, P.J., Perrone-Bizzozero, N.I., Finklestein, S.P. and Zwiers, H. (1988) Anatomical distribution of the growth-associated protein GAP-43/B-50 in the adult rat brain. *J. Neurosci.*, 8: 339–352.

Benowitz L.I., Perrone-Bizzozero N.I., Finklestein S.P. and Bird, E.D. (1989a) Localization of the growth-associated phosphoprotein GAP-43 in the human cerebral cortex. *J. Neurosci.*, 9: 990–995.

Benowitz, L.I., Rodriguez, W.R., Prusky, G.T. and Cynader, M.S. (1989b) GAP-43 levels in cat striate cortex peak during the critical period. *Neuroscience Abstr.*, 15: 796; ms. submitted.

Benowitz, L.I., Rodriguez, W.R. and Neve, R.L. (1990) The pattern of GAP-43 immunostaining changes in the rat hippocampal formation during reactive synaptogenesis. *Mol. Brain Res.*, 8: 17–23.

Bernstein, P. and Ross, J. (1989) Poly(A), poly(A) binding

protein and the regulation of mRNA stability. *Trends Biochem.*, 14: 373–377.

Bisby, M.A. (1988) Dependence of GAP43 (B50, F1) transport on axonal regeneration in rat dorsal root ganglion neurons. *Brain Res.*, 458: 157–161.

Cleveland, D.W. and Havercroft, J.C. (1983) Is apparent autoregulatory control of tubulin synthesis nontranscriptionally regulated? *J. Cell. Biol.*, 97: 919–924.

Costello, B., Meymandi, A. and Freeman, J.A. (1990) Factors influencing GAP-43 gene expression in PC-12 pheochromocytoma cells. *J. Neurosci.*, 10: 1398–1406.

Cotman, C.W., Matthews, D.A., Taylor, D. and Lynch, G. (1973) Synaptic rearrangement in the dentate gyrus: histochemical evidence of adjustments after lesions in immature and adult rats. *Proc. Natl. Acad. Sci. USA*, 70: 3473–3477.

Dani, J.W., Armstrong, D.M. and Benowitz, L.I. (1990) Mapping the development of the rat brain by GAP-43 immunocytochemistry. *Neuroscience*, 40: 277–288.

De Graan, P.N.E., Heemskerk, F.M.J., Dekker, L.V., Melchers, B.P.C., Gianotti, C. and Schrama, L.H. (1987) Phorbol esters induce long- and short-term enhancement of B-50/GAP-43 phosphorylation in rat hippocampal slices. *Neurosci. Res. Commun.*, 3: 175–182.

Dekker, L.V., De Graan, P.N.E., Oestreicher, A.B., Versteeg, D.H.G. and Gispen, W.H. (1989) Inhibition of noradrenaline release by antibodies to B-50 (GAP-43). *Nature*, 34: 274–276.

De la Monte, S.M., Federoff, H.J., Ng, S.-C., Grabczyk, E. and Fishman, M.C. (1989) GAP-43 gene expression during development: persistence in a distinctive set of neurons in the mature central nervous system. *Dev. Brain Res.*, 46: 161–168.

DiFiglia, M., Roberts, R.C. and Benowitz, L.I. (1990) Immunoreactive GAP-43 in adult rat caudate neuropil: localization in unmyelinated fibers, axon terminals and dendric spines. *J. Comp. Neurol.*, 302: 992–1001.

Erzurumlu, R.S., Jhaveri, S., Moya, K.L. and Benowitz, L.I. (1989) Peripheral nerve regeneration induces elevated expression of GAP-43 in the brainstem trigeminal complex of adult hamster. *Brain Res.*, 498: 135–139.

Erzurumlu, R.S., Jhaveri, S. and Benowitz, L.I. (1990) Transient patterns of GAP-43 expression during the formation of barrels in the rat somatosensory cortex. *J. Comp. Neurol.*, 292: 443–456.

Federoff, H.J., Grabczyk, E. and Fishman, M.C. (1988) Dual regulation of GAP-43 gene expression by nerve growth factor and glucocorticoids. *J. Biol. Chem.*, 263: 19290–19295.

Fernyhough, P., Mill, J.F., Roberts, J.L. and Ishii, D.N. (1989) Stabilization of tubulin mRNAs by insulin and insulin-like growth factor I during neurite formation. *Mol. Brain Res.*, 6: 109–120.

Fitzgerald, M., Reynolds, M.L. and Benowitz, L.I. (1991) The development of the rat lumbar spinal cord: a GAP-43 immunocytochemical study. *Neuroscience*, 41: 187–189.

Freeman, J.J., Lettes, A.A. and Costello, B. (1988) Possible role of GAP-43 in calcium regulation/transmitter release. *Neurosci. Abstr.*, 14: 1126.

Gispen, W.H., Leunissen, L.M., Oestreicher, A.B., Verkleij, A.J. and Zwiers, H. (1985) Presynaptic localization of B-50 phosphoprotein: the ACTH-sensitive protein kinase substrate involved in rat brain phosphoinositide metabolism. *Brain Res.*, 328: 381–385.

Gorgels, T.G.M.F, van Lookeren Campagne, M., Oestreicher, A.B., Gribnau, A.A.M. and Gispen, W.H. (1989) Ultrastructural localization of B-50/GAP-43 in developing and mature pyramidal tract in the rat: predominant localization at the cytoplasmic side of the plasma membrane. *J. Neurosci.*, 9: 3861–3869.

Goslin, K., Schreyer, D.J., Skene, J.H.P. and Banker, G. (1988) Development of neuronal polarity: GAP-43 distinguishes axonal from dendritic growth cones. *Nature*, 336: 672–674.

Grabczyk, E., Zuber, M.X., Federoff, H.,.Ng, S.C., Pack, A. and Fishman, M.C. (1989) GAP-43 gene structure. *Neurosci. Abst.*, 15: 957.

Hoffman PN. (1989) Expression of GAP-43, a rapidiy transported growth-associated protein, and class II beta tubulin, a slowly transported cytoskeletal protein, are coordinated in regenerating neurons. *J. Neurosci.*, 9: 893–897.

Jacobson, R.D., Virag, I. and Skene, J.H.P. (1986) A protein associated with axon growth, GAP-43, is widely distributed and developmentally regulated in rat CNS. *J. Neurosci.*, 6: 1843–1855.

Jolles, J., Zwiers, H., van Dongen, C., Schotman, P., Wirtz, K.W.A. and Gispen, W.H. (1980) Modulation of brain polyphosphoinositide metabolism by ACTH-sensitive protein phosphorylation. *Nature*, 286: 623–625.

Karns, L.R., Ng, S.-C., Freeman, J.A., and Fishman, M.C. (1987) Cloning of complementary DNA for GAP-43, a neuronal growth-related protein. *Science*, 236: 597–600.

Kater, S.B., Mattson, M.P., Cohan, C. and Connor, J. (1988) Calcium regulation of the neuronal growth cone. *Trends Neurosci.*,, 11: 315–321.

LaBate, M.E. and Skene, J.H.P. (1989) Selective conservation of GAP-43 structure in vertebrate evolution. *Neuron*, 3: 299–310.

Lindenbaum, M.H., Carbonetto, S., Grosveld, F., Flavell, D. and Mushynski, W.E. (1988) Transcriptional and posttranscriptonal effects of nerve growth factor on expression of three neurofilaments subunits in PC-12 cells. *J. Biol. Chem.*, 263: 5662–5667.

Lynch, G., Gall, C., Rose, G. and Cotman, C. (1976) Changes in the distribution of the dentate gyrus associational system following unilateral or bilateral entorhinal lesion in the adult rat. *Brain Res.*, 110: 57–71.

86

Malter, J. (1989) Identification of an AUUUA-specific messenger RNA binding protein. *Science*, 246: 664–666.

Martz, D., Garner, J., and Lasek, R.J. (1989) Protein changes during anterograde-to-retrograde conversion of axonally transported vesicles. *Brain Res.*, 476: 199–203.

McGuire, C.B., Snipes, G.J. and Norden, J.J. (1988) Light-microscopic immunolocalization of the growth-associated protein GAP-43 in the developing brain. *Dev. Brain Res.*, 41: 277–291.

Meiri, K.F. and Gordon-Weeks, P.R. (1990) GAP-43 in growth cones is associated with areas of membrane that are tightly bound to substrate and is a component of a membrane skeleton subcellular fraction. *J. Neurosci.*, 10: 256–266.

Meiri, K., Pfenninger, K.H. and Willard, M. (1986) Growth-associated protein, GAP-43, a polypeptide that is induced when neurons extend axons, is a component of growth cones and corresponds to pp46, a major polypeptide of a subcellular fraction enriched in growth cones. *Proc. Natl. Acad. Sci. USA*, 83: 3537–3541.

Meiri, K.F., Willard, M. and Johnson, M.I. (1988) Distribution and phosphorylation of the growth-associated protein GAP-43 in regenerating sympathetic neurons in culture. *J. Neurosci.*, 8: 2571–2581.

Messina, J.L. (1989) Insulin and dexamethasone regulation of a rat hepatoma messenger ribonucleic acid: insulin has a transcriptional and a post-transcriptional effect. *Endocrinology*, 124: 754–761.

Moss, D.J., Fernyhough, P., Chapman, K., Baizer, L., Bray, D., and Allsopp, T. (1990) Chicken growth-associated protein GAP-43 is tightly bound to the actin-rich neuronal membrane skeleton. *J. Neurochem.*, 54: 729–736.

Moya, K.L., Benowitz, L.I., Jhaveri, S. and Schneider, G.E. (1988) Changes in rapidly transported proteins in developing hamster retinofugal axons. *J. Neurosci.*, 8: 4445–4454.

Moya, K.L., Jhaveri, S., Schneider, G.E. and Benowitz, L.I. (1989) Immunohistochemical localization of GAP-43 in the developing hamster retinofugal pathway. *J. Comp. Neurol.*, 288: 51–58.

Nedivi, E., Basi, G.S. and Skene, I.H.P. (1989) Structural and potential regulatory domains of a rat GAP-43 gene. *Neurosci. Abstr.*, 15: 1269.

Nelson, R.B. and Routtenberg, A., (1985) Characterization of protein F1 (47 kDa, 4.5 pI): a kinase C substrate directly related to neural plasticity. *Exp. Neurol.*, 89: 213–224.

Neve, R.L., Perrone-Bizzozero, N.I., Finklestein, S.P., Zwiers, H., Bird, E., Kurnit, D.M. and Benowitz, L.I. (1987) The neuronal growth-associated protein GAP-43 (B-50, F1): neuronal specificity, developmental regulation and regional distribution of the human and rat mRNAs. *Mol. Brain Res.*, 2: 177–183.

Neve, R.L., Finch, E.A., Bird, E.D. and Benowitz, L.I. (1988) The growth-associated protein GAP-43 (B-50, F1) is expressed selectively in associative regions of the adult human brain. *Proc. Natl. Acad. Sci. USA*, 85: 3638–3642.

Ng, S.-C., de la Monte, S.M., Conboy, G.L., Karns, L.R., and Fishman, M.C. (1988) Cloning of human GAP-43: growth association and ischemic resurgence. *Neuron*, 1: 133–139.

Norden, J.J., Costello, B. and Freeman, J.A. (1987) The growth-associated protein, GAP43, is associated synaptic vesicles and with plasma membrane in presynaptic terminals. *Neurosci. Abstr.*, 13: 1480.

Norden, J.J., Woltjer, R. and Steward, O. (1988) Changes in the immunolocalization of the growth and plasticity-associated protein GAP-43 during lesion-induced sprouting in the rat dentate gyrus, *Neurosci. Abstr.,,* 14: 116.

Oestreicher, A.B. and Gispen, W.H. (1986) Comparison of the immunocytochemical distribution of the phosphoprotein B-50 in the cerebellum and hippocampus of immature and adult rat brain. *Brain Res.*, 375: 267–279.

Pandey, N.B. and Marzluff, W.F.,(1987) The stem-loop structure at the 3′ end of the histone mRNA is necessary and sufficient for regulation of histone mRNA stability. *Mol. Cell. Biol.*, 7: 4557–4559.

Perrone-Bizzozero, N.I., Finklestein, S.P. and Benowitz, L.I. (1986) Synthesis of a growth-associated protein by embryonic rat cerebrocortical neurons in vitro. *J. Neurosci.*, 6: 3721–3730.

Perrone-Bizzozero, N.I., Neve, R.L Franck, E, Gossels, J., Irwin, N. and Benowitz, L.I. (1990a) Regulation of GAP-43 mRNA levels in the regenerating goldfish optic pathway, submitted.

Perrone-Bizzozero, N.I., Irwin, N., Lewis, S.E., Fischer, I., Neve, R.L.and Benowitz, L.I. (1990b) Posttranscriptional regulation of GAP-43 mRNA levels during process outgrowth. *Neurosci. Abstr.*, 16: 814.

Perry, G.W., Burmeister, D.W. and Grafstein, B. (1987) Fast axonally transported proteins in regenerating goldfish optic axons. *J. Neurosci.*, 7: 792–806.

Pfenninger, K.H. (1986) Of nerve growth cones, leukocytes and memory: second messenger systems and growth-regulated proteins. *Trends Neurosci.*, 9: 562–565.

Reh, T.A., Redshaw, J.D., and Bisby, M.A. (1987) Axons of the pyramidal tract do not increase their transport of growth associated proteins after axotomy. *Mol. Brain Res.*, 2: 1–6.

Reynolds, M.L., Fitzgerald, M. and Benowitz, L.I. (1991) The development of cutaneous and muscle innervation in the rat hind limb: a study of peripheral axon growth using GAP-43 immunocytochemistry. *Neuroscience*, 41: 201–211.

Rosenthal, A., Chan, S.Y., Henzel, W., Haskell, C., Kuang, W.-I., Chen, E., Wilcox, J.N., Ullrich, A., Goeddel, D.V. and Routtenberg, A. (1987) Primary structure and mRNA localization of protein F1, a growth-related protein kinase C substrate associated with synaptic plasticity. *EMBO J.*, 6: 3641–3646.

Schrama, L.H., Eggen, B.J.C., Nielander, H.B., Schotman, P., Oestreicher, A.B., Spruijt, B.M. and Gispen, W.H. (1989) Increased expression of the growth-associated protein B-50

(GAP-43) in unilateral fimbria-fornix lesioned rats. *Neurosci. Abstr.*, 15: 319.

Shaw, G. and Kamen, R. (1986) Regulation of GM-CSF mRNA mediates selective mRNA degradation. *Cell*, 46: 659–667.

Shea, T.B., Perrone-Bizzozero, N.I., Beermann, M.L. and Benowitz, L.I. (1991) Phospholipid-mediated delivery of anti-GAP-43 antibodies into neuroblastoma cells prevents neuritogenesis, *J. Neurosci.*, 11: 1685–1690.

Skene, J.H.P. (1989) Axonal growth-associated proteins, *Annu. Rev. Neurosci.*, 12: 127–156

Skene, J.H.P. and Willard, M. (1981a) Changes in axonally transported proteins during axon regeneration in toad ganglion cells. *J. Cell Biol.*, 89: 86–95.

Skene, J.H.P. and Willard, M. (1981b) Axonally transported proteins associated with axon growth in rabbit central and peripheral nervous system. *J. Cell Biol.*, 89: 96–103.

Snipes, G.J., McGuire, C.B., Chan, S., Costello, B.R., Norden, J.J., Freeman, J.A. and Routtenberg, A. (1987) Evidence for the coidentification of GAP-43, a growth-associated protein, and F1, a plasticity-associated protein. *J. Neurosci.*, 7: 4066–4075.

Stein, R., Orit, S. and Anderson, D. (1988) The induction of a neural-specific gene, SCG10, by nerve growth factor is transcriptional, protein synthesis dependent, and glucocorticoid inhibitable. *Dev. Biol.*, 127: 316–325.

Strittmatter, S.M., Valenzuela, D., Kennedy, T.E., Neer, E.J., and Fishman, M.C. (1990) G_o is a major growth cone protein subject to regulation by GAP-43. *Nature*, 344: 836–841.

Van der Zee, C.E.E.M., Nielander, H.B., Vos, J.P., da Silva, S.L., Verhaagen, J., Oestreicher, A.B., Schrama, L.H., Schotman, P. and Gispen, W.H. (1989) Expression of growth-associated protein B-50 (GAP43) in dorsal root ganglia and sciatic nerve during regenerative sprouting. *J. Neurosci.*, 9: 3505–3512.

Van Lookeren Campagne, M., Oestreicher, A.B., van Bergen en Henegouwen, P.M.P. and Gispen, W.H. (1989) Ultrastructural immunocytochemical localization of B-50/GAP-43, a protein kinase C substrate, in isolated presynaptic nerve terminals and neuronal growth cones. *J. Neurocytol.*, 18: 479–489.

Verhaagen, J., Oestreicher, A.B., Edwards, H., Veldman, F., Iennekens, G.I. and Gispen, W.H. (1988) Light and electromicroscopic study of phosphoprotein B-50 following denervation and reinnervation of the rat soleus muscle. *J. Neurosci.*, 8: 1759–1766.

van Hooff, C.O.M., Holthius, J., Oestreicher, A.B., Boonstra, J., De Graan, P.N.E., and Gispen, W.H. (1989) Nerve growth factor-induced changes in the intracellular localization of the protein kinase C substrate B-50 in pheochromocytoma PC-12 cells. *J. Cell Biol.*, 108: 1115–1125.

Wells, R.D., Collier, D.A., Hanvey, J.C., Shimizu, M.and Wohlrab, F. (1988) The chemistry and biology of unusual DNA structures adopted by oligopurine-oligopyrimidine sequences. *FASEB J.*, 2: 2939–2949.

Woolf, C.J., Molander, C., Reynolds, M. and Benowitz, L.I. (1990) GAP-43 appears in the rat dorsal horn following peripheral nerve injury. *Neuroscience*, 34: 465–478.

Yankner, B.A., Benowitz, L.I., Villa-Komaroff, L. and Neve, R.L. (1990) Transfection of PC-12 cells with the human GAP-43 gene: effects on neurite outgrowth and regeneration. *Mol. Brain Res.*, 7: 39–44.

Zuber, M.X., Goodman, D.W., Karns, L.R. and Fishman, M.C. (1989) The neuronal growth-associated protein GAP-43 induces filopodia in non-neuronal cells. *Science*, 244: 1193–1195.

W.H. Gispen and A. Routtenberg (Eds.)
Progress in Brain Research, Vol. 89
© 1991 Elsevier Science Publishers B.V.

CHAPTER 7

GAP-43 and neuronal remodeling

Mark C. Fishman and Dario Valenzuela

Developmental Biology Laboratory, Massachusetts General Hospital, Boston, MA 02114, U.S.A.

One of the roles postulated for the protein GAP-43 is in the fashioning of the shape of nerve cells. This could be accomplished either by its causing growth (e.g. Skene and Willard, 1981; Benowitz and Lewis, 1983) or by its permitting a variety of plastic changes in the cell, especially in the nerve terminal (Lovinger et at., 1985; De-Graan et al., 1986), changes including but not limited to growth. That a protein might function both in growth cones and synapses is reasonable, since nerve terminal plasticity is a feature not only of development (Fishman, 1984; Purves and Lichtman, 1980), but also of learning and memory (Greenough et al., 1986). This model proposes, in its simplest form, that expression of the GAP-43 gene causes or permits remodeling (Fishman and Zuber, 1990).

Our interest in GAP-43 derived from experiments designed to determine which components of growth cones distinguish them from synapses. Our approach was to determine which nerve terminal components are down-regulated when growth cones become synapses (Sonderegger et al., 1983; Baizer and Fishman, 1987). We identified several growth cone proteins, one of which was GAP-43, that were diminished by specific target contact in cell culture, suggesting that these proteins played roles more important to growth cones than to synapses.

We hence cloned the cDNA for GAP-43 (Karns et al., 1987; Ng et al., 1988), as did others (e.g. Basi et al., 1987; Cimler et al., 1987; Neve et al., 1987; Nielander et al., 1987; Rosenthal et al., 1987), in order to pursue several questions. (1) Which growth cone functions might GAP-43 affect? (2) Is it sufficient by itself to cause any of the structure of the growth cone? (3) How can it be restricted to particular membrane domains of the cells where growth must occur? (4) How is it regulated at the genetic level? (5) Is there a homologous gene in invertebrates? (6) Most importantly, how could an intrinsic GAP-43-induced propensity to remodeling be coordinated with morphological responses induced by the host of known extracellular growth signals?

Presented at the Third International Workshop on Brain Phosphoproteins, University of Utrecht, The Netherlands, August 24–27, 1990.
Correspondence: Mark C. Fishman, M.D., Chief, Developmental Biology Laboratory, Massachusetts General Hospital, Boston, MA 02114, U.S.A. Tel.: (617) 726-3738

GAP-43 and filopodia

GAP-43 becomes associated with the plasma membrane when it is expressed in cultured cells (Zuber et al., 1989a). Interestingly, high levels of GAP-43 cause non-neuronal cells, such as CHO or COS cells, to begin to extend long filamentous processes that resemble filopodia (Zuber et al., 1989a). The number of these processes and their complexity all correlate well with GAP-43 levels within the cell. Since even non-neuronal cells have the capability to extend filopodia, which serve as important components of motility and cell division, it is likely that in these experiments GAP-43 co-opts a more universal cytoskeletal machinery related to filopodial extension than that it donates a uniquely neuronal phenotype. However, given the important role of such filopodia in growth cones, it seems reasonable to speculate that GAP-43's function may be related to them in nerves. This type of experiment also provides a bioassay for effects of GAP-43 and for their dissection by mutational analysis.

Growth cone targeting

Many cells in the body, including both neurons and epithelia, are asymmetrical in the distribution of membrane proteins. For example, in epithelial cells certain proteins are concentrated in the basolateral surface and others in the apical surface of the cell (Matlin, 1986). How this is accomplished is not clear, although it has been speculated that proteins may bear within their primary sequence "sorting signals" which are recognized by a transport machinery that directs them through the cell to the appropriate location. Since GAP-43 is enriched in growth cones (although not restricted there) and since its function may relate to filopodial extension from growth cones, it is of interest to know how the protein might be restricted to particular domains of the cell.

GAP-43 is a membrane protein, and associated with the membrane in a manner resembling integral membrane proteins rather than peripheral membrane proteins. However, the predicted amino acid sequence of GAP-43 does not predict a hydrophobic stretch long enough to insert into the membrane. It was therefore of interest to ask how GAP-43 binds to the membrane and specifically how it becomes enriched in particular membrane domains. GAP-43 expressed in COS cells is associated with the membrane (Zuber et al., 1989a,b). Mutational analysis suggests that GAP-43 depends upon its amino terminus for this membrane association (Zuber et al., 1989b). Most specifically required are two cysteines at positions 3 and 4, cysteines that have been shown to be palmitoylated by Skene and Virag (1989). In fact, the amino terminal 10 amino acids are sufficient to cause a normally cytosolic protein, chloramphenicol acetyl transferase (CAT), to become membrane-associated in exactly the same manner as is native GAP-43. Interestingly, this 10 amino acid stretch constitutes the coding sequence of the first exon of GAP-43 (see below). This suggests that the amino terminus includes a cell membrane binding region. This same stretch is sufficient to cause accumulation of chimeric proteins in the growth cone of PC12 cells. In other words, GAP-43 1-10 fused to CAT is enriched in PC12 growth cones, whereas, CAT itself remains cytosolic (Zuber et al., 1989b). This suggests that GAP-43 bears at its amino terminus a targeting sequence responsible for partitioning in particular membrane domains. It is an open question as to whether there is a particular GAP-43 "receptor" in the growth cone. It is also an interesting question how GAP-43 traffic through the cell itself is guided. However, taken together this evidence suggests that GAP-43 may be targeted for particular regions of the cell where growth will occur and once there can participate in causing filopodial extension.

GAP-43 gene and its regulation

GAP-43 is expressed solely in the brain and is mainly restricted to neurons, although some experiments suggest that there is expression in some glia (da Cunha and Vitkovic, 1990). As development proceeds GAP-43 levels diminish, although adult expression continues in selected groups of neurons in both the central and peripheral nervous systems, in neurons that have been speculated to continue synaptic remodeling in the adult (De la Monte et al., 1989; Neve et al., 1988; Ng et al., 1988). Skene and Willard (1981) demonstrated differences in the axon transport of GAP-43 in regenerating nerves in the central nervous system and peripheral nervous system, suggesting that expression might differ between those two regions.

In the search for the origin of the cell specificity of expression, we cloned the genomic GAP-43 (Grabczyk et al., 1990). Encoded in three exons, the first of which includes only the membrane targeting sequence and the second of which includes the calmodulin binding domain (Alexander et al., 1988) and the phosphorylation site (Coggins and Zwiers, 1989), the GAP-43 gene is unusual in several regards. The most unexpected finding concerns the GAP-43 promoter, which lacks TATA and CAAT boxes and utilizes multiple start sites for gene transcription initiation. Other genes that lack TATA boxes can also have multiple transcription initiation sites, but most such genes are constitutive in their expression rather than tissue-specific, although there are some exceptions such as the Thy-1 gene (Spanopoulou et al., 1988). Some of the upstream start sites are utilized to a greater degree in the central nervous system than in the peripheral, although the level of expression from these sites is always low. Included in the promoter and within the RNA itself are long homopurine stretches. Such GA stretches on one strand have been noted in promoter elements of other genes and are believed to be capable of forming unusual triple-stranded DNA, so called, "H-DNA" (Htun and Dahlberg, 1989). The role of such structures in regulating transcription is speculative.

Since GAP-43 expression seems to correlate with expression of the neuronal cell fate, it was of interest to study cells in which bimodal cell fate decisions are made. PC12 cells can be induced to assume a more neuronal phenotype under the influence of nerve growth factor and to retain a more chromaffin-life phenotype under the influence of corticosteroids. GAP-43 expression is enhanced by nerve growth factor and diminished by corticosteroids (Federoff et al., 1988). Both effects are direct, in other words do not require intervening protein synthesis. There is no evidence for a significant portion of the nerve growth factor effect being at the transcriptional level, although the steroid repression does appear to be transcriptional in nature (Federoff et al., 1988). One possibility is that the GAP-43 gene is expressed at a relatively steady constitutive level and that nerve growth factor enhances RNA stability, thereby causing increased expression. We have begun to isolate the elements responsible for transcription suppression by corticosteroids.

An evolutionary view of neuronal plasticity

What is the importance of neuronal plasticity to the whole animal? Remodeling of nerves, especially at their terminals, is a life-long process. Unlike the response of the lymphocytic system of cells to specific challenges, which are met by selective clonal expansion, nerves adapt by strengthening or weakening connections. With only the rarest of exceptions, such as in the brain regions responsible for song in certain birds (Nottebohm, 1986), cell division has not been adopted by the nervous system as a mode to meet external challenges. In mammals, changes in synaptic structure, visible microscopically (Purves et al., 1987), are ongoing throughout life. Accumulated evidence suggests that selective nerve terminal expansion underlies learning and long-term memory (Greenough et al., 1986). This plasticity is a major way by which the nervous system helps the animal adapt to its ecological niche during the lifetime of the animal. Thus, unlike responses of many other organs – heart, liver, kidney – which can hypertrophy, or expand their existing repertoire of responses, the nervous system can generate whole new patterns of responses by combinatorial rearrangement of neuronal circuits. Therefore, plasticity of the nervous system may well be viewed as its most important evolutionary function as it helps the brain to adapt to its specific eco-niche (Fishman, 1984). At a cellular level neuronal shape is controlled by both its local environment, and by that of the whole animal, the latter reflected in patterns of neuronal activity. It is not surprising that similar cellular behaviors, such as growth cone collapse, can be brought about by either electrical stimulation (Cohan and Kater, 1986) or contact with specific targets (Keynes and Cook, 1990).

Since plasticity is a feature even of invertebrate organisms, it is reasonable to predict that some of its molecular constituency might also be shared. Indeed, low stringency hybridization has revealed GAP-43 cross-hybridizing genes in *Drosophila melanogaster* and *C. elegans*, but these genes include only a very short stretch of sequence closely related to mammalian GAP-43 (Ng et al., 1989). In *Drosophila* this gene, referred to as KZ30, is localized not in neurons, but rather in specific glial cells that form part of the scaffolding of the nervous system (Bastiani and Goodman, 1986). Although there has been a recent suggestion that certain glia, at least in cul-

ture, may use GAP-43 (Vitkovic et al., 1988), thereby suggesting a role in glial "growth", it still seems unlikely that the described gene serves a purpose directly analogous to vertebrate GAP-43.

GAP-43 associated proteins

It seems quite likely that GAP-43 should participate in protein-protein interactions. Its transport to and enrichment at the cytosolic face of restricted plasma domains suggest a need for physical association with other proteins. They might include for example, integral membrane proteins, cytoskeletal components, or vesicular proteins. Furthermore, GAP-43 might serve to localize proteins in addition to calmodulin to specific regions of the cell or submembranous face. In order to identify candidate proteins which may bind GAP-43, affinity chromatography was used. Purified GAP-43 was covalently coupled to a cross-linked agarose support, through which we passed a soluble fraction from 10-day old rat brain. Loosely bound proteins were removed by washing with the loading buffer (which included $CaCl_2$ and $MgCl_2$), and elution was then performed in the absence of divalent cations and the presence of EDTA.

As shown in Fig. 1, SDS-gel electrophoresis and protein staining of the EDTA eluate shows a single protein band of a molecular weight of about 39 kd. Figure 1 also shows that no protein is present in the EDTA eluate of a control column which contains the agarose support and no bound GAP-43. As a further control for nonspecific binding, a column containing the GTP-binding protein, G_o, was run under identical conditions and found not to bind the 39 kDa polypeptide (not shown). Taken together, these results indicate that the 39 kDa polypeptide interacts specifically with GAP-43 in vitro.

To obtain sequence information, the 39 kDa polypeptide band was electroblotted onto Immo-

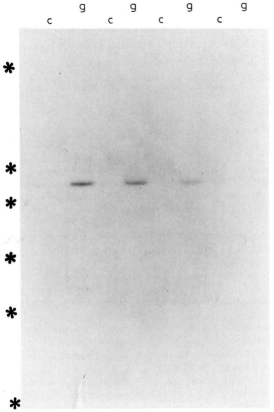

Fig. 1. Elution of a 39 kDa cytoplasmic polypeptide bound to a GAP-43 column. The figure compares the polypeptide profiles of 4 successive fractions of the EDTA eluate of control (c) and GAP-43 (g)-linked Sepharose columns to which a soluble fraction from 10-day-old rat brain has been passed. The 39 kDa polypeptide has been identified as a mixture of the two isozymes of brain aldolase (see text for further details). Asterisks indicate the position of polypeptide molecular weight markers which are from top to bottom: bovine albumin (66 kDa), ovalbumin (45 kDa), pepsin (35 kDa), trypsinogen (24 kDa), β-lactoglobulin (18 kDa), and lysozyme (14 kDa). The only major eluted band, in all 4 fractions, is the 39 kd protein that is eluted from the GAP-43-linked column. Methods: Aliquots of an S100 fraction from 10-day-old rat brain in buffer A (50 mM Tris, pH 7.5, 1 mM $CaCl_2$, 1 mM $MgCl_2$, 1 mM DTT, 1 mM PMSF) were passed through a Sepharose or a GAP-43-linked Sepharose column. The columns were washed with 13.5 column volumes of buffer A, followed by elution in the same buffer, except that 5 mM EDTA was substituted for the divalent cations. The fractions were run on a 10% acrylamide SDS gel and the polypeptides stained with Coomassie blue.

bilon-P Transfer Membrane and subjected to 30 cycles of Edman degradation in the gas phase sequencer. A database search of the resulting sequence revealed that the polypeptide is a mixture of two highly homologous sequences which correspond to the N-terminus of rat fructose-diphosphate aldolase isozymes A and C, each polypeptide having a molecular weight of 39 kDa. It has been previously shown that rat brain contains both the "muscle-specific" or A isozyme, and the "brain-specific" or C isozyme. It is also known that in addition to the homotetramers A_4 and C_4, brain aldolase can form tetrameric hybrids of both isozymes, i.e. A_3C_1, A_2C_2, and A_1C_3.

Besides its crucial regulatory role in glucose metabolism, little is known about brain aldolase and the subcellular distribution and physiological role of its isozymes. Further studies need to be done in order to determine whether the isozymes differ in their interaction with GAP-43, and to assess the in vivo significance of this interaction.

GAP-43 and G_o in the growth cone

An important question is how can growth control from within the cell by proteins such as GAP-43 be coordinated with signals from without? Cell surface molecules, extracellular matrix, and soluble signals all are known to have dramatic effects on nerve growth. In order to address this question, we have isolated other proteins from the growth cone. The major non-cytoskeletal protein in growth cone membranes turns out to be G_o, a member of the heterotrimeric G protein family that transduces between receptors and intracellular second messengers. G_o is largely concentrated in the nervous system, where it has been speculated to link receptors to potassium and calcium channels, and to phospholipase C activation.

G proteins are known to be activated by the displacement of GDP from the alpha subunit and the subsequent binding of GTP. Receptor-ligand complexes activate G proteins in this manner. GAP-43 also is capable of enhancing GTP binding to G_o (Strittmatter et al., 1990). This was surprising because, to date, the heterotrimeric G proteins have not been shown to be subject to intracellular regulation.

Since previously only receptor-ligand complexes had been demonstrated to activate G proteins, it was of interest to investigate whether GAP-43 bears a resemblance to such receptors. The only region in which there is a relationship is in the first 7 amino acids of GAP-43, which are homologous to the amino terminal segment of the cytoplasmic tail of a number of G-linked receptors. The consensus sequence is hydrophobic-Leu-Cys-Cys-X-basic-basic. In the cases where it has been studied, the cysteines are palmitoylated in both these receptors and in GAP-43, enhancing the similarity. A peptide constituted of the first 24 amino acids of GAP-43 can, in fact, activate G_o, although less effectively than can the intact protein (Strittmatter et al., 1990). So the region of GAP-43 that interacts with G_o is the same region that resembles the G protein-linked receptors. Thus, one notion is that GAP-43 acts as an intracellular mimic of the receptors.

If the activation of intracellular second messenger systems and the increase in calcium play a role in nerve growth, as many investigators have suggested, GAP-43 and extracellular ligands might be coordinated in their action by funneling their effects through G_o and thereby regulating second messenger systems. In theory, GAP-43 could lower the threshold for stimulation of G proteins by extracellular ligands. Alternatively, if GAP-43 stimulates G_o to its maximal degree, and that is related to growth cone activity, the presence of GAP-43 could persistently stimulate G_o and thereby permit constitutive growth, until GAP-43 was turned off. At that time appropriate stop signals would be acknowledged by the grow-

ing axon, as G_o again became susceptible. Whether this effect of GAP-43 can be modulated by the degree of palmitoylation or phosphorylation or by calmodulin remains to be determined, but these effects could serve to fine tune the system.

References

Alexander, K.A., Wakim, B.T., Doyle, G.S., Walsh, K.A. and Storm, D.R. (1988) Identification and characterization of the calmodulin-binding domain of neuromodulin, a neurospecific calmodulin-binding protein. *J. Biol. Chem.*, 263: 7544–7549.

Baizer, L. and Fishman, M.C. (1987) Recognition of specific targets by cultured dorsal root ganglion neurons. *J. Neurosci.*, 7: 2305–2311.

Basi, G.S., Jacobson, R.D., Virag, I., Schilling, J. and Skene, J.H.P. (1987) Primary structure and transcriptional regulation of GAP-43, a protein associated with nerve growth. *Cell*, 49: 785–791.

Benowitz, L.I. and Lewis, E.R. (1983) Increased transport of 44,000–49,000 dalton acidic proteins during regeneration of the goldfish optic nerve: A two dimensional gel analysis. *J. Neurosci.*, 3: 2153–2163.

Cimler, B.M., Giebelhaus, D.H., Wakim, B.T., Storm, D.R. and Moon, R.T. (1987) Characterization of murine cDNAs encoding P-57, a neural-specific calmodulin-binding protein. *J. Biol. Chem.*, 262: 12158–12163.

Coggins, P.J. and Zwiers, H. (1989) Evidence for a single protein kinase C-mediated phosphorylation site in rat brain protein B-50. *J. Neurochem.*, 53: 1895–1901.

Cohan, C.S. and Kater, S.B. (1986) Suppression of neurite elongation and growth cone motility by electrical activity. *Science*, 232: 1638–1640.

Da Cunha, A. and Vitkovic, L. (1990) Regulation of immunoreative GAP-43 expression in rat cortical macroglia is cell type specific. *J. Cell Biol.*, 111: 209–215.

De Graan, P.N.E., Oestreicher, A.B., Schrama, L.H. and Gispen, W.H. (1986) Phosphoprotein B-50: localization and function. *Prog. Brain Res.*, 69: 37–50.

De la Monte, S.M., Federoff, H.J., Ng, S.-C., Grabczyk, E. and Fishman, M.C. (1989) GAP-43 gene expression during development: persistence in a distinctive set of neurons in the mature central nervous system. Dev. Brain Res., 46: 161–168.

Federoff, H.J., Grabczyk, E. and Fishman, M.C. (1988) Dual regulation of GAP-43 gene expression by nerve growth factor and glucocorticoids. *J. Biol. Chem.*, 263: 19290–19295.

Fishman, M.C. (1984) Plasticity of the developing synapse. In J.M. Lauder and P.G. Nelson (Eds.), *Gene Expression and Cell-Cell Interactions in the Developing Nervous System*, Plenum Publishing Corporation, New York, pp. 241–246.

Fishman, M.C. and Zuber, M.X. (1990) GAP-43: a gene for neuronal remodeling. In A. Bjorklund, A.J. Aguayo and D. Ottoson (Eds.), *Brain Repair*, The Macmillan Press Ltd., London, pp. 175–184.

Grabczyk, E., Zuber, M.X., Federoff, H.J., Ng, S.-C., Pack, A. and Fishman, M.C. (1990) Cloning and characterization of the rat gene encoding GAP-43. *Eur. J. Neurosci.*, In Press.

Greenough, W.T., McDonald, J.W., Parnisari, R.M. and Camel, J.E. (1986) Environmental conditions modulate degeneration and new dendrite growth in cerebellum of senescent rats. *Brain Res.*, 380: 136–143.

Htun, H. and Dahlberg, J.E. (1989) Topology and formation of triple-stranded H-DNA. *Science*, 243: 1571–1576.

Karns, L.R., Ng, S.-C., Freeman, J.A. and Fishman, M.C. (1987) Cloning of complementary DNA for GAP-43, a neuronal growth-related protein. *Science*, 236: 597–600.

Keynes, R. and Cook, G. (1990a) Cell-cell repulsion. *Cell*, 62: 609–610.

Lovinger, D.M., Akers, R.F., Nelson, R.B., Barnes, C.A., McNaughton, B.L. and Routtenberg, A. (1985) A selective increase in phosphorylation of protein F1, a protein kinase C substrate, directly related to three-day growth of long-term synaptic enhancement. *Brain Res.*, 343: 137–143.

Matlin, K.S. (1986) The sorting of proteins to the plasma membrane in epithelial cells. *J. Cell Biol.*, 103: 2565–2568.

Neve, R.L., Perrone-Bizzozero, N.I., Finklestein, S., Zwiers, H., Bird, E., Kurnit, D.M. and Benowitz, L.I. (1987) The neuronal growth-associated protein GAP-43 (B-50, F1: neuronal specificity, developmental regulation and regional distribution of the human and rat mRNAs. *Mol. Brain Res.*, 2: 177–183.

Ng, S.-C., De la Monte, S.M., Conboy, G.L., Karns, L.R. and Fishman, M.C. (1988) Cloning of human GAP-43: Growth association and ischemic resurgence. *Neuron*, 1: 133–139.

Ng, S.-C., Perkins, L.A., Conboy, G., Perrimon, N. and Fishman, M.C. (1989) A *Drosophila* gene expressed in the embryonic CNS shares one conserved domain with the mammalian GAP-43. *Development*, 105: 629–638.

Nielander, H.B., Schrama, L.H., Van Rozen, A.J., Kasperaitis, M. Oestreicher, A.B., De Graan, P.N.E., Gispen, W.H. and Schotman, P. (1987) Primary structure of the neuro-specific phosphoprotein B-50 is identical to growth-associated protein GAP-43. *Neurosci. Res. Commun.*, 1: 163–172.

Purves, D. and Lichtman, J.W. (1980) Elimination of synapses in the developing nervous system. *Science*, 210: 153–157.

Purves, D., Voyvodic, J.T., Magrassi, L. and Yawo, H. (1987) Nerve terminal remodeling visualized in living mice by repeated examination of the same neuron. *Science*, 238: 1122–1126.

Rosenthal, A., Chan, S.Y., Henzel, W., Haskell, C., Kuang, W.-J., Chen, E., Wilcox, J.N., Ullrich, A., Goeddel, D.V. and Routtenberg, A. (1987) Primary structure and mRNA localization of protein F1, a growth-related protein kinase C substrate associated with synaptic plasticity. *EMBO J.*, 6: 3641–3646.

Skene, J.H.P. and Virag, I. (1989) Posttranslational membrane attachment and dynamic fatty acylation of a neuronal growth cone protein, GAP-43. *J. Cell Biol.*, 108: 613–624.

Skene, J.H.P. and Willard, M. (1981) Axonally transported proteins associated with axon growth in rabbit central and peripheral nervous systems. *J. Cell Biol.*, 89: 96–103.

Sonderegger, P., Fishman, M.C., Bokoum, M., Bauer, H.C. and Nelson, P.G. (1983) Axonal proteins of presynaptic neurons during synaptogenesis. *Science*, 221: 1294–1297.

Spanopoulou, E., Giguere, V. and Grosveld, F. (1988) Transcriptional unit of the murine Thy-1 gene: different distribution of transcription initiation sites in brain. *Mol. Cell. Biol.*, 8: 3847–3856.

Strittmatter, S.M., Valenzuela, D., Kennedy, T.E., Neer, E.J. and Fishman, M.C. (1990) G_o is a major growth cone protein subject to regulation by GAP-43. *Nature*, 344: 836–841.

Vitkovic, L., Steisslinger, H.W., Aloyo, V.J. and Mersel, M. (1988) The 43-kDa neuronal growth-associated protein (GAP-43) is present in plasma membranes of rat astrocytes. *Proc. Natl. Acad. Sci. USA*, 85: 8296–8300.

Zuber, M.X., Goodman, D.W., Karns, L.R. and Fishman, M.C. (1989) The neuronal growth-associated protein Gap-43 induces filopodia in non-neuronal cells. *Science*, 244: 1193–1195.

Zuber, M.X., Strittmatter, S.M. and Fishman, M.C. (1989) A membrane-targeting signal in the amino terminus of the neuronal protein GAP-43. *Nature*, 341: 345–348.

W.H. Gispen and A. Routtenberg (Eds.)
Progress in Brain Research, Vol. 89
© 1991 Elsevier Science Publishers B.V.

CHAPTER 8

Regulation of gene expression in the olfactory neuroepithelium: a neurogenetic matrix

F.L. Margolis [1], J. Verhaagen [2], S. Biffo [3], F.L. Huang [4] and M. Grillo [1]

[1] *Department of Neurosciences, Roche Institute of Molecular Biology, Nutley, NJ 07110, U.S.A.,* [2] *Division of Molecular Biology, Rudolf Magnus Institute for Pharmacology and Institute of Molecular Biology and Medical Biotechnology, Utrecht, The Netherlands,* [3] *Dipartimento di Biologia Animale, Università di Torino, Albertina 17, Turin, Italy,* [4] *National Institute of Child Health and Human Development, NIH, Bethesda, MD, U.S.A.*

Introduction

In adult mammals neuronal death following trauma or disease is essentially an irreversible event. The ability of damaged CNS axons to correctly reinnervate their appropriate targets is also compromised. In contrast, although lesion, infection, environmental insult or trauma can result in the death of mature olfactory neurons the neuroepithelium manifests ongoing neurogenesis, axon outgrowth, and target reinnervation.

The olfactory neuroepithelium of adult vertebrates (shown schematically in Fig. 1) contains a population of stem cells that are capable of dividing and differentiating into mature olfactory neurons (reviewed in Graziadei and Monti-Graziadei, 1978). This process of neurogenesis can occur throughout life in many species and is essential for the replacement of mature olfactory neurons

that degenerate in response to noxious environmental stimuli entering the nasal cavity (Graziadei and Monti-Graziadei, 1978; Hinds et al., 1984) or recovery from trauma damaging the first cranial nerve (Schechter and Henkin, 1974). Similarly, experimental surgical or chemical damage to the mature receptor neurons results in their death and subsequent replacement from the progenitor cells (Graziadei, 1973; Harding et al., 1977, 1978; Graziadei and Monti-Graziadei, 1978; Nadi et al., 1981; Simmons and Getchell, 1981; Constanzo and Graziadei, 1983; Samanen and Forbes, 1984). Thus, the exceptional ability of the vertebrate olfactory neuroepithelium to replace lost receptor neurons from a population of stem cells present in the basal region of the epithelium makes it an important neural system for the study of mechanisms operative in neuronal development, differentiation, degeneration, and postlesion plasticity.

The morphological and electrophysiological events associated with postlesion neuronal reconstitution of the olfactory epithelium and its ability to reform its terminal synaptic field have been studied previously in several species (Graziadei,

---

Correspondence: Dr. F. Margolis, Department of Neurosciences, Roche Institute of Molecular Biology, Nutley, NJ 07110, U.S.A.

1973; Harding et al., 1977; Graziadei et al., 1980; Graziadei and Monti-Graziadei, 1980; Simmons and Getchell, 1981; Samanen and Forbes, 1984; Baker, 1988; Ehrlich et al., 1990). The replacement and initial differentiation of newly formed olfactory neurons can occur independently of the olfactory bulb. However, the importance of the target olfactory bulb in successful functional maturation and maintenance of newly formed olfactory neurons could be inferred from lesion (Constanzo and Graziadei, 1983), organ culture (reviewed in Farbman, 1988), cell culture (Calof and

Chikaraishi, 1989) and transplantation studies (Barber and Jensen, 1988: Monti-Graziadei and Morrison, 1988) in several species.

The major afferent synaptic input to the rat olfactory bulb is provided by the primary olfactory receptor neurons located in the olfactory neuroepithelium in the nasal cavity. Several lines of evidence suggest that these neurons have the capacity to induce morphological and biochemical changes in the developing and adult olfactory bulb (reviewed in Farbman, 1988 and Baker, 1988). If primary olfactory nerve fibers are pre-

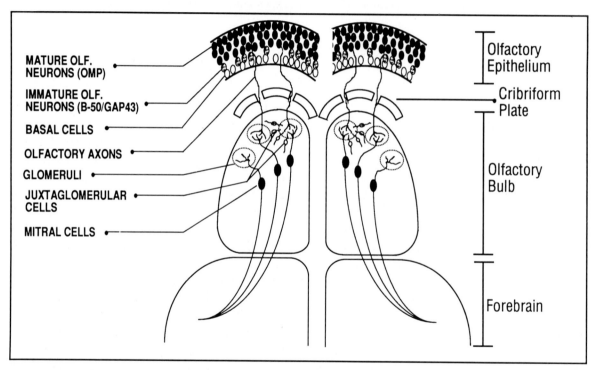

Fig. 1. Schematic representation of the anatomical relationships in the primary olfactory pathway. The cell bodies of the olfactory neurons, localized in the olfactory epithelium, project their axons into the olfactory bulb glomeruli where they form synapses on the processes of their target cells, the juxtaglomerular, mitral, and tufted cells. In addition to monitoring expression of specific gene products during normal ontogeny their response to olfactory neural lesions was evaluated. The recovery process in the primary olfactory pathway was investigated following two lesioning procedures: (1) peripheral lesioning with solutions of TX-100; and (2) olfactory bulbectomy. Peripheral lesioning through intranasal irrigation with TX-100 destroys both the mature and immature olfactory neurons in the neuroepithelium, but does not directly affect the olfactory bulb. Thus, following this lesioning procedure the remaining progenitor cells give rise to new olfactory neurons. These newly formed neurons are able to reinnervate their target cells in the olfactory bulb. Removal of the olfactory bulb (olfactory bulbectomy) results in the degeneration of the damaged olfactory neurons. The newly formed olfactory nerve cells do elaborate axons but, in contrast to the situation after TX-100 lesioning, these axons cannot form synaptic contacts as their target cells have been ablated (Verhaagen et al., 1990b).

vented from reaching the CNS during prenatal ontogeny of the olfactory system, the olfactory bulb does not develop (Giroud et al., 1965). In addition, odor stimulus deprivation subsequent to closure of the naris in neonatal or adult animals results in the arrested development of the olfactory bulb (Baker, 1990; Maruniak et al., 1989; Meisami, 1976). During adulthood, chronic respiratory infection (Smith, 1935) or de-afferentation results in shrinkage of the olfactory bulb (Margolis et al., 1974). Furthermore, transplantation of olfactory epithelium to ectopic brain regions induces the formation of glomerular structures in the olfactory bulb (Graziadei and Monti-Graziadei, 1986). These morphological changes are accompanied by biochemical alterations in subclasses of olfactory bulb neurons (Nadi et al., 1981). Chemical or surgical lesions of the olfactory mucosa or neonatal closure of the nares caused diminished expression of the enzyme tyrosine hydroxylase, its mRNA, and a reduction in dopamine concentrations in the olfactory bulb (Nadi et al., 1981; Kawano and Margolis, 1982; Baker et al., 1983, 1984; Ehrlich et al., 1990). Similar effects were noted for the neuropeptides substance P and CCK in juxtaglomerular neurons of the bulb (Kream et al., 1984; Scarisbrick and Gall, 1988). If reconstitution of the olfactory neuroepithelium occurs, the expression of tyrosine hydroxylase mRNA and protein is reestablished according to a time course that coincides with the reinnervation of the olfactory bulb by primary olfactory nerve fibers (Ehrlich et al., 1990).

These reports demonstrate that gene expression and morphology of the olfactory bulb can be profoundly influenced by events occurring in the olfactory neuroepithelium. Following denervation, the subsequent reinnervation of olfactory bulb neurons by new primary olfactory receptor neurons deriving from stem cells in the basal region of the olfactory epithelium is a phenomenon that can occur throughout adult life

(reviewed in Graziadei and Monti-Graziadei, 1978).

The specific molecular events that underlie the exceptional neurogenic capacity of the olfactory receptor cell precursors are largely unknown. Maturation of the olfactory neuroepithelium is paralleled by a progressive increase in the number of neurons containing the olfactory marker protein (OMP; Margolis, 1972, 1988) and expressing olfactory cilia (Chuah et al., 1985). The OMP-containing cells are absent from the basal cell layer of the olfactory epithelium in the adult animal and OMP is expressed only 7–8 d after the last [^{3}H]thymidine incorporation (Farbman and Margolis, 1980; Miragall and Monti-Graziadei, 1982). These observations indicate that OMP is only expressed in olfactory neurons that have reached a relatively advanced state of maturation. The phenotypic diversity of olfactory receptor cells at various stages of maturation also became apparent in studies with monoclonal antibodies raised in mice using crude epithelial homogenates as the immunogen (Allen and Akeson, 1985a,b; Fujita et al., 1985; Hempstead and Morgan, 1985). With some of these antibodies, it was possible to discriminate between receptor cells in the basal and neuronal cell layers of the mature neuroepithelium. However, the antigens recognized by these antibodies are only beginning to be characterized.

B50/GAP43 is a well-characterized neuron-specific phosphoprotein. Recent cloning experiments have demonstrated that this 24 kDa membrane-associated protein has been studied from different perspectives in several laboratories using a variety of designations (Basi et al., 1987; Cimler et al., 1987; Karns et al., 1987; Nielander et al., 1987; Rosenthal et al., 1987; reviewed by Benowitz and Routtenberg, 1987). Thus, B50, a protein kinase C-substrate implicated in polyphosphoinositide turnover; GAP43 and GAP48 proteins with an axonal growth-related expression pattern; protein F_1, a phosphoprotein associated with long-term potentiation in hippocampal neurons; pp46, a phosphoprotein of growth cone membranes; and P57, a neuron-specific calmodulin binding protein are identical. For convenience we will refer to the protein as B50/GAP43 in this paper.

The growth-related appearance of B50/GAP-43 in neurons has previously been established using ^{35}S-methionine pulse-labeling (Benowitz et al., 1981; Skene and Willard, 1981) and immunochemical methods (De Graan et al., 1985; Jacobson et al., 1986; Meiri et al., 1986; Oestreicher and Gispen, 1986; Skene et al., 1986; Van Hooff et al., 1986; Verhaagen et al., 1986, 1988a). It has been suggested that the developmental regulation of B50/GAP43 expression takes place largely at the level of gene transcription (Basi et al., 1987; Karns et al., 1987; Nielander et al., 1987; Rosenthal et al., 1987). However, more recent studies (this volume) have strongly implicated the role of post transcriptional events.

We have evaluated the response of B50/GAP43 protein and mRNA in both the olfactory neuroepithelium and bulb (Fig. 1) during development (Verhaagen et al., 1989, 1990a; Biffo et al., 1990) and in response to decentralizing and deafferenting lesions (Verhaagen et al, 1990b). In the olfactory neuroepithelium of mature animals B50/GAP43 expression is restricted to a subset of olfactory neurons while most of the olfactory neurons lack B50/GAP43 and express olfactory marker protein (OMP), a protein associated with terminal differentiation of these neurons. In immature animals the converse is true and most of the neurons are B50/GAP43 positive and OMP negative. Following lesion the mature pattern is lost and is replaced by that seen in immature animals. If the synaptic target is intact the neurons proceed through a process similar to that observed during ontogeny and the high level of B50/GAP43 declines and is replaced by OMP.

However, in the absence of target, B50/GAP43 expression is maintained for a prolonged period of time without the onset of OMP expression. This demonstrates the critical role of the target in the maturation of these sensory neurons. The mechanism by which the target stabilizes the maturing neurons, down regulates B50/GAP43 expression and/or triggers OMP gene expression is unknown. However, the lesions permit the demonstration of two different states of the olfactory neuron as evidenced by expression of B50/GAP43 or OMP. Related studies have characterized the distribution and responses of B50/GAP43 in bulbar neurons (Verhaagen et al., 1989, 1990a,b).

B50/GAP43 is phosphorylated in olfactory mucosa in vitro and in vivo. The neurospecific γ-type PKC appears to be absent from the olfactory epithelium while α, β, δ and ζ are clearly expressed in mature intact epithelium. Interestingly, PKC-ϵ expression is induced following bulbectomy suggesting a growth associated pattern of expression for this PKC subtype very similar to that seen for B50/GAP43. These observations may facilitate identification of the role of specific PKC isozymes in olfactory function. Furthermore, the role of B50/GAP43 as a calmodulin (CaM) binding protein in olfactory neurons is unclear as CaM expression is highest in these neurons when B50/GAP43 is lowest (Biffo et al., 1991). These observations offer insights into the interrelated roles of these several proteins in olfactory neuron function, development and regeneration. Furthermore they illustrate the distinct, yet complementary, programs regulating expression of the various genes for CaM, PKC, OMP and B50/GAP43.

Ontogeny

Olfactory epithelium

In the mouse the primary olfactory neuroepithelium is formed from the olfactory placode at embryonic day 10. At embryonic day 11.5 the nasal chamber has formed and axons from the olfactory neurons in the neuroepithelium grow in the direction of the presumptive olfactory lobes. At this age numerous olfactory neurons express B50/GAP43 and elaborate immunoreactive dendrites and axons coincident with the onset of the formation of the olfactory nerve (Biffo et al., 1990). Elevated levels of B50/GAP43 continue to be expressed concurrently with the outgrowth of new processes.

On postnatal days 1, 3, and 9 B50/GAP43-immunoreactive neurons were present throughout the entire olfactory epithelium (Fig. 2A, C). B50/GAP43 immunoreactivity was observed in the cell bodies, apical dendrites, and descending neuritic processes of the nerve cells. Olfactory knobs and proximal portions of olfactory cilia were intensely stained during this stage of development. The axonal processes formed small nerve bundles in the lamina propria mucosae adjacent to the neuroepithelium, and these nerve bundles collected into heavily immunostained fascicles. On postnatal day 1 OMP was present in a small number of neurons occupying a well-defined superficial region of the neuroepithelium. The OMP-positive neurons constituted a discontinuous band of single cells at this age (Fig. 2B). The numbers of OMP-immunoreactive cells increased progressively with age, and the OMP-containing cell layer increased in thickness from about 1 cell at postnatal day 1 to 3–4 cells at postnatal day 9 (Fig. 2B, D). A patchy pattern of OMP immunoreactivity could be seen in the nerve bundles in the lamina propria mucosae (Fig. 3B).

In 3.5- and 5-week-old rats the most intensely stained B50/GAP43-positive cells were located in the lower half of the epithelium, while only faint B50/GAP43 immunoreactivity occurred in neurons in the upper compartment of the epithelium. In 7-week-old rats the olfactory epithelium was virtually devoid of B50/GAP43-positive re-

ceptors cells (Fig. 2G). During the 3.5–7 week period OMP immunoreactivity was present in several layers of receptor neurons, but the deepest region of the epithelium, which contained a gradually decreasing number of B50/GAP43 immunoreactive cells, remained nearly devoid of OMP staining.

In the adult rats (3.5 and 6 months of age) B50/GAP43 immunoreactive cell bodies are observed adjacent to the basal cell layer of the epithelium exclusively. These B50/GAP43-containing cells appeared either in small groups or as individual cells so that some areas of the basal cell region were devoid of B50/GAP43-positive cells (Fig. 2I). A considerable number of B50/GAP43-positive cells in the basal cell region of adult rats formed dendritic and neuritic processes. Often the dendritic processes reached the surface of the epithelium and formed a dendritic knob (Fig. 2I). However, in contrast to the olfactory knobs in young rats, the knobs of the 3.5- and 6-month-old rats never bore B50/GAP43-positive olfactory cilia.

In the 3.5- to 7-week-old rats the pattern of B50/GAP43 immunoreactivity in the nerve bundles tended to change from a homogeneous staining of the total cross section of the nerve bundles to a patchy staining pattern. In the 3.5- and 6-month-old animals this typical patchy distribution of B50/GAP43 immunoreactivity was evident in the majority of nerve bundles and was clearly distinct from the nearly homogeneous staining obtained with OMP antibodies.

In two day old animals B50/GAP43 mRNA is expressed in olfactory neurons throughout the neuroepithelium while OMP mRNA is only expressed in a few cells located in the upper part of the epithelium (Verhaagen et al., 1990b). As ontogeny proceeds olfactory neurons expressing B50/GAP43 mRNA become progressively restricted to the lower portion of the epithelium and are virtually absent from the upper part of

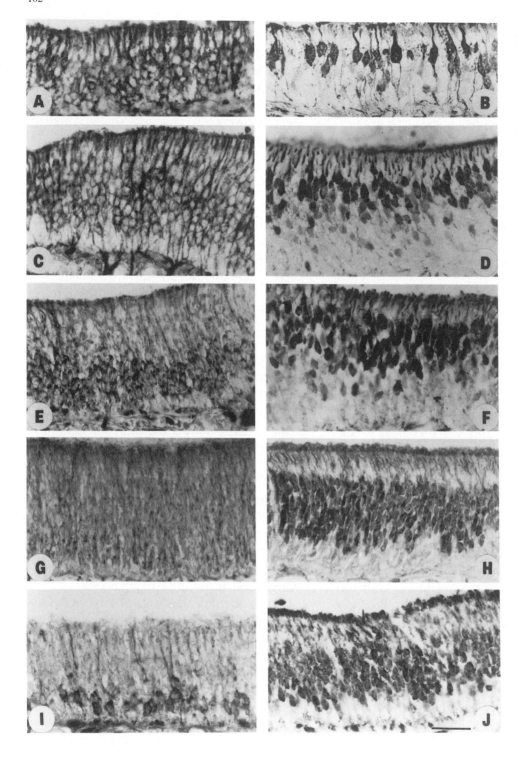

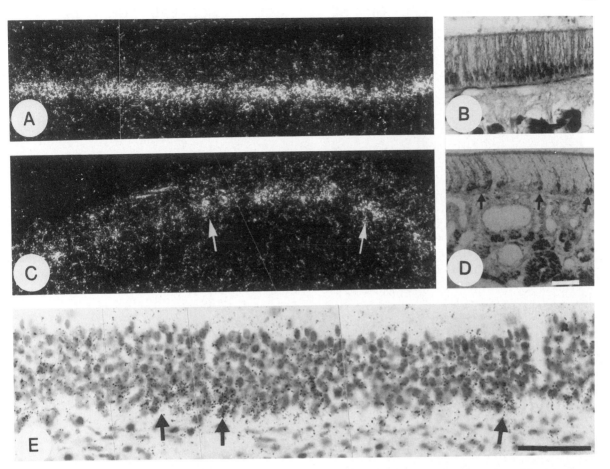

Fig. 3. Expression patterns of B50/GAP43-mRNA and protein in 7.5 week and 18-month-old rats. In 7.5-week-old rats B50/GAP43 mRNA (A) and protein (B) are present in a virtually continuous strip of cells in the basal region of the olfactory epithelium. In 18-month-old rats B50/GAP43 mRNA (C, E) and protein (D) expression has become discontinuous and occurs in small groups (arrows) of olfactory neurons in the basal region of the epithelium. Note the patchy immunostaining in the nerve bundles in the lamina propria mucosae in 18-month-old rats. Scale bars are 50 μm (Verhaagen et al., 1990a).

the epithelium. Concurrently the number of cells expressing OMP-mRNA increases and the autoradiographic signal remains confined to the upper region of the epithelium at short times after birth. The number of OMP expression cells in the upper three fourths of the epithelium increased substantially in 1–2-month-old rats (Verhaagen et al., 1990b).

←

Fig. 2. Immunohistochemical localization of B50/GAP43 and OMP in rat olfactory epithelium. B50/GAP43 (A, C, E, G, I) and OMP (B, D, F, H, J) in rats at the following ages: 1 d (A, B), 9 d (C, D), 3.5 weeks (E, F), 7 weeks (G, H), and 6 months (I, J). The nearly reciprocal relationship between B50/GAP43 and OMP expression is evident. The different character of the staining between OMP and B50/GAP43 as seen in the mucosa is related to the fact that OMP is strictly cytoplasmic, while B50/GAP43 is largely membrane associated. Scale bar = 50 μm (Verhaagen et al., 1989).

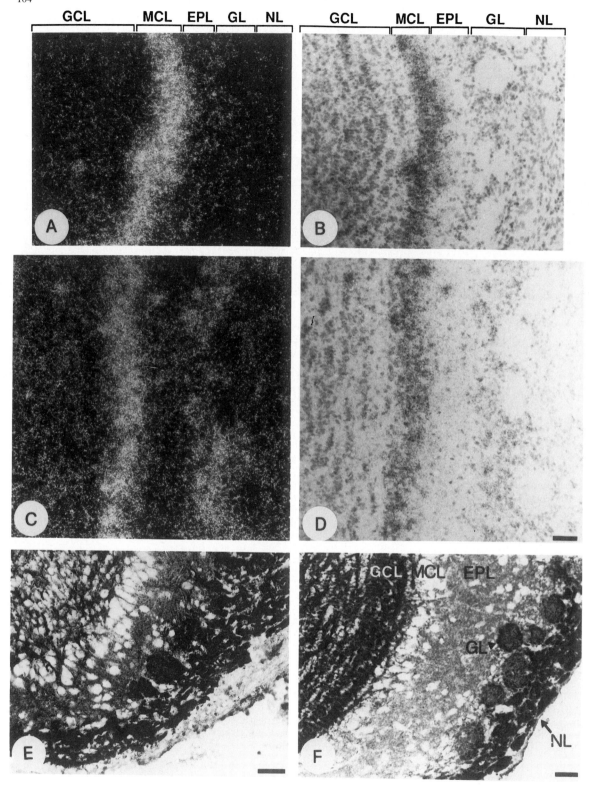

The expression of B50/GAP43 mRNA and protein becomes progressively restricted with increasing age (Fig. 3). Thus, in rats 1–2 months of age the in situ signal is confined to a relatively narrow, nearly continuous, band of cells located in the basal region of the olfactory epithelium (Fig. 3A, B). As the animals reach 6–18 months of age the expression of B50/GAP43 mRNA and protein in the basal region of the epithelium has become discontinuous and appears to be restricted to either small groups of cells or individual cells (Fig. 3C, D, E).

Olfactory bulb

In 2-day-old rats B50/GAP43 mRNA expression is very abundant in the mitral cell layer, while expression in the granule and juxtaglomerular cells is generally low (Fig. 4A, B). During the second postnatal week numerous B50/GAP43 positive juxtaglomerular cells are evident between glomeruli and at the inner border of the glomerular layer (Fig. 4C, D). The labeling in the granule cells has increased slightly but remains much lower than the robust signal in mitral- and juxtaglomerular cells. In the external plexiform layer occasional labeled cells can be observed, representing the tufted cells (Fig. 4C, D).

Between 12 days and 7.5 weeks after birth the expression of B50/GAP43 decreases substantially in all cell types in the olfactory bulb. However, in 7.5-week-old animals virtually all mitral cells and juxtaglomerular cells still express B50/GAP43-mRNA. In older animals (6 and 18 months of age) B50/GAP43 mRNA levels de-

creased further but some mitral cells and juxtaglomerular cells still clearly express B50/GAP43 mRNA even 18 months after birth (Verhaagen et al., 1990a).

To investigate whether the expression of B50/GAP43 mRNA in older animals is due to mitral cells, in situ hybridization was performed on sections of olfactory bulb obtained from 4-month-old mice of the pcd strain. Homozygous affected animals (pcd/pcd) lose a large number of cerebellar Purkinje cells in the first 4 weeks after birth. In addition, by 3 months of age 75% of the mitral cells have been lost from the olfactory bulb (Greer and Shepherd, 1982; Greer and Halasz, 1987) but the tufted cell population is unchanged (Baker and Greer, 1990). A striking difference in B50/GAP43 mRNA expression was observed in homozygous affected mice (pcd/pcd) as compared to their phenotypically normal littermates (pcd/+). B50/GAP43 mRNA was clearly expressed in the mitral cell layer of olfactory bulbs from pcd/+ animals, but was virtually absent at the equivalent anatomical location in bulbs from the pcd/pcd mice. This clearly indicates that the persistent in situ signal in this region of the olfactory bulb in older normal animals is due to B50/GAP43 mRNA expression in mitral cells.

Response to lesions

Effect of intranasal irrigation with TX-100

Quantitative assessment of B50/GAP43 and OMP levels in the olfactory epithelium following TX-100 treatment demonstrated a sharp initial

Fig. 4. Expression of B50/GAP43-mRNA and protein in the olfactory bulb of 2-day-old and 12-day-old rats. Dark field (A, C) and corresponding bright field (B, D) photomicrographs demonstrate abundant B50/GAP43-mRNA expression in mitral cells and juxtaglomerular cells of 2-day-old (A, B) and 12-day-old rats (C, D). Immunohistochemistry in 2-day-old (E) and 12-day-old (F) rats reveals highest B50/GAP43 immunostaining in the nerve layer, glomeruli and in neuronal processes among granule cells. Mitral cell layer (MCL), nerve layer (NL), glomerular layer (GL), external plexiform layer (EPL) and granule cell layer (GCL). Scale bar = 50 μm (Verhaagen et al., 1990a)

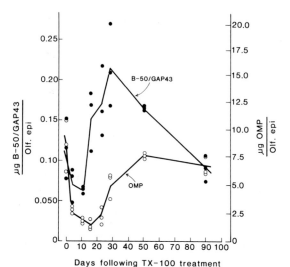

Fig. 5. Time course of the changes in the levels of B50/GAP43 (closed circles) and OMP (open circles) following TX-100 lesioning as measured by radioimmunoassay. At each time point B50/GAP43 and OMP levels were determined in three samples. The amounts of B50/GAP43 and OMP in individual samples are shown. Triton X-100 treatment resulted in an initial sharp decline in the levels of both B50/GAP43 and OMP. Subsequently the expression of B50/GAP43 increases rapidly and exceeds control levels by the fourth week postlesion. OMP expression remains reduced during the first 3 weeks following the lesion and then gradually increases. Three months after the lesion OMP expression has been restored to original levels, while B50/GAP43 levels have decreased to control values. Olf. epi., olfactory epithelium (Verhaagen et al., 1990b).

decline in the amounts of both proteins (Fig. 5). After 10 days the expression of B50/GAP43 increases rapidly and by 4 weeks is nearly double pretreatment levels. OMP levels remain reduced until 3 weeks following the lesion and then increase gradually. Three months after TX-100 treatment B50/GAP43 levels have declined to control values, while OMP expression has been restored to pretreatment levels (Fig. 5).

Immunohistochemical and in situ hybridization studies showed that in the control 5–6-month-old unlesioned mice B50/GAP43 mRNA and immunoreactivity are present in cell bodies localized in the basal cell region of the epithelium

(Fig. 6A). In the olfactory epithelium of control animals OMP-immunoreactive neurons are present in several layers of receptor cells, but the deepest region of the epithelium that contained the B50/GAP43-expressing neurons is virtually devoid of OMP-immunoreactive cell bodies (Fig. 6B). Chemical deafferentation of the olfactory epithelium with TX-100 results in the degeneration of the B50/GAP43- and OMP-immunoreactive cells (Fig. 6C and D). The nerve bundles in the lamina propria did shrink noticeably and contain diminished B50/GAP43 and OMP immunoreactivity. Between 10 days and 4 weeks following the lesion a pronounced increase in the number of B50/GAP43-expressing cells is evident (Fig. 6E). The first OMP-positive cells are observed at around 3 weeks following the lesion. These cells are only present in the upper compartment of the epithelium (Fig. 6F). Between 7 and 8 weeks following the lesion the epithelium appears to contain the full complement of OMP-positive neurons (Fig. 6H). At this postlesion time B50/GAP43-containing cell bodies in the epithelium are once again confined to a thin discontinuous band of cells in the basal region (Fig. 6G). In the nerve bundles in the lamina propria mucosae, the typical patchy staining as observed in the majority of nerve bundles in control tissue, begins to return. However, some nerve bundles in the lamina propria mucosae still contain more B50/GAP43 immunoreactivity than controls (not shown).

Changes in expression of B50/GAP43 and OMP during reinnervation of the olfactory bulb after TX-100 treatment were also studied (Verhaagen et al., 1990b). In control, nondenervated olfactory bulb, the nerve layer and the glomeruli display robust OMP immunoreactivity. In contrast, a patchy pattern of B50/GAP43 immunoreactivity was evident in these areas of the olfactory bulb. As in the olfactory epithelium by 5 days after the lesion, the axons and terminals of the primary olfactory neurons in the nerve- and glomerular layers of the olfactory bulb have degenerated, and the B50/GAP43 and OMP immunoreactivity in the nerve layer has diminished.

The glomerular layer shrinks, and the glomeruli lose their typical globular appearance. During the third week after the lesion, the first B50/GAP43-containing nerve fibers reach the nerve

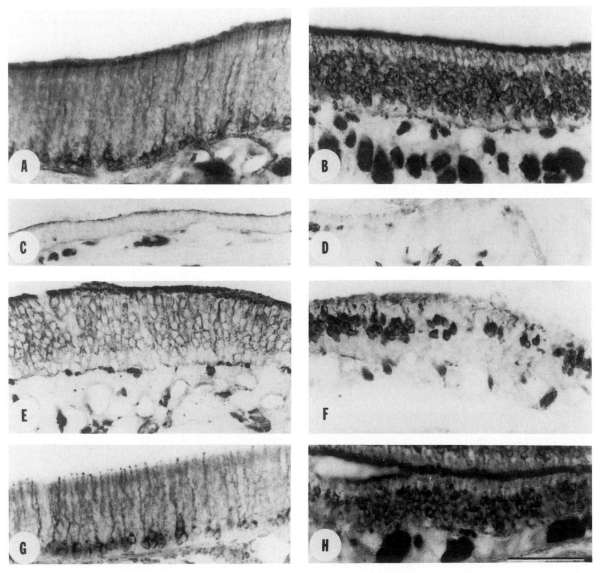

Fig. 6. Immunohistochemical localization of B50/GAP43 and OMP during reconstitution of the olfactory epithelium following TX-100 lesioning. B50/GAP43 (A, C, E, G) and OMP (B, D, F, H) in control unlesioned epithelium (A, B) and at 5 days (C, D), 3 weeks (E, F), and 7.5 weeks (G, H) after the lesion. B50/GAP43 is expressed in virtually all olfactory neurons at 3 weeks after the lesion, while OMP expression is confined to cells in the upper compartment of the epithelium. These expression patterns for B50/GAP43 and OMP are reminiscent of those seen in early postnatal animals. At 7.5 weeks following the lesion B50/GAP43 and OMP staining patterns are very similar to those seen in control tissues. Scale bar = 50 μm (Verhaagen et al., 1990b).

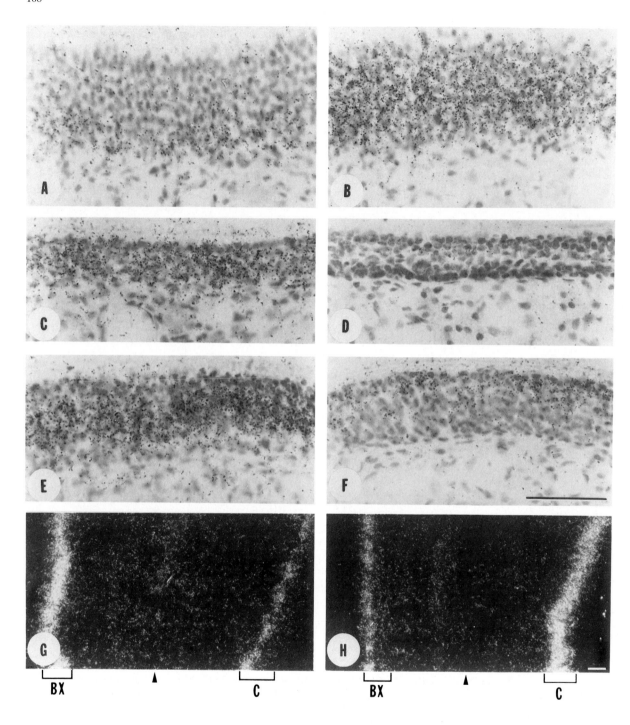

layer of the olfactory bulb, and by 3 weeks following the lesion bundles of B50/GAP43 -containing fibers have penetrated the glomeruli. Although OMP is expressed in the nerve layer at this time, the glomeruli still virtually lack OMP immunoreactivity. In addition to an increase in B50/GAP43 immunoreactivity in the nerve and glomerular layers of the olfactory bulb as a result of ingrowth of new primary olfactory nerve fibers, an increase in B50/GAP43 immunoreactivity is observed in the juxtaglomerular cell, mitral cell, external plexiform and granule cell layers. At about 2 months following the lesion, B50/GAP43 immunoreactivity in the olfactory bulb has diminished noticeably and is comparable to immunostaining patterns observed in control animals (Fig. 5G and H). At this time all glomeruli display OMP immunoreactivity equivalent to the pattern of OMP expression seen in control animals, indicating a reconstitution of the mature organization of the olfactory bulb.

Response to unilateral bulbectomy

The changes in the expression of B50/GAP43- and OMP-mRNA following unilateral bulbectomy in the rat are illustrated in Fig. 7. On the contralateral control side, B50/GAP43 mRNA is only present in a subset of olfactory neurons in the basal region of the olfactory epithelium (Fig. 7A). OMP expression is specifically observed in the mature olfactory neurons constituting the up-

per three-fourths of the epithelium (Fig. 7B). The nerve bundles in the lamina propria mucosae display a typical patchy B50/GAP43 immunostaining pattern that is present in the control nerve bundles of 2-month-old rats but is more evident in older animals (5 months of age). The expression patterns in the epithelium and nerve bundles in untreated age-matched control animals are indistinguishable from those seen in the ipsilateral control side of lesioned animals.

Six days after unilateral olfactory bulbectomy, the thickness of the olfactory epithelium on the lesioned side is substantially reduced, and expression of OMP mRNA is virtually completely lost (Fig. 7D). In contrast, abundant B50/GAP43 mRNA is evident in the thin remaining layer of olfactory epithelium on the lesioned side (Fig. 7C). In Northern blot analyses B50/GAP43 mRNA is barely visible in controls but is clearly elevated by 3 days after lesion and remains high for several weeks thereafter. OMP mRNA declines dramatically by three days after lesion and remains low thereafter.

At 1 and 3 months postlesion the thickness of the olfactory epithelium on the lesioned side has increased substantially, although it has not fully regained the thickness of the control epithelium. At these time points a large number of cell bodies in the neuroepithelium express B50/GAP43 mRNA (Fig. 7E and G) and protein and nerve bundles in the lamina propria mucosae are heav-

Fig. 7. Expression of B50/GAP43- and OMP-mRNA in rat olfactory epithelium following bulbectomy. B50/GAP43 (A, C, E, G) and OMP (B, D, F, H) mRNA localization 6 days after unilateral olfactory bulbectomy on the contralateral control side (A, B), the ipsilateral lesioned side (C, D), and the lesioned side (E, F) in an animal 3 months after bulbectomy. Dark-field photomicrographs of sections taken through the nasal septum of a rat 1 month after bulbectomy showing the expression of mRNA for B50/GAP43 (G) and OMP (H) in the control (C) side (right) and bulbectomized (BX) side (left) in the same section. The middle of the septum is indicated by an arrowhead and the width of the neuroepithelium is indicated by brackets. Note that B50/GAP43 and OMP are expressed in a virtually reciprocal fashion. On the control side B50/GAP43 is expressed in a discontinuous band of cells in the lower region of the epithelium, while OMP-mRNA is observed in the mature neurons forming a broad band of labeled cells. On the lesioned side the number of cells expressing B50/GAP43-mRNA is substantially increased, while OMP-mRNA is only expressed in a thin band of neurons in the upper part of the epithelium. Scale bar = 50 μm (Verhaagen et al., 1990b).

ily immunoreactive for B50/GAP43. In the upper compartment of the newly formed epithelium a thin band of OMP-expressing neurons is visible (Fig. 7F and H), but the number of OMP-containing cells at 1 and 3 months postlesion is only a fraction of the number of OMP-positive cells in the control unlesioned side.

The olfactory neuroepithelium of mature vertebrates possesses a functional stem cell compartment that endows this tissue with the unusual ability to replace mature neurons lost as a result of trauma or environmental insult. In part, this process of olfactory neurogenesis, maturation, synaptic reconnection and activation of transduction recapitulates the sequence of events occurring during the normal ontogeny of the olfactory neuroepithelium. It is characterized by an organized program of cellular and molecular events leading to the reexpression of the terminally differentiated functional cellular phenotype. An example of this is the demonstration here of at least two biochemically distinct populations of olfactory neurons. These are (1) immature neurons that express B50/GAP43, a neuronal specific, calmodulin binding, phosphoprotein, and (2) mature neurons that express olfactory marker protein (OMP), a cytoplasmic protein associated with olfactory neuron maturation. Immature neurons predominate during the fetal and early postnatal development of the olfactory epithelium, as well as after surgical removal of the olfactory bulb and during the early phases of regeneration of the neuroepithelium following reversible lesion. Conversely, mature neurons constitute virtually the entire neuroepithelium in adult animals and in the late phases of post-lesion recovery.

This system offers a model in which to study neurogenesis under normal conditions, as well as following lesions that generate different endpoints, and the associated responses of a set of neuronal genes whose protein products form an interactive set.

Phosphorylation and PKC-isozyme expression

B50/GAP43 can be phosphorylated in vivo and in vitro by PKC (Van Hooff et al, 1988; Coggins and Zwiers, 1989; Baudier et al., 1989) and in vitro by casein kinase II (Pisano et al., 1988). Phosphorylation by PKC reduces the affinity of B50/GAP43 for CaM (Alexander et al., 1989) potentially regulating the availability of CaM for interaction with a variety of CaM-modulated enzymes and proteins (Liu and Storm, 1990). This process could become self limiting as free CaM activates calcineurin, a phosphoprotein phosphatase that can dephosphorylate B50/GAP43 (Liu and Storm, 1989; Schrama et al., 1989). Since the function of this protein is at least partially dependent on its state of phosphorylation it was important to test whether it can be phosphorylated in the immature olfactory receptor neurons and if so what enzyme is responsible. This was evaluated both in vivo and in vitro.

Tissue homogenates of CNS and olfactory mucosa incubated with ^{32}P-ATP in vitro phosphorylate many endogenous proteins (Fig. 8). The Ca^{2+} dependence of phosphorylation varies for different proteins in both brain and olfactory mucosa. However, in both tissues the in vitro phosphorylation of B50/GAP43 is Ca^{2+} dependent (Fig. 8). The ability of broken cell preparations from olfactory mucosa to phosphorylate endogenous B50/GAP43 is encouraging but does not prove that this reaction can occur within the intact immature olfactory neurons. To evaluate this young rats received ^{32}P$_i$ by either the intranasal or intracerebral route. Immunoprecipitable ^{32}P-labeled B50/GAP43 was obtained from extracts of brain following intracerebral injection of ^{32}P$_i$ as well as from extracts of olfactory mucosa following intranasal administration of ^{32}P$_i$ (data not shown). In the case of the olfactory mucosa the signal was weaker, but could be clearly demonstrated. In part, the weaker signal was due to the much lower level of $^{32}P_i$ uptake and incorporation into protein by the olfactory tissue even when much more $^{32}P_i$ was administered in-

tranasally than intracerebrally. Both the in vitro and in vivo studies clearly demonstrated the ability of olfactory tissue to phosphorylate

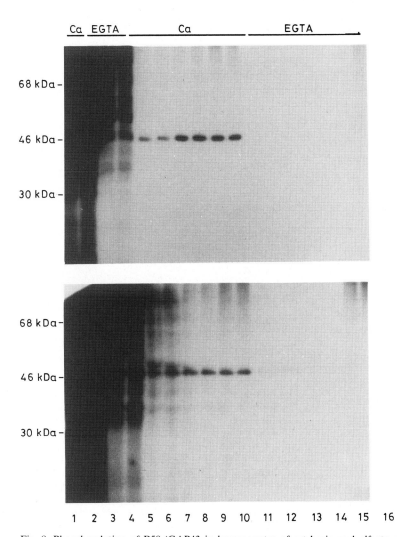

Fig. 8. Phosphorylation of B50/GAP43 in homogenates of rat brain and olfactory mucosa. Olfactory mucosa (lower panel) and CNS (upper panel) from 9 day postnatal rats were homogenized in 10 mM Tris-HCl, pH 7.4, 10mM MgCl$_2$, 0.1 mM CaCl$_2$ and 1 mM PMSF. Aliquots were preincubated in the presence or absence of 5 mM EGTA at 30°C for 5 min. [^{32}P]ATP was added to 100 μM final concentration (2 μCi to brain, 10 μCi for olfactory mucosa). After 2 min, the samples were adjusted to 1% SDS, held at 80°C for 3 min. and diluted. ^{32}P labeled B50/GAP43 was immunoprecipitated with a rabbit antiserum (#6377) directed against the C-terminal trisdeka-peptide of B50/GAP43 and Ig-Sorb. The immunoprecipitate was dissociated, electrophoresed on 11% SDS-PAGE and exposed on X-ray film. Lanes were loaded in duplicate and contain whole homogenate (1–4), 5 μl (5–6, 11–12) 10 μl (7–8, 13–14) and 20 μl (9–10, 15–16) of antiserum. Note that the phosphorylation is Ca^{2+} dependent and EGTA inhibited in both tissues. Furthermore, the antipeptide antisera (#6377 and 6378) exhibit high specificity and saturability (Biffo et al. (1990) and unpublished observations).

B50/GAP43. PKC has been reported to be present in olfactory cilia of amphibia (Anholt et al., 1987) although nothing is known about its presence in mammalian olfactory tissue. Therefore, it was important to evaluate the isozyme distribution of PKC in rodent olfactory tissue to determine whether the occurrence of a specific isozyme might be functionally related to the presence of B50/GAP43 in immature olfactory neurons.

We first tested extracts of olfactory bulb and mucosa for the presence of PKC immunoreactivity on immunoblots. A polyclonal antiserum (Huang and Huang, 1986) capable of reacting with PKC isozymes type I, II and III (γ, β, α, respectively) demonstrated the presence of PKC immunoreactivity in extracts of olfactory bulb in young and adult rats (Fig. 9). Further studies with type specific PKC antibodies (Huang et al., 1987) demonstrated that these three isozymes are all expressed in olfactory bulb and that I (α) and III (γ) but not II (β) manifest significant ontogenetic increases (Huang, Verhaagen and Margolis, unpublished observations). In contrast, no immunoreactivity was observed with extracts of olfactory mucosa prepared from the same animals (Fig. 9). However, B50/GAP43 is a PKC substrate and is phosphorylated in olfactory homogenates by a Ca^{2+} dependent mechanism (Fig. 8). This, coupled with our prior demonstration (Fig. 7; Verhaagen et al., 1990b) that B50/GAP43 protein and mRNA increases in olfactory mucosa following olfactory bulbectomy encouraged us to consider that a different PKC isozyme might be operative in this tissue.

RNA was isolated from olfactory mucosa of control and bulbectomized rats and analyzed for the presence of mRNA for each of the known PKC isozymes (Kosaka et al, 1988; Ono et al., 1988 and Nishizuka, 1988) by RT-PCR (Grillo and Margolis, 1990). To verify the method, amplicons of the correct sizes were observed to be generated for each of the isozymes in brain RNA

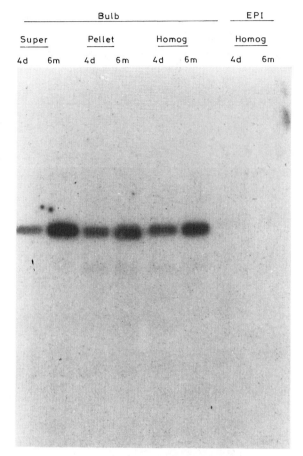

Fig. 9. Protein kinase-C (PKC) immunoreactivity in rat olfactory tissue. Homogenates were prepared from olfactory bulbs (Bulb) and olfactory mucosa (Epi) of 4 day and 6 month rats. Aliquots (100μg protein) of total homogenate from each, as well as supernatant and resuspended pellet from olfactory bulb, were subjected to SDS-PAGE and transferred to nitrocellulose. An unfractionated polyclonal antiserum directed against PKC (Huang and Huang, 1986) was used to probe the blots. Olfactory bulb extracts are clearly immunoreactive at both ages while extracts of olfactory mucosa are devoid of significant PKC immunoreactivity.

using selected oligonucleotide pairs (Table I). In olfactory mucosa the decline in the OMP amplicon and the elevation in the B50/GAP43 amplicon following bulbectomy (Fig. 10) serve as internal controls to demonstrate the efficacy of the surgery and to confirm our earlier observations

TABLE I

Tissue distribution of PKC isozyme mRNAs by RT-PCR

Isozyme	Oligo pair	Amplicon	CNS	Olfactory mucosa	
				Control	Bulbx
δ	3 + 9	354	+	+	+
ϵ	4 + 8	366	+	−	↑
γ	2 + 10	376	+	−	−
$\beta, \beta_1, \beta_{II}$	1 + 6	490	+	+	↓
$\alpha, \beta, \beta_1, \beta_2$	1 + 11	284	+	+	+
ζ	5 + 7	466	+	+	↓

The presence of mRNAs for various PKC isozymes was determined by the coupled reverse transcriptase-polymerase chain reaction (RT-PCR) as described in Grillo and Margolis (1990). Oligonucleotides were selected by comparison of the nucleotide sequences for the various PKC isozymes listed in Genbank and analyzed using Intelligenetics. The predicted lengths of the amplicons are given as base pairs. The apparent presence (+) or absence (−) of the various mRNAs was evaluated by ethidium bromide staining of the amplicons following agarose gel electrophoresis. Arrows illustrate the change in perceived intensity of the ethidium bromide stained amplicon at 26 days following olfactory bulbectomy (bulbx) compared to unoperated control.

on the responses of these two mRNAs (Verhaagen et al., 1990b). The amplicon for each PKC isozyme mRNA responds to bulbectomy in a characteristic manner. No amplicon is seen for either the ϵ or the γ mRNAs in control olfactory mucosa. However, the α, β, δ and ζ mRNAs are present. After bulbectomy the most dramatic change is the appearance of ϵ mRNA. This is accompanied by little or no change in α or δ while ζ declines as does the low level of β. The presence of mRNAs for α and β is unexpected in view of the immunoblot (Fig. 10) but may reflect differences in sensitivity or specificity of the two techniques.

The most intriguing observation after bulbectomy is the appearance of the mRNA for the ϵ isozyme in concert with that for B50/GAP43 at a time when the olfactory mucosa is abundantly populated with immature olfactory neurons. The expression of this PKC isozyme is reported to decline when a neuronal cell line is induced to differentiate in vitro (Wada et al., 1988) consistent with our observation of the ϵ isozyme in tissue with immature neurons. Our observations suggest the possibility that the expression of B50/GAP43 in immature olfactory neurons and PKC-ϵ may be functionally linked. It will be of interest to characterize the expression of PKC-ϵ during normal ontogeny and in response to lesion for comparison with our prior studies on

Expression of mRNA for PKC isozymes influence of olfactory bulbectomy

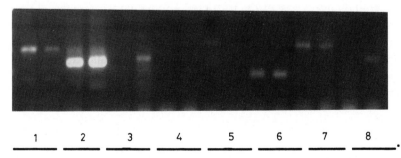

Fig. 10. Effect of olfactory bulbectomy on mRNA of protein kinase C (PKC)-isozymes determined by RT-PCR. RNA isolated from olfactory mucosa of rats 26 days after olfactory bulbectomy and from age matched controls was analyzed as described in the legend to Table I. For each pair of lanes the left represents control and the right 26 days after surgery. OMP (1), PKC-δ (2), PKC-ϵ (3), PKC-γ (4), PKC-β (5), PKC-β,α (6), PKC-ζ (7) and B50/GAP43 (8).

114

B50/GAP43 to determine if the expression of these two gene products is coordinately regulated and is of functional significance.

Calmodulin expression

Calmodulin (CaM) is a ubiquitous calcium binding protein that modulates many enzyme activities associated with signal transduction (Cheung, 1988; Klee et al., 1980; Rasmussen, 1986) as well as progression through the cell cycle (Rasmussen and Means, 1989). Both of these processes are intimately associated with the behavior of the primary olfactory neurons. Furthermore, the demonstration (summarized here) of the presence, and response to lesion, of the neuron specific, calmodulin binding protein B50/GAP43 (Cimler et al., 1987; Gispen et al., 1990; Alexander et al., 1987) in these neurons, prompted us to characterize the expression of calmodulin in the olfactory system during ontogeny and in response to lesion. A CaM cDNA clone that contains the open reading frame of the rat calmodulin gene CaMI (Nojima and Sokabe, 1987; Nojima, 1989; Nojima et al., 1987) was isolated from rat olfactory mucosa (Biffo et al., 1989; Biffo et al., 1991).

In the adult olfactory mucosa CaM mRNA is localized in a wide band of cells spanning from the basal lamina to the sustentacular cell layer (Fig. 12A, B). Even after an extended exposure time, the submucosa, the respiratory epithelium and the nasal glands were devoid of significant labelling. The extent of labelling is grossly similar in all the regions of the neuroepithelium (anterior vs. posterior, dorsal vs. ventral), although occasionally a few scattered zones manifest lower

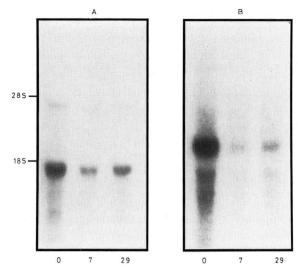

Fig. 11. Influence of olfactory bulbectomy on expression of CaM and OMP mRNAs. Each lane contains 10μg of total RNA isolated from rat olfactory mucosa at 0, 7 and 29 days after surgery. After being probed for CaM mRNA (panel A) the blot was stripped and reprobed for OMP mRNA (panel B) (Biffo et al., 1991).

levels of labelling. The absence of labelling when using the sense transcript (Fig. 12C, D) testifies to the specificity of the antisense CaM probe.

During ontogeny CaM mRNA is detectable in the olfactory epithelium from at least postnatal day 1. Interestingly, in neonates the labelling spans the full thickness of the neuroepithelium between the basal and sustentacular cell layers but only in limited zones. Otherwise, the labelling is often restricted to the upper regions of the neuroepithelium. By three weeks postnatal the pattern of labelling is indistinguishable from that observed in the adult animal. In all cases the cells in the basal cell region are virtually devoid of label (Fig. 12A). High levels of CaM mRNA are

Fig. 12. Specificity of in situ hybridization for CaM mRNA in adult rat olfactory mucosa. CaM antisense RNA (A, B) or CaM sense RNA (C, D). A and C, bright field; B and D, dark field. Scale bar is 30μm. A strong labelling is seen in a broad band of cells in the olfactory neuroepithelium (A, B). The sustentacular and the basal layer are relatively devoid of grains. No specific labelling is seen in sections probed with CaM sense probes (Biffo et al., 1991).

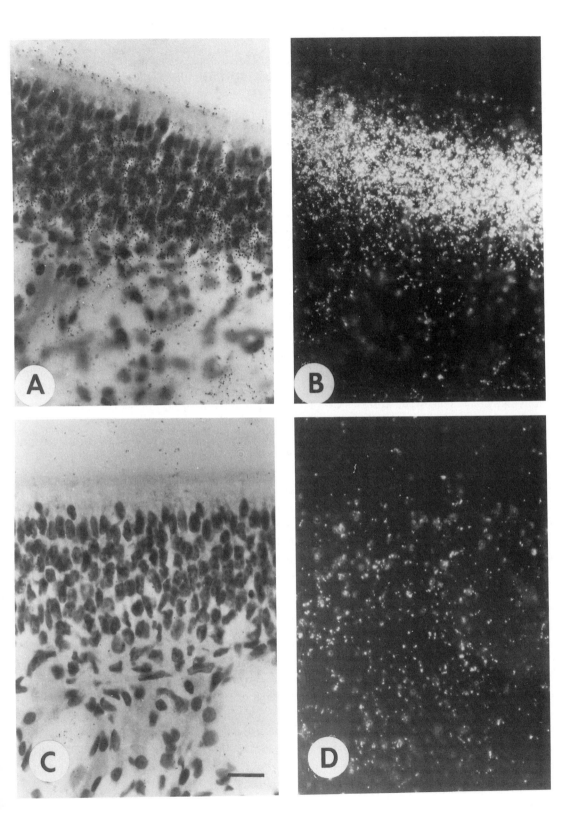

present in several neuronal cells of the adult olfactory bulb particularly in the mitral cell layer (Biffo et al., 1991). This is especially noteworthy given the prolonged expression of B50/GAP43 in mitral cells (Fig. 4 and Verhaagen et al., 1989; 1990a, b; Rosenthal et al., 1987). Nevertheless as already described for other Ca binding proteins (Celio, 1990) not all the neurons appear to express CaM mRNA.

The rapid decline in steady state levels of both CaM and OMP mRNAs after bulbectomy followed by an increase in CaM mRNA but not OMP mRNA in olfactory mucosa at 29 days after surgery (Fig. 11) illustrates the differential regulation of these two genes. Seven days after olfactory bulbectomy, when the mature olfactory neurons are fully degenerated, both OMP mRNA and CaM mRNA have declined dramatically (Fig. 11). Twenty-nine days after lesion CaM mRNA levels are partially restored while OMP mRNA remains barely detectable consistent with the virtual absence of mature olfactory neurons. Since these data suggest that CaM mRNA is expressed in both immature and mature populations of olfactory neurons the cellular localization of CaM mRNA was analyzed by in situ hybridization in control and bulbectomized rats. Six days after bulbectomy the thickness of the olfactory neuroepithelium is dramatically reduced due to the degeneration of virtually all of the mature olfactory receptor neurons (Fig. 7) and a decline of OMP mRNA and protein. At this time little or no expression of CaM mRNA could be seen.

Subsequently, by one month after surgery, new replacement neurons form, the thickness of the neuroepithelium is largely restored (Fig. 7) and CaM mRNA is reexpressed in a relatively broad band of cells (Biffo et al., 1991). However, in the absence of target these neurons remain immature and express B50/GAP43 but do not achieve the fully differentiated state as indicated by the absence of ciliated olfactory knobs or expression of

OMP (Fig. 7 and Verhaagen et al., 1990b). Following olfactory bulbectomy the steady state level of CaM mRNA declines indicating that it, like OMP, is expressed in mature olfactory neurons. However, the subsequent partial recovery in steady state level of CaM mRNA, but not OMP mRNA, indicated that it is expressed in immature neurons as well.

By in situ hybridization we have been able to demonstrate four striking features of CaM expression in the olfactory mucosa (Biffo et al., 1991). First, abundant expression of CaM mRNA is highly restricted to the olfactory neurons. Second, CaM is virtually absent from the basal layer of the olfactory neuroepithelium where the neuronal precursor cells are located. Third, CaM mRNA expression is very low in non-neuronal cells (i.e. sustentacular cells, submucosal glandular cells and adjacent respiratory epithelium). Fourth, CaM mRNA is expressed both in mature and immature neurons throughout the neuroepithelium.

The pattern of expression of CaM mRNA differs significantly from those observed earlier for the mRNAs of B50/GAP43 and OMP. CaM mRNA is abundant in the olfactory epithelium of neonates (when OMP expressing neurons are in the minority and dispersed) as well as in adults (when B50/GAP43 containing neurons occur infrequently and are located near the basal region of the neuroepithelium). The partial recovery of steady state levels of CaM mRNA following bulbectomy is similar to the increased expression of B50/GAP43 following this lesion. However, in contrast to B50/GAP43 and OMP, CaM mRNA is expressed in *both* mature and immature olfactory neurons. These observations, coupled with the pattern of CaM mRNA expression in adult and neonatal animals demonstrates that CaM gene expression is subject to different regulatory constraints than are those for either B50/GAP43 or OMP. Nevertheless, as observed for OMP

(Farbman, 1988; Farbman and Margolis, 1980; Verhaagen et al., 1989, 1990b), contact of olfactory neurons with their olfactory bulb target seems to be essential for full expression of CaM mRNA.

The highly restricted localization of CaM mRNA to the olfactory neurons suggests it may fulfill one or more distinct roles in these sensory neurons. Thus, CaM may play one role in immature olfactory neurons where both CaM mRNA and the mRNA and protein for B50/GAP43 are coexpressed. However, in the mature olfactory neurons, that are active in chemosensory transduction, CaM may play a very different role as it is expressed abundantly in the absence of B50/GAP43.

Summary and conclusions

The olfactory neuroepithelium exhibits neurogenesis throughout adult life, and in response to lesions, a phenomenon that distinguishes this neural tissue from the rest of the mammalian brain. The newly formed primary olfactory neurons elaborate axons into the olfactory bulb. Thus, denervation and subsequent reinnervation of olfactory bulb neurons may occur throughout life. This unique ability of the olfactory neuroepithelium to generate new neurons from a population of precursor cells present in the basal cell layer of this tissue makes it a valuable model in the study of neural development and regeneration.

The molecular processes underlying the neurogenic properties of the olfactory neuroepithelium are poorly understood. Here we have reviewed our studies on the expression of B50/GAP43 during ontogeny of the olfactory system and following lesioning. This analysis includes the characterization of the expression of OMP, a protein expressed in mature olfactory neurons, as well as PKC and calmodulin. The latter two molecules are of particular interest to the function of

B50/GAP43 since the degree of phosphorylation of B50/GAP43 appears to determine B50/GAP43's ability to bind calmodulin (see also Storm, chapter 4, this volume).

In the mature olfactory epithelium B50/GAP-43 expression is restricted to a subset of cells located in the basal region. Since the expression of B50/GAP43 is high in developing and regenerating nerve cells we are confident that the B50/GAP43 positive cells are new neurons derived from the stem cells in the basal region of the epithelium. B50/GAP43 is absent from the stem cells themselves and also from the mature OMP-expressing neurons.

On the basis of the patterns of B50/GAP43 and OMP expression two stages could be discriminated in the regeneration of the olfactory epithelium. First, as an immediate response to lesioning a large population of B50/GAP43 positive, OMP negative neurons are formed. Subsequently, during the second stage, these newly formed differentiating neurons mature as evidenced by a decrease in B50/GAP43 and an increase in OMP expression. The second stage in the regeneration process is only manifested if the regenerating neurons can reach their target cells in the olfactory bulb. Hence, bulbectomy results in the arrest of the reconstituted olfactory epithelium in an immature state. The differential patterns of B50/GAP43 expression following peripheral lesioning and bulbectomy suggest the existence of a target derived signal molecule involved in the down-regulation of B50/GAP43 expression in olfactory neurons that have established synaptic contacts in the olfactory bulb (see also Willard, chapter 2, this volume, "the suppressor hypothesis").

In the olfactory bulb a robust signal for B50/GAP43 mRNA is evident in mitral cells of olfactory bulbs of neonatal rats. However, most intriguing is the observation that significant B50/GAP43 expression persists in some of the mitral-and juxtaglomerular cells even in 18-month-old rats. Three possible explanations of this may be considered. First, it is possible that B50/GAP43 expression in these neurons in 18-month-old rats reflects rearrangements of their dendrites in the glomeruli in response to the arrival of new primary olfactory nerve fibers. Second, "reinnervation" of mitral- and juxtaglomerular cells by ingrowing primary olfactory neurons may result in enhanced transduction of signals requiring an increase in the levels of B50/GAP43 in these cells. The ability of primary olfactory neurons to affect the synthesis of tyrosine hydroxylase in juxtaglomerular cells demonstrates that neurotransmitter synthesis in these cells is, at least partially, under the control of the primary olfactory neurons. Finally, in pyriform cortex, high levels of B50/GAP43 immunoreactivity are present in layer I, the site of mitral cell synapses on the dendrites of neurons located in layer II. Thus, the persistent expression of B50/GAP43 mRNA in mitral cells may be related to the modulation or the continuing plasticity of the olfactory system and reflect its involvement in the processing of constantly changing input from the external volatile chemical environment. Curiously, within the olfactory bulb, mitral cells are the primary neuronal population expressing calmodulin.

Despite this similarity, the pattern of expression of CaM mRNA differs significantly from those observed for the mRNAs of B50/GAP43 and OMP. CaM mRNA is abundant in the olfactory epithelium of neonates (when OMP expressing neurons are in the minority and dispersed) as well as in adults (when B50/GAP43 containing neurons occur infrequently and are located near the basal region of the neuroepithelium). Following bulbectomy CaM mRNA declines in the olfactory neuroepithelium as does OMP mRNA. In contrast to the latter, CaM mRNA makes a partial recovery by one month after surgery. These results, coupled with those from in situ hybridiza-tion, indicate that CaM mRNA is expressed in both mature and immature olfactory neurons.

A model has been proposed (Liu and Storm, 1990) suggesting that the role of CaM in synaptic plasticity and growth could be regulated by its interaction with B50/GAP43 (neuromodulin). Phosphorylation of neuromodulin (B50/GAP43) by protein kinase-C reduces its ability to bind CaM. Thus, phosphorylation of the CaM-neuromodulin complex could release CaM to interact with a variety of CaM-modulated enzymes and cytoskeletal proteins. This process could become self limiting since free CaM also activates the phosphoprotein phosphatase calcineurin that can dephosphorylate neuromodulin facilitating its re-association with CaM.

However, CaM may play a different role in the mature olfactory neurons where its mRNA is expressed abundantly in the absence of any B50/GAP43. Mature olfactory neurons are active in chemosensory transduction and express molecules and organelles associated with this process. CaM is present in the olfactory cilia where transduction is thought to take place.

Possibly mature olfactory neurons contain a CaM binding protein that fulfills the role played by B50/GAP43 (neuromodulin) in immature olfactory neurons. Thus, although the role of CaM in the olfactory chemoreceptor neurons is uncertain, its abundant expression, cellular localization and stringent regulation imply that it plays a critical role in their maturation and function.

Finally, the ability of PKC to phosphorylate B50/GAP43 and regulate its affinity for CaM coupled with the presence of only certain isozymes of PKC in the olfactory mucosa suggested that modulation of PKC isozyme expression may be a critical step in this process. Thus, it was especially intriguing to observe selective responses of PKC isozymes in olfactory mucosa in response to olfactory bulbectomy. In particular, the increase of PKC ϵ concomitant with declines in OMP and increases in B50/GAP43 and CaM as the olfactory epithelium becomes populated by immature olfactory neurons suggests that a window of opportunity has been opened. The ability to selectively modulate these various components in the olfactory neuroepithelium suggests that we will now be able to explore the regulatory interrelationships of these molecules during normal neurogenesis and in response to lesion induced neural regeneration.

Note added in proof

Our recent results using RT-PCR to study the effect of bulbectomy on expression of mRNAs for PKC isozymes in rat olfactory mucosa confirm the absence of the g isozyme and virtual absence of the b isozyme. The mRNAs for the other isozymes are clearly present but the variability in their apparent levels and temporal response to bulbectomy indicate that it is essential to determine their cellular localization in the olfactory neuroepithelium before detailed interpretations can be offered.

References

Alexander, K.A., Cimler, B.M., Meier, K.E. and Storm, D.R. (1987) Regulation of calmodulin binding to P-57. A neurospecific calmodulin binding protein. *J. Biol. Chem.*, 262: 6108–6113.

Allen, W.K. and Akeson R. (1985a) Identification of a cell surface glycoprotein family of olfactory receptors neurons with a monoclonal antibody. *J. Neurosci.*, 5: 284–296.

Allen, W.K. and Akeson R. (1985b) Identification of an olfactory receptor neuron subclass: Cellular and molecular analysis during development. *Dev. Biol.*, 109: 393–401.

Anholt, R.R.H. and Rivers, A.M. (1990) Olfactory transduction: Crosstalk between second messenger systems. *Biochemistry*, in press.

Anholt, R.R.H., Mumby, S.M., Stoffers, D.A., Girard, P.R., Kuo, J.F. and Snyder, S.H. (1987) Transduction proteins of olfactory receptor cells: Identification of guanine nucleotide binding proteins and protein kinase C. *Biochem.*, 26: 788–795.

Baker, H. (1988) Neurotransmitter plasticity in the juxta-glomerular cells of the olfactory bulb. In F.L. Margolis and T.V. Getchell (Eds.), *Molecular Neurobiology of the Olfactory System*, Plenum Publishing Corporation, New York, pp. 185–216.

Baker, H. (1990) Unilateral, neonatal olfactory deprivation alters tyrosine hydroxylase expression but not aromatic amino acid decarboxylase or GABA immunoreactivity. *Neuroscience*, in press.

Baker, H., Kawano, T., Albert, V., Joh, T.H., Reis, D.J. and Margolis, F.L. (1984) Olfactory bulb dopamine neurons survive deafferentation induced loss of tyrosine hydroxylase. *Neuroscience*, 11: 605–615.

Baker, H., Kawano, T., Margolis, F.L. and Joh, T.H. (1983) Transneuronal regulation of tyrosine hydroxylase expression in olfactory bulb of mouse and rat. *J. Neurosci.*, 3: 69–78.

Barber, P.C. and Jensen S. (1988) Olfactory tissue interaction studied by intraocular transplantation. In F.L. Margolis and T.V. Getchell (Eds.), *Molecular Neurobiology of the Olfactory System*, Plenum Publishing Corporation, New York, pp. 333–352.

Basi, G.S., Jacobson, R.D., Virag, I., Schilling, J. and Skene, J.H.P. (1987) Primary structure and regulation of GAP43, a protein associated with nerve growth. *Cell*, 49: 785–791.

Baudier, J., Bronner, C., Kligman, D. and Cole R.D. (1989) Protein kinase C substrates from bovine brain. *J. Biol. Chem.*, 264: 1824–1828.

Benowitz, L.I. and Routtenberg, A. (1987) A membrane phosphoprotein associated with neural development, axonal regeneration, phospholipid metabolism and synaptic plasticity. *Trends Neurosci.*, 10: 527–532.

Benowitz, L.I., Yoon, M.G. and Lewis, E.R. (1981) Specific changes in rapidly transported proteins during regeneration of the goldfish optic nerve. *J. Neurosci.*, 1: 300–307.

Biffo, S., Goren, Khew-Goodall, Y.S., Miara, J. and Margolis, F.L. (1991) Expression of calmodulin mRNA in rat olfactory neuroepithelium. *Mol. Brain Res.*, submitted.

Biffo, S., Goren, T., Miara-Donnerer, J., Khew-Goodall, Y.-S. and Margolis, F.L. (1989) Calmodulin mRNA in the olfactory neuroepithelium is associated with both mature and immature receptor neurons, *Proceedings of the European Journal of Cell Biology, Ital. Assoc. Cell Biology and Differentiation Soc.*, Abstr., 49

Biffo, S., Verhaagen, J., Schrama, L.H., Schotman, P., Danho, W. and Margolis, F.L. (1990) B50/GAP43 expression correlates with process outgrowth in the embryonic mouse nervous system. *Eur. J. Neurosci.*, 2: 487–499.

Calof, A.L. Chikaraishi. D.M. (1989) Analysis of neurogenesis in a mammalian neuroepithelium: Proliferation and differentiation of an olfactory neuron precursor in vitro. *Neuron*, 3: 115–127.

Chuah, M.I., Farbman, A.L., Menco, P.Ph.M. (1985) Influence of olfactory bulb on dendritic knob density of rat olfactory receptor neurons in vitro. *Brain Res.*, 338: 259–266.

Celio, M.R. (1990) Calbindin D-28K and Parvalbumin in the rat nervous system. *Neuroscience,*, 35: 375–475.

Cimler, B.M., Giebelhaus, D.H., Wakim, B.T., Storm, D.R. and Moon, R.T. (1987) Characterization of murine cDNAs encoding P-57, a neural-specific calmodulin-binding protein. *J. Biol. Chem.*, 262: 12158–12163.

Coggins, P.J. and Zwiers, H. (1989) Evidence for a single protein kinase C-mediated phosphorylation site in rat brain protein B50. *J. Neurochem.*, 53: 1895–1901.

Constanzo, R.M., Graziadei, P.P.C. (1983) A quantitative analysis of changes in the olfactory epithelium following bublectomy in the hamster. *J. Comp. Neurol.*, 215: 370–381.

De Graan, P.N.E., Van Hooff, C.O.M., Tilly, B.C., Oestreicher, A.B., Schotman, P. and Gispen, W.H. (1985) Phosphoprotein B-50 in nerve growth cones from fetal rat brain. *Neurosci. Lett.*, 61: 235–241.

Ehrlich, M.E., Grillo, M., Joh, T.H., Margolis, F.L. and Baker H. (1990) Transneuronal regulation of neuronal specific gene expression in the mouse olfactory bulb. *Mol. Brain Res.*, 7: 115–122.

Farbman, A.I. (1988) Cellular interactions in the development of the vertebrate olfactory system. In F.L. Margolis and T.V. Getchell (Eds.), *Molecular Neurobiology of the Olfactory System*, Plenum Publishing Company, New York, pp. 319–332.

Farbman, A.I. and Margolis, F.L. (1980) Olfactory marker protein ontogeny: Immunohistochemical localization. *Dev. Biol.*, 74: 205–215.

Fujita, S.C., Mori, K., Imamura, K. and Okata, K. (1985) Subclass of olfactory receptor cells and their segregated central projections demonstrated by a monoclonal antibody. *Brain Res.*, 326: 192–196.

Giroud, A., Martinet, M. and Deluchat, C. (1965) Mechanisme de developpement du bulbe olfactif. *Arch. Anat. Histol. Embryol. (Strassb.)*, 48: 203–217.

Gispen, W.H., Boonstra, A., De Graan, P.N.E., Jennekens, F.G.I., Oestreicher, A.B., Schotman, P., Schrama, L.H., Verhaagen, J. and Margolis, F.L. (1990) B-50/GAP43 in neuronal development and repair. *J. Restorat. Neurol. Neurosci.*, 1: 237–244.

Graziadei, P.P.C. (1973) Cell dynamics in the olfactory mucosa. *Tissue Cell*, 5: 113–131.

Graziadei, P.P.C. and Monti-Graziadei, G.A. (1978) The olfactory system: a model for the study of neurogenesis and axon regeneration in mammals In: C.W. Cotman (Ed.), *Neuronal Plasticity*, Raven Press, New York.

Graziadei, P.P.C. and Monti-Graziadei, G.A. (1980) Neurogenesis and neuron regeneration in the olfactory system of mammals: III. Deafferentation and reinnervation of the olfactory bulb following section of the fila olfactoria in rat. *J. Neurocytol.*, 9: 145–165.

Graziadei, P.P.C. and Monti-Graziadei, G.A. (1986) Princi-

ples of organization of the vertebrate olfactory glomerulus: an hypothesis. *Neuroscience,* 19: 1025–1035.

Graziadei, P.P.C., Karlan, M.S., Monti-Graziadei, G.A. and Bernstein, J.J. (1980) Neurogenesis of sensory neurons in the primate olfactory system after section of the filia olfactoria. *Brain Res.*, 186: 289–300.

Greer, C.A. and Halasz, N. (1987) Plasticity of dendrodendritic microcircuits following mitral cell loss in the olfactory bulb of the murine mutant Purkinje cell degeneration. *J. Comp. Neurol.* 250: 284–298.

Greer, C.A. and Shepherd, G.M. (1982) Mitral cell degeneration and sensory function in the neurological mutant mouse Purkinje cell degeneration (pcd). *Brain Res.*, 235: 150–161.

Grillo, M. and Margolis, F.L. (1990) Use of reverse transcriptase polymerase chain reaction to monitor expression of intronless genes. *Biotechniques*, 9: 262–268.

Harding, J., Graziadei, P.P.C., Monti-Graziadei, G.A. and Margolis, F.L. (1977) Denervation in the primary olfactory pathway of mice. IV. Biochemical and morphological evidence for neuronal replacement following nerve section. *Brain Res.*, 132: 11–28.

Harding, J.W., Getchell, T.V. and Margolis, F.L. (1978) Denervation of the primary olfactory pathway in mice. V. Long-term effect of intranasal $ZnSO_4$ irrigation on behavior, biochemistry and morphology. *Brain Res.*, 140: 271–285.

Hempstead, J.L. and Morgan, J.I. (1985) A panel of monoclonal antibodies to the rat olfactory epithelium. *J. Neurosci.*, 5: 438–449.

Hinds, J.W., Hinds, P.L. and McNelly, N.A. (1984) An autoradiography study of the mouse olfactory epithelium: Evidence for long-lived receptors. *Anat. Rec.*, 210: 375–383.

Huang, F.L., Yoshida, Y., Nakabayashi, H. and Huang, K.P. (1987) Differential distribution of protein kinase C isozymes in the various regions of brain. *J. Biol. Chem.*, 262: 15714–15720.

Huang, K.P. and Huang F.L. (1986) Immunochemical characterization of rat brain protein kinase C. *J. Biol. Chem.*, 261: 14781–14787.

Jacobson, R.D., Virag, I. and Skene, J.H.P. (1986) A protein associated with axon growth, GAP43 is widely distributed and developmentally regulated in rat CNS. *J. Neurosci.*, 6: 1843–1855.

Karns, L.R., Ng, S.-C., Freeman, J.A. and Fishman, M.C. (1987) Cloning of complementary DNA for GAP-43, a neuronal growth related protein. *Science*, 236: 597–600.

Kawano, T. and Margolis, F.L. (1982) Transsynaptic regulation of olfactory bulb catecholamines in mice and rats. *J. Neurochem.*, 39: 342–348.

Kosaka, Y., Ogita, K., Ase, K., Nomura, H., Kikkawa, U. and Nishizuka, Y. (1988) The heterogeneity of protein kinase C in various rat tissues. *Biochem. Biophys. Res. Commun.*, 151: 973–981.

Kream, R.M., Davis, B.J., Kawano, T., Margolis, F.L. and

Macrides, F. (1984) Substance P and catecholaminergic expression in neurons of the hamster main olfactory bulb. *J. Comp. Neurol.* 222: 140–154.

Liu, Y. and Storm, D. (1990) Regulation of free calmodulin levels by neuromodulin: neuron growth and regeneration. *Trends Pharmacol. Sci.*, 11: 107–111.

Liu, Y.C. and Storm, D.R. (1989) Dephosphorylation of neuromodulin by calcineurin. *J. Biol. Chem.*, 264: 12800–12804.

Margolis, F.L. (1972) A brain protein unique to the olfactory bulb. *Proc. Natl. Acad. Sci. USA*, 69: 1221–1224.

Margolis, F.L. (1988) Molecular cloning of olfactory specific gene products. In F.L. Margolis and T.V. Getchell (Eds.), *Molecular Neurobiology of the Olfactory System,* Plenum Press, New York, pp. 237–265.

Margolis, F.L., Roberts, N., Ferriero, D. and Feldman, J. (1974) Denervation in the primary olfactory pathway of mice: biochemical and morphological effects. *Brain Res.,* 81: 469–483.

Maruniak, J.A., Taylor, J.A., Henegar, J.R. and Williams, M.B. (1989) Unilateral naris closure in adult mice: atrophy of the deprived-side olfactory bulbs. *Brain Res.,* 47: 27–33.

Meiri, K., Pfenninger, K.. and Willard, M. (1986) Growth-associated protein GAP43, a polypeptide that is induced when neurons extend axons, is a component of growth cones and corresponds to pp46, a major polypeptide of a subcellular fraction enriched in growth cones. *Proc. Natl. Acad. Sci. USA*, 83: 3537–3541.

Meisami, E. (1976) Effects of olfactory deprivation on postnatal growth of rat olfactory bulb utilizing a new method for production of neonatal unilateral anosmia. *Brain Res.,* 107: 437–444.

Miragall, F. and Monti-Graziadei, G.A. (1982) Experimental studies on the olfactory marker protein II. Appearance of the olfactory marker protein during differentiation of the olfactory sensory neurons of mouse: an immunohistochemical and autoradiographic study. *Brain Res.,* 329: 245–250.

Monti-Graziadei, G.A. and Morrison G.E. (1988) Experimental studies on the olfactory marker protein. IV. Olfactory marker protein in the olfactory neurons transplanted within the brain. *Brain Res.,* 455: 401–406.

Nadi, N.S., Head, R., Grillo, M., Hempstead, J., Grannot, N., Reisfeld, C. and Margolis, F.L. (1981) Chemical deafferentiation of the olfactory bulb: plasticity of the levels of tyrosine hydroxylase, dopamine and norepinephrine. *Brain Res.,* 213: 365–377.

Nielander, H.B., Schrama, L.M., Van Rozen, A.J., Kasperaitis, M., Oestreicher, A.B., De Graan, P.N.E., Gispen, W.H. and Schotman, P. (1987) Primary structure of the neuron-specific phosphoprotein B-50 is identical to growth-associated protein GAP43. *Neurosci. Res. Commun.*, 1: 163–172.

Nishizuka, Y. (1988) The molecular heterogeneity of protein kinase C and its implications for cellular regulation. *Nature*, 334: 661–665.

Nojima, H. (1989) Structural organization of multiple rat calmodulin genes. *J. Mol. Biol.*, 208: 269–282.

Nojima, H. and Sokabe, H. (1987) Structure of a gene for rat calmodulin. *J. Mol. Biol.*, 193: 439–445.

Nojima, H., Kishi, K. and Sokabe, H. (1987) Multiple calmodulin mRNA species are derived from two distinct genes. *Mol. Cell Biol.*, 7: 1873–1880.

Oestreicher, A.B. and Gispen, W.H. (1986) Comparison of the immunocytochemical distribution of the phosphoprotein B-50 in the cerebellum and hippocampus of immature and adult rat brain. *Brain Res.*, 375: 267–279.

Ono, Y., Fujii, T., Ogita, K., Kikkawa, U., Igarashi, K. and Nishizuka, Y. (1988) The structure, expression, and properties of additional members of the protein kinase C family. *J. Biol. Chem.*, 265: 6927–6932.

Pisano, M.R., Hegazy, M.G., Reimann E.M. and Dokas, L.A. (1988) Phosphorylation of protein B50 (GAP43) from adult rat brain cortex by casein kinase II. *Biochem. Biophys. Res. Commun.*, 155: 1207–1212.

Rasmussen, C.D. and Means, A.R. (1989) Calmodulin, cell growth and gene expression. *Trends Neurol. Sci.*, 12: 433–438.

Rasmussen, H. (1986) The calcium messenger system. *New Engl. J. Med.*, 314: 1164–1170.

Rosenthal, A., Chan, S.Y., Menzel, W., Haskell, C., Kuang, W.-J., Chen, E., Wilcox, J.N., Ullrich, A., Goeddel, D.V. and Routtenberg, A. (1987) Primary structure and mRNA localization of protein F₁, a growth related protein kinase C substrate associated with synaptic plasticity. *EMBO J.,* 6: 3641–3646.

Samanen D.W. and Forbes W.B. (1984) Replication and differentiation of olfactory receptor neurons following axotomy in the adult hamster: A morphometric analysis of post-natal neurogenesis. *J. Comp. Neurol.,* 225: 201–211.

Scarisbrick, I.A. and Gall, C.M. (1988) Regulation of CCK and NPY immunoreactivity within the adult rat olfactory bulb by primary afferent input. *Soc. Neurosci. Abst.*, 14: 447.

Schechter P.J. and Henkin R.I. (1974) Abnormalities of taste and smell after head trauma. *J. Neurol. Neurosurg. Psychiat.*, 37: 802–810.

Schrama, L.H., Heemskerk, F.M.J. and De Graan, P.N.E. (1989) Dephosphorylation of protein kinase C phosphorylated B-50/GAP-43 by the calmodulin-dependent. *Neurosci. Res. Commun.*, 5 141–147.

Simmons P.A., and Getchell T.V. (1981) Physiological activity of newly differentiated olfactory receptor neurons correlated with morphological recovery from olfactory nerve section in the salamander. *J. Neurophys.*, 45: 529–549.

Skene, J.H.P. and Willard, M. (1981) Changes in axonally transported proteins during regeneration in toad retinal ganglion cells. *J. Cell Biol.*, 89: 96–105.

Skene, J.H.P., Jacobson, R.O., Snipes, W.J., McGuire, C.B., Norden, J.J. and Freeman, J.A. (1986) A protein induced during nerve growth (GAP43) is a major component of growth cone membranes. *Science*, 233: 763–768.

Smith, C.G. (1935) The change in volume of the olfactory and accessory olfactory bulbs of the albino rat during postnatal life. *J. Comp. Neurol.*, 61: 477–508.

Van Hooff, C.O.M., De Graan, P.N.E., Boonstra, J., Oestreicher, A.B., Schmidt-Michels, M.V. and Gispen, W.H. (1986) Nerve growth factor enhances the level of the protein kinase C substrate B-50 in pheochromocytoma PC12 cells. *Biochem. Biophys. Res. Commun.*, 139: 644–651.

Van Hooff, C.O.M., De Graan, P.N.E., Oestreicher, A.B. and Gispen, W.H. (1988) B50 phosphorylation and polyphosphoinositide metabolism in nerve growth cone membranes. *J. Neurosci.*, 8: 1789–1795.

Verhaagen, J., Van Hooff, C.O.M., Edwards, P.M., De Graan, P.N.E., Oestreicher, A.B., Schotman, P., Jennekens, F.G.I. and Gispen, W.H. (1986) The kinase C substrate protein B50 and axonal regeneration. *Brain Res. Bull.*, 17: 737–741.

Verhaagen, J., Oestreicher, A.B., Edwards, P.M., Veldman, H., Jennekens, F.G.I. and Gispen, W.H. (1988) Light and electromicroscopical study of phosphoprotein B50 following denervation and reinnervation of the rat soleus muscle. *J. Neurosci.*, 8: 1759–1766.

Verhaagen J., Oestreicher A.B., Gispen W.H. and Margolis F.L. (1989) The expression of the growth associated protein B-50/GAP43 in the olfactory system of neonatal and adult rats. *J. Neurosci.*, 9: 683–691.

Verhaagen, J., Greer, C.A. and Margolis, F.L. (1990a) B50/GAP43 gene expression in the rat olfactory system during postnatal development and aging. *Eur. J. Neurosci.*, 2: 397–407.

Verhaagen, J., Oestreicher, A.B., Grillo, M., Khew-Goodall, Y.-S., Gispen, W.H. and Margolis, F.L. (1990b) Neuroplasticity in the olfactory system: differential effects of bulbectomy and peripheral lesion of the primary olfactory pathway on the expression of the growth-associated protein B-50/GAP43 and olfactory marker protein. *J. Neurosci. Res.*, 26: 31–44.

Wada, H., Ohno, S., Kubo, K., Taya, C., Tsuji, S., Yonehara, S. and Suzuki K. (1989) Cell type-specific expression of the genes for the protein kinase C family: Down regulation of mRNAs for PKCα and nPKCε upon in vitro differentiation of a mouse neuroblastoma cell line neuro. *Biochem. Biophys. Res. Commun.*, 165: 533–538.

SECTION III

Structure and Characteristics of Protein Kinases

W.H. Gispen and A. Routtenberg (Eds.)
Progress in Brain Research, Vol. 89
© 1991 Elsevier Science Publishers B.V.

CHAPTER 9

Protein kinase C family and nervous function

Y. Nishizuka, M.S. Shearman, T. Oda, N. Berry, T. Shinomura, Y. Asaoka, K. Ogita, H. Koide, U. Kikkawa, A. Kishimoto, A. Kose, N. Saito and C. Tanaka

Departments of Biochemistry and Pharmacology, Kobe University School of Medicine, Kobe 650, Japan, and Biosignal Reseach Center, Kobe University, Kobe 657, Japan

Introduction

The hydrolysis of inositol phospholipids is now generally accepted as a signal transduction pathway that has been shown to be utilized in many neuronal circuits to convey information from the presynaptic terminal to the postsynaptic cell. The immediate products of this pathway, inositol 1,4,5-trisphosphate (IP_3) and a 1,2-*sn*-diacylglycerol (DG), act as the second messengers at the beginning of a bifurcating signal pathway involving Ca^{2+} mobilisation from intracellular storage pools, and the activation of protein kinase C (PKC) (Berridge, 1984; Nishizuka, 1984). This appears to be the primary mechanism for initiating PKC-mediated effects, but recent evidence has indicated that signal-dependent hydrolysis of additional phospholipids such as phosphatidylcholine (PC) may also provide DG which is needed for PKC activation, and Ca^{2+} may also

come from an extraneuronal source, by the activation of plasma membrane ion channels. Physiological activation of PKC by DG can be mimicked by tumour-promoting phorbol esters, such as 12-*O*-tetradecanoylphorbol 13-acetate (TPA), for which it appears to be the cell surface receptor (Castagna et al., 1982). The mammalian PKC is well known to exist as a family of multiple subspecies (for review, see Nishizuka, 1988). The members of the family are differently distributed in brain tissues and limited intracellular localisations.

PKC is involved in many aspects of neuronal function, such as the short-term modulation of membrane excitability and neurotransmitter release. In areas such as the hippocampus, it appears also, together with other protein kinases and possibly proteases, to play a part in the long-term, use-dependent potentiation of synaptic transmission (for review, see Kennedy, 1989). While the extent of the diversity of PKC family is not fully clarified at present, some aspects of current studies and prospective of this enzyme will be briefly discussed in this article.

Correspondence: Y. Nishizuka, Department of Biochemistry, Kobe University School of Medicinie, Kobe 650, Japan.

Signal route for PKC activation

Although the receptor-mediated hydrolysis of inositol phospholipids was once thought to be the sole mechanism leading to the activation of PKC, recent studies suggest that there are several additional routes to provide DG that is needed for enzyme activation as schematically given in Fig. 1. For instance, PC may also be hydrolysed to produce DG at a relatively later phase of cellular responses, particularly to those to long-acting signals such as some growth factors (for review, see Exton, 1990). However, several isoforms of Ca^{2+}-dependent phospholipase C so far identified do not react with PC, and it has been postulated that phospholipase D followed by phosphatidic acid phosphatase may be involved in such receptor-mediated phospholipid degradation. In fact, agonist-stimulated activation of phospholipase D has been shown in several tissues and cell types, but precise mechanism of this activation remains to be explored. It is possible that PKC itself may take part in this phospholipid degradation cascade.

In addition, both receptor-mediated and voltage-dependent Ca^{2+}-gate opening may cause membrane phospholipid degradation, that is probably initiated by the activation of phospholipase C and also phospholipases D and A_2 due to the Ca^{2+}-influx. As briefly described below, arachidonic acid and its metabolites are able to activate some subspecies of PKC under certain conditions. Thus, the signal routes leading to the activation of PKC may greatly vary with cell types,

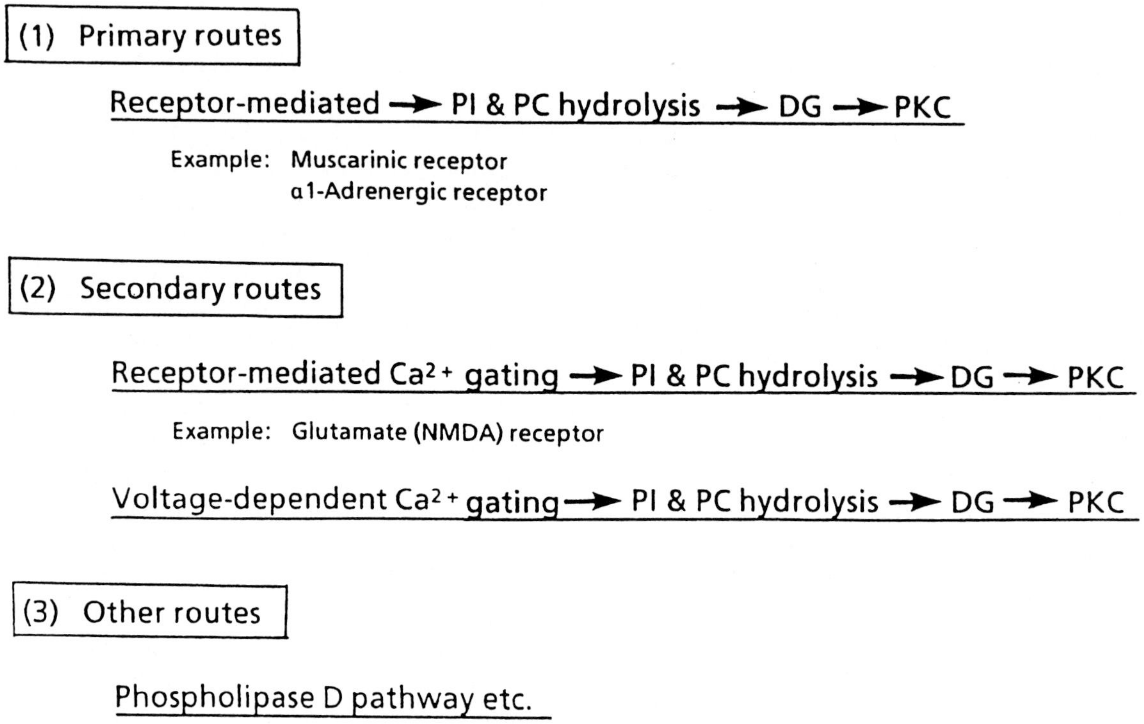

(1) **Primary routes**

Receptor-mediated → PI & PC hydrolysis → DG → PKC

    Example: Muscarinic receptor
                  α1-Adrenergic receptor

(2) **Secondary routes**

Receptor-mediated Ca^{2+} gating → PI & PC hydrolysis → DG → PKC

    Example: Glutamate (NMDA) receptor

Voltage-dependent Ca^{2+} gating → PI & PC hydrolysis → DG → PKC

(3) **Other routes**

Phospholipase D pathway etc.

Fig. 1. Signal routes for PKC activation. The detailed explanation is given in the text. PI, phosphatidylinositol and its mono- and bisphosphate; PC, phosphatidylcholine; DG, diacylglycerol; and NMDA, N-methyl-D-aspartate. This figure is taken from Farago and Nishizuka (1990).

extracellular signals, and perhaps with the time after receptor stimulation.

Molecular heterogeneity and properties

Molecular cloning and biochemical analysis has revealed PKC to exist as a family of multiple subspecies having closely related structures as schematically given in Fig. 2 (for reviews, see Nishizuka, 1988). The α-, βI-, βII- and γ-subspecies appear to form a subgroup within the family, as their primary structures are highly conserved, and distinguish them from the δ-, ϵ- and ζ-subspecies. The α-, β- and γ-subspecies exhibit

sufficient differences, however, to allow them to be resolved by hydroxyapatite column chromatography. The PKC activity in brain tissues can be separated clearly into three peaks (types I, II and III), corresponding to the γ-, β (βI plus βII)- and α-subspecies, respectively (Kikkawa et al., 1987). The βI- and βII-subspecies are almost identical, being derived from a single gene transcript by alternative splicing (Ono et al., 1987), and have yet to be resolved by conventional chromatographic procedures. The ability to purify to homogeneity and resolve the different PKC subspecies has allowed detailed examination of their kinetic and catalytic properties (Sekiguchi et al.,

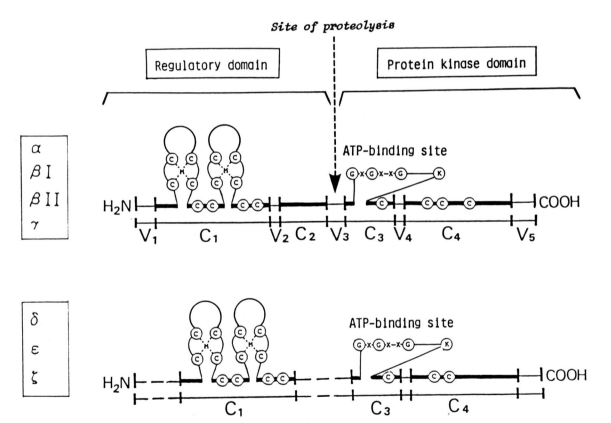

Fig. 2. Common structure of protein kinase C subspecies. C, G, K, X and M represent cystein, glycine, lysine, any amino acid and metal ion, respectively. The numbers indicate the amino acid residues of the PKC subspecies. Other details are described in the text. This figure is adapted from Nishizuka (1988).

1988; Huang et al., 1988a), and has revealed that they exhibit subtle, but identifiable, differences.

The enzymes with γ- (type I) and α-sequence (type III) show much less activation by DG than the enzymes with βI- and βII-sequence (type II). The βI- and βII-subspecies show substantial activity without added Ca^{2+} in the presence of DG and phospholipid. A characterisitic feature of the enzyme having γ-sequence is that it can be activated at a micromolar range of arachidonic acid in the absence of phospholipid and DG (Sekiguchi et al., 1988). Lipoxin A is also active but other unsaturated free fatty acids such as linoleic and oleic acids are less active. The α- and β-subspecies respond to higher concentrations of arachidonic acid, but only at elevated levels of Ca^{2+}. It is worth noting that, the α- and β-subspecies can be activated by the synergistic action of DG and arachidonic acid. Various PKC subspecies, therefore, may be activated at different phases of cellular responses by a series of phospholipid metabolites, such as DG, arachidonic acid and lipoxin A, that successively appear subsequent to stimulation of the receptor.

In addition to these well defined PKC subspecies, structurally undefined enzymes, which are present in some tissues and cell types such as HL-60 promyelocytic leukemic cells, mammary gland and platelets, respond to phospholipid, DG and Ca^{2+} in different ways (Hashimoto et al., 1990). Probably, more subspecies will be found in brain tissues. However, the dependency of PKC on Ca^{2+}, DG and phospholipid in in vitro systems varies markedly with the phosphate acceptor protein employed, and, in the most extreme case,

when protamine is used as phosphate acceptor, neither Ca^{2+} nor phospholipid is needed for the enzymatic activity. Thus, comparison of the members of the family is possible only under limited, defined conditions. In addition, following disruption of the cell, most protein kinases show an activity to phosphorylate many proteins that may or may not represent physiological substrates. Therefore, the fact that various PKC subspecies probably display different responsiveness to activators within the cell, and that various PKC subspecies are differentially distributed both in different cell types and within the same cell as described below, greatly complicates the search for physiological substrates for each member of the PKC family. Presumably, the PKC subspecies show different preferences for substrate proteins that are located in specific intracellular compartments.

Differential regional expression

Using a combination of immunohistochemical and biochemical procedures with subspecies-specific antibodies, the relative activity and individual pattern of expression of multiple PKC subspecies in brain tissues have recently been examined extensively and clarified in more detail (for review, see Nishizuka, 1988).

The enzyme with γ-sequence is apparently expressed solely in the brain and spinal cord, and is not found in any other tissues and cell types. Immunohistochemical studies with antibodies specific to this subspecies indicate that the highest activity is associated with the hippocampus, cerebral cortex, amygdaloid complex, and cere-

→

Fig. 3. A typical electron-microscopic picture of the synapse in rat hippocampus pyramidal cell. The dense immunostaining with the antibodies specific to the γ-subspecies is seen in the post-synaptic pyramidal cell of the rat hippocampus CA1 area, whereas this PKC subspecies is apparently absent in the presynaptic terminal. The γ-subspecies is associated with the membranous structure of the pyramidal cell body. The detailed experimental conditions are described elsewhere (Saito et al., 1988; Kose et al., 1988, 1990).

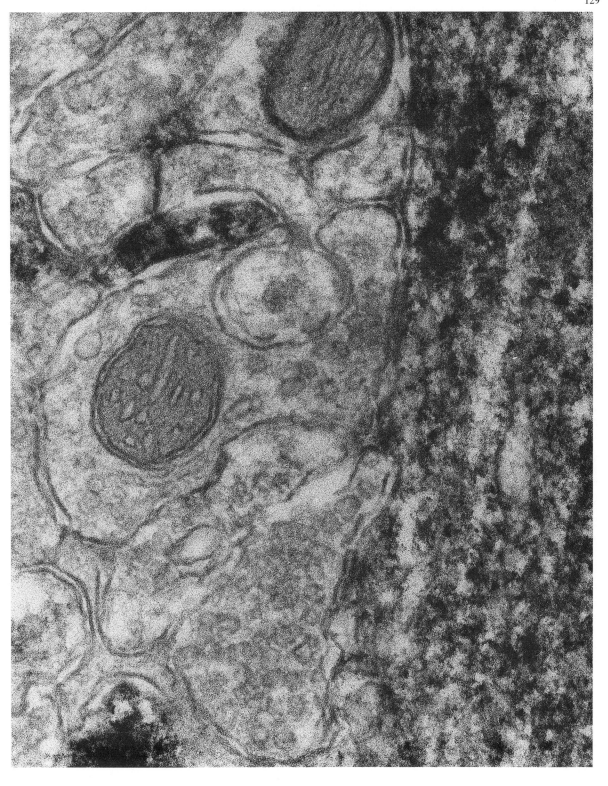

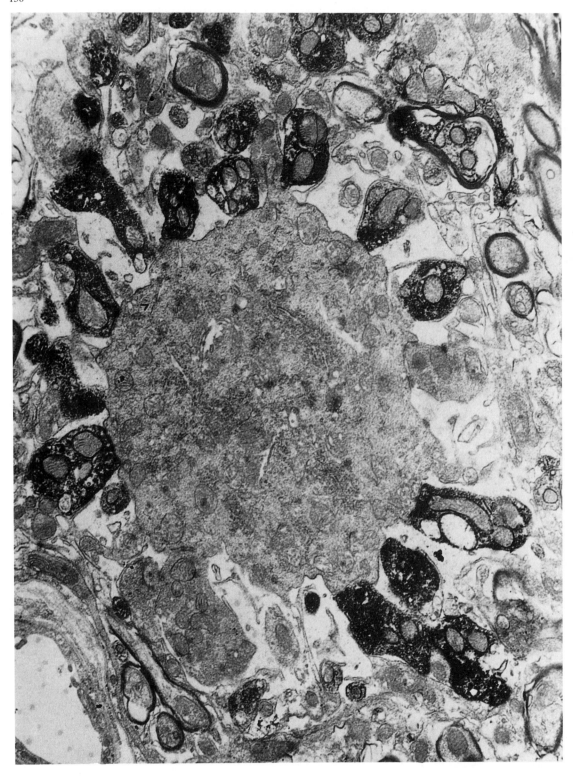

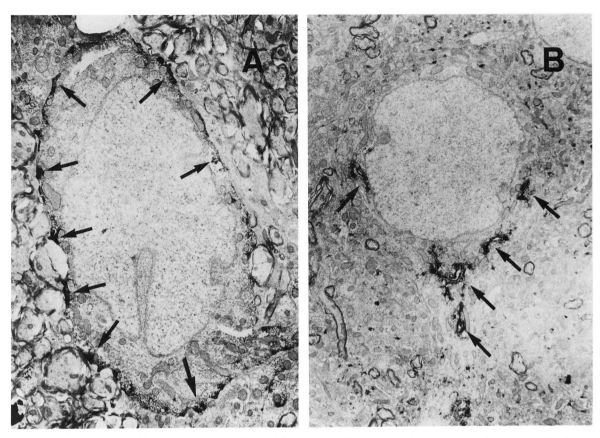

Fig. 5. Distinct intracellular localization of protein kinase C subspecies. (A) PKC with βI-sequence. Dense immunoreaction is seen in the periphery of a neuronal cell in the rat pontine nucleus. (B). PKC with βII-sequence. Dense immunoreaction is located around the Golgi complex of a neuronal cell in the rat cerebral cortex. The detailed experimental conditions are described elsewhere (Saito et al., 1989; Hosoda et al., 1989; Kose et al., 1988, 1990).

bellar cortex (Saito et al., 1988; Kose et al., 1988, 1990). In the rat hippocampus this subspecies is present predominantly in the post-synaptic py-ramidal cell bodies, but not pre-synaptic termi-nals (Fig. 3). On the other hand, in the deep cerebellar nucleus it is localized exclusively in pre-synaptic terminals but not post-synaptic cells (Fig. 4). In the cerebellum, this PKC subspecies is present in the cell bodies, dendrites and axons of Purkinje cells. Immunoelectron microscopic anal-ysis has revealed that the γ-subspecies is associ-ated with most membranous structures present

←

Fig. 4. A typical electron-microscopic picture of the rat deep cerebellar nuclear cell. The dense immunostaining with the antibodies specific to the γ-subspecies is seen in the presynaptic terminals of Purkinje axon surround the deep cerebellar nuclear cell, but not in the postsynaptic cell body. The detailed experimental conditions are described elsewhere (Saito et al., 1988; Kose et al., 1988, 1990).

throughout the cell, probably endoplasmic reticulum. The nucleus and mitochondria poorly express PKC.

On the other hand, the enzymes with βI- and βII-sequence are expressed in different ratios. Normally, the activity of βII-subspecies far exceeds that of βI-subspecies. Cytochemical analysis with the antibodies specific to each subspecies has indicated a clearly distinct intracellular localisation (Ase et al., 1988b; Hosoda et al., 1989; Saito et al., 1989). In neuronal cells the βI-subspecies is sometimes associated with plasma membranes, whereas the βII-subspecies is often localized around the Golgi complex as shown in Fig. 5.

In contrast, PKC with α-sequence is ubiquitously distributed in all tissues and cell types so far examined, implying that this subspecies may play a role in the control of more fundamental cellular processes such as gene expression. Most brain regions contain α-, βI- and βII-subspecies in variable ratios.

It has been established that the level of PKC activity in the mammalian brain progressively increases during development, reaching a peak concomitant with functional maturation (Turner et al., 1984). Several studies examining the ontogenic expression of the enzyme subspecies either by biochemical (Hashimoto et al., 1988), radioimmunocytochemical (Yoshida et al., 1988) or in situ hybridisation techniques (Sposi et al., 1989), have concluded that the γ-subspecies is either absent or expressed at very low levels at birth, but increases rapidly afterward to reach its steady-state level (at around 3–4 weeks in rat), whereas both the α- and β-subspecies are detectable prenatally, and incease steadily to their final levels.

At present, the distribution and biochemical properties of the enzymes encoded by γ-, ϵ- and ζ-sequence have not been defined, since the substrate of this class of enzymes is not known. Histone appears to be a poor phosphate acceptor. Northern blot analysis suggests that these members of the PKC family are expressed in several brain regions in different ratios (Ono et al., 1988). Presumably, a distinct regional pattern of the members of the PKC family per se may be an important factor in determing their different functions.

Modulation of neuronal functions

It is well recognized that the synergistic interaction between PKC and Ca^{2+} pathways underlies a variety of cellular responses to external stimuli (for reviews, see Nishizuka, 1984, 1986). Activation of PKC in neuronal tissues, using phorbol esters such as TPA which mimic the physiological activator, DG, has been associated with changes in neuronal development (Påhlman et al., 1981; Montz et al., 1985), the modulation of ion channels (for review, see Shearman et al., 1989), enhancement of neurotransmitter release (Zurgil and Zisapel, 1985; Nichols et al., 1987; Versteeg and Ulenkate, 1987; Allgaier et al., 1988; Shuntoh et al., 1988) and synaptic plasticity and long-term potentiation (for review, see Kennedy, 1989). These diverse functions may well involve activation of different PKC subspecies.

TPA can cause the translocation of the PKC activity from the cytosolic fraction to the membrane, which may involve insertion of the enzyme into the lipid bilayer, resulting initially in activation, but followed by down-regulation (Rodriguez-Pena and Rozengurt, 1984; Young et al., 1987; Stabel et al., 1987). Several studies have investigated the relationship between cell function and PKC activity by using PKC-depleted cell lines, induced by chronic treatment with TPA (Blackshear et al., 1985; Vyas et al., 1990). Recent analysis in this laboratory has suggested that TPA-induced down-regulation of PKC may involve proteolysis of the membrane-bound, activated form of PKC by calpain, particularly by calpain I (Kishimoto et al., 1983), and also, that the α- and β- subspecies co-expressed in a single cell type disappear at different rates upon treatment with TPA (Ase et al., 1988a).

Consistent with the findings of immunoelectron microscopic studies described above, biochemical analysis of synaptosomes prepared from various brain regions has clearly indicated that the γ-subspecies is undetectable in the pre-synaptic nerve terminals which end at the hippocampal pyramidal cell and Purkinje cell of adult rats (Shearman et al., 1991). The synaptosomal PKC expression, however, exhibits significant regional differences. For instance, the synaptosomes prepared from the cerebrum contain the γ-subspecies in addition to the α- and β-subspecies,

whereas the synaptosomes prepared from the deep cerebellar nuclei contain a large quantity of the γ-subspecies with relatively small amounts of the α- and β-subspecies (Shearman et al., 1990). Treatment of these synaptosomes with TPA prompted the translocation to membranes and subsequent depletion of both the α- and β- subspecies within a min, but does not change the γ-subspecies in the synaptosomes for at least 15 min (Oda et al., 1991). The γ-subspecies is gradually decreased at a slower rate. It is possible, therefore, to prepare various synaptosomes which contain different PKC subspecies in different ratios, as exemplified in Fig. 6. The experiments given in this figure have shown, in the synaptosomes thus having only the γ-subspecies, a subse-

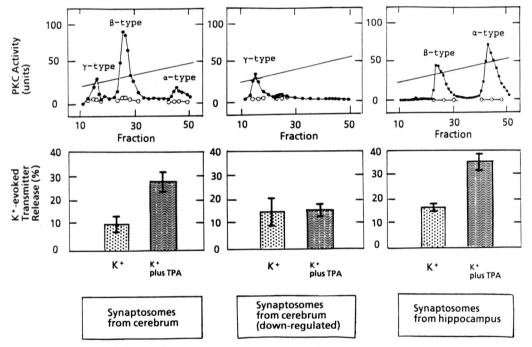

Fig. 6. TPA-enhanced K^+-evoked release of noradrenalin from various synaptosomes having PKC subspecies in different ratios. Synaptosomes were prepared from the rat cerebrum and hippocampus by the methods of Dunkley et al. (1988). The synaptosomes from the rat cerebrum were treated with TPA for 10 min, and the α- and β-subspecies were depleted. The PKC subspecies in these synaptosomes were analysed biochemically by chromatography on a hydroxyapatite column as described (Shearman et al., 1991). These synaptosomes were pre-labeled with [^{3}H]noradrenalin, and its release reaction was determined by challenging with 41 mM K^+ with or without TPA. The detailed experimental conditions are described elsewhere (Oda et al., 1991). ●, PKC activity assayed in the presence of Ca^{2+}, PS and DG; and ○, PKC activity assayed in the presence of EGTA instead of Ca^{2+}, PS and DO.

134

quent dose of TPA can not enhance K⁺-evoked noradrenalin release, although, in original synaptosomes, TPA is able to enhance K⁺-evoked noradrenalin release (Oda et al., 1991). Pretreatment with TPA does not alter the K⁺-evoked noradrenalin release itself. TPA is also found to enhance the K⁺-evoked noradrenalin release from synaptosomes prepared from both hippocampus, that express the γ-subspecies of PKC at a negligible level, and cerebral cortex, that have a significant level of the γ-subspecies, to the same degree. The results suggest that the γ-subspecies of PKC is not essential for the modulation of evoked neurotransmitter release. This conclusion agrees with the previous findings that, in PC 12 neuroblastoma cells and rat sympathetic neurons which contain the α- and β-subspecies, the enzyme depletion results in a loss of the ability to enhance transmitter release and in a reduction in the extent of depolarisation-evoked transmitter release (Matthies et al., 1987).

The enzyme with the γ-sequence is, however, expressed transiently in synaptosomes in the hippocampus at postnatal day 7, reaches its peak levels at around postnatal day 14, and then steadily declines to the undetectable level as given in Fig. 7 (Shearman et al., 1991). The transient nature of the expression of the synaptosomal γ-subspecies suggests that it has some function at an early stage of development. It is interesting to note that expression of this enzyme subspecies coincides, to some extent, with levels of a growth-associated protein, so-called GAP-43, B-50 or F1 (Benowitz and Routtenberg, 1987; Gispen et al., 1985; Meiri et al., 1986; Skene et al., 1986), which are highest during the first postnatal week, and decline thereafter (Jacobson et al., 1986).

The poor expression of the γ-subspecies in the pre-synaptic terminals of the adult hippocampus appears striking because the major post-synaptic components, i.e. hippocampus pyramidal cells,

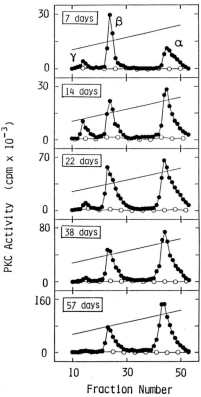

Fig. 7. Ontogenic changes in the pattern of expression of rat hippocampus synaptosomal PKC subspecies. PKC activity present in the cytosolic and membrane extracts of the hippocampal synaptosomal fraction, prepared from rats of the postnatal ages indicated, was partially purified by anion exchange column chromatography and then subjected to hydroxyapatite column chromatography to resolve the enzyme subspecies. The detailed experimental conditions are described elsewhere (Shearman et al., 1991). ●, PKC activity assayed in the presence of Ca^{2+}, PS and DG; and ○, PKC activity assayed in the presenc of EGTA instead of Ca^{2+}, PS and DO.

exhibit dense immunoreactivity toward antibodies against this subspecies, in both the cell soma and the dendritic tree (Saito et al., 1988; Kose et al., 1990; Huang et al., 1988b). This strongly suggests that the primary function of this subspecies is related to the modulation of the response of these neurons to released neurotransmitter, particularly glutamic acid. Studies directed toward delineating the involvement of individual PKC subspecies in these areas are still at their outset, but electrophysiological and biochemical approaches have suggested that in the adult hippocampus the modulation of voltage-dependent Ca^{2+} and K^+ channels (Doerner et al., 1988) and feedback inhibition of quisqualate/kainate receptor-mediated responses (Canonico et al., 1988; Crepel and Krupa, 1988) may be candidates for the action of the γ-subspecies of PKC as schematically shown in Fig. 8. It is attractive to surmise, however, that this subspecies may have some function in synaptogenesis at an early stage of development since GAP-43 is an excellent substrate of PKC, and that arachidonic acid and its metabolites may serve as retrograde signals for this enzyme activation. It is equally attractive to speculate that in the adult hippocampus the α- or β-subspecies in the pre-synaptic terminals may be regulated in a retrograde fashion, because these subspecies can be activated by synergistic action

of DG and arachidonic acid as briefly mentioned above.

Modulation and cross-talk of various signalling pathways

A large body of evidence indicates that PKC exerts negative feedback control over, as well as cross-talk with, various steps of cell signalling pathways (for review, see Nishizuka, 1986). In biological systems, a positive signal is frequently followed by an immediate negative feedback control. In short-term responses, for instance, a major role of PKC appears to lie in decreasing Ca^{2+} concentrations in a manner given in Fig. 9. The appearance of physiological second messengers is normally very transient. A number of reports have suggested that, in various cell types, PKC has a function to activate the Ca^{2+}-transport ATPase and Na^+/Ca^{2+} exchange protein, both of

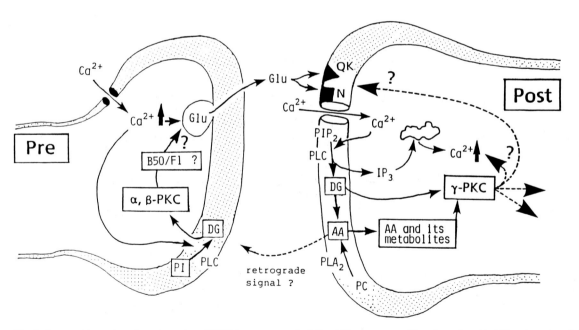

Fig. 8. Presumptive roles of γ-subspecies of PKC in synapses of adult rat hippocampus. Glu, glutamic acid; QK, quisqualate/kinate receptor; N, NMDA receptor; IP$_3$, inositol trisphosphate; DG, diacylglycerol; PIP$_2$, phosphatidylinositol bisphosphate; AA, arachidonic acid; PC, phosphatidylcholine; PI, inositol phospholipids; PLC, phospholipase C; and PLA$_2$, phospholipase A$_2$.

136

which remove Ca^{2+} from the cytosol (for review, see Nishizuka, 1986). PKC often inhibits the receptor-mediated hydrolysis of inositol phospholipids, thereby blocking the activation of the Ca^{2+}-signalling pathway.

Although it sounds paradoxical, such a negative feedback role of PKC is not confined to the receptor functions for short-term responses, but may also be extended, in some tissues and cell types, to those for long-term responses such as cell growth and differentiation. The receptor for epidermal growth factor and insulin has repeatedly been shown to be phosphorylated by PKC to give a functionally down-regulated receptor (for review, see Nishizuka, 1986). Under physiological conditions the activation of PKC is probably transient, since DG once produced in membranes disappears within a few seconds or at most several minutes after its formation. On the other hand, the treatment of cells with TPA causes sustained activation of PKC but frequently depletes the enzyme from the cell as described above. Thus, DG and TPA cause entirely different cellular responses, especially in long-term as described below.

It is also becoming clear that PKC plays a crucial role in the cross-talk of cell signalling pathways which can be seen between various levels of receptors, coupling factors such as G-proteins, effectors such as adenylate cyclase and phospholipases, second messengers such as cyclic AMP, DG and IP_3, Ca^{2+} and many protein kinases including protein kinase cascade, as well as between various types of ion channels. At present, such intra- and inter-cellular events have not yet been fully substantiated on a firm biochmical basis, but it is extremely important to clarify the molecular mechanism of such cross-talk among cell signalling pathways for better understanding the role of PKC in nervous tissues.

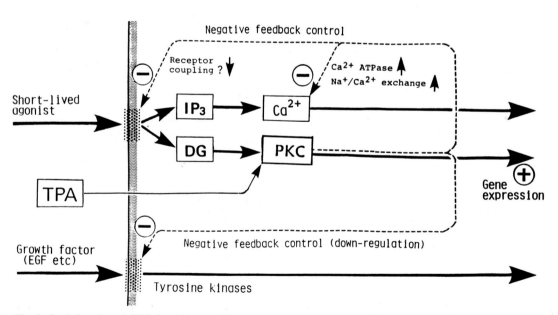

Fig. 9. Dual function of PKC furnishing positive and negative actions on cellular regulation. IP_3, inositol trisphosphate; DG, diacylglycerol.

Role in gene expression

Plausible evidence seems to indicate the involvement of PKC in gene expression and some proto-oncogene activation. The c-*fos* protein is present at low basal levels in most cells, but the c-*fos* mRNA and the protein therefrom are both induced rapidly by various external signals including TPA. It has been proposed that, upon stimulation of the cell, this protein undergoes more extensive post-translational modification, and produces a homodimer, or heterodimer complex with other nuclear phosphoproteins, such as c-*jun* and CREB (for review, see Hoeffler et al., 1989).

Apparently, sustained activation of PKC is needed to cause gene activation, eventually leading to cell proliferation. It has been recently demonstrated that, with purified T-cells as a model system, multiple and repeated additions of membrane-permeable DG result in both interleukin-2 receptor expression and cell proliferation as shown in Fig. 10 (Berry et al., 1990; Berry and Nishizuka, 1990). Unlike phorbol esters which activate PKC persistently, a single dose of permeable DG, which causes only transient activation of PKC, cannot elicit T-cell activation (Berry et al., 1989). Cell activation and proliferation process is obviously the result of interaction of a number of signalling pathways. One of these is PKC activation, which is required for a prolonged length of time, brought about, physiologically, by a combination of various signals involving phospholipid degradation cascade discussed above. This may be needed for maintenance of phosphorylation of a protein whose activity is critical in the regulation of gene activation, such as transcription factors. Presumably, the signalling pathways, including tyrosine kinases and Ca^{2+}-dependent protein kinases may also play critical roles in this process. Recently, considerable interest has centered on protein kinase cascade in direct or indirect fashion as given in Fig. 11 (for review,

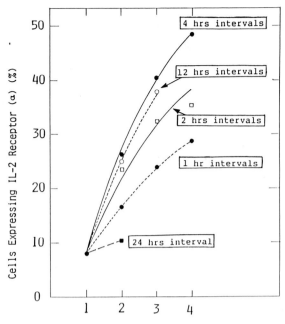

Fig. 10. Effect of increasing both number and interval between multiple additions of permeable diacylglycerol and ionomycin on T-cell responses. The expression of interleukin-2 receptor α was measured. The detailed experimental conditions are descried elsewhere (Berry et al., 1990). DG, diacylglycerol; Ca^{2+}, ionomycin; and Il-2, interleukin-2.

see Weiel et al., 1990). It has been proposed that integration of two signalling pathways involving tyrosine and serine/threonine kinases is needed for activation of a protein kinase, MAP kinase (Anderson et al., 1990). It is likely, then, that gene activation and cell proliferation may result from dynamic interactions between PKC and many other signalling pathways.

Implications for brain research

Following the discovery of PKC in 1977, our knowledge of this enzyme has expanded explosively and diverse roles of this enzyme in cell-to-cell communication appear to be firmly established. PKC is most abundant both in quantity

138

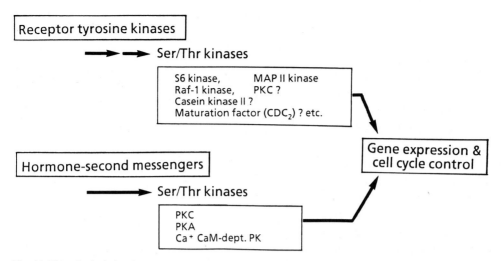

Fig. 11. Hypothetical signal cascades involving various protein kinases, eventually leading to gene activation and cell cycle control. Search has been made in recent years to elucidate a possible link between various growth factor receptors having tyrosine kinases and serine/threonine kinases in direct or indirect fashions. MAP II kinase, microtubulus-associated protein II kinase; PKA, cyclic AMP-dependent protein kinase; and Ca^{2+} CaM-dept. PK, Ca^{2+} and calmodulin-dependent protein kinase. This figure is taken from Farago and Nishizuka (1990).

and in variety in the brain tissues having trillions of neuron networks. Available evidence suggests that the members of the PKC family each play discretely different, specific functions in the processing and modulation of neuronal signals. Extensive electrophysiological studies have shown that the phosphorylation of proteins catalyzed by PKC is most likely related to the induction of synaptic plasticity, particularly of hippocampus long-term potentiation. It is also suggested that PKC may take part in the induction of neuronal cell damage during anoxia and over-stimulation of glutamate receptors. Attempts have thus been made to identify any roles of this enzyme in developing various neuronal disorders.

The involvement of PKC in the growth control and cell differentiation is now evident. In fact, it is becoming clearer that several growth factors, including nerve growth factor, may well involve PKC in their signal transduction processes, and

that abnormal activity of this enzyme may result in malfunction of nervous tissues. The evidence to discuss the detailed functions of the individual PKC subspecies is still premature, but it is hoped that precise molecular basis of the role of each member of the PKC family will be clarified further in the next few years.

Acknowledgements

The skillful secretarial assistance of Mrs. S. Nishiyama and Miss Y. Kimura is cordially acknowledged. The investigation was supported in part by research grants from the Special Research fund of the Ministry of Education, Science and Culture, Japan; Muscular Dystrophy Association, U.S.A.; Juvenile Diabetes Foundation International, U.S.A.; Yamanouchi Foundation for Research on Metabolic Disorders; Merck Sharp & Dohme Research Laboratories; Biotechnology

Laboratories of Takeda Chemical Industories; and New Lead Research Laboratories of Sankyo Company.

References

Allgaier, C., Daschmann, B., Huang, H.Y. and Hertting, G. (1988) Protein kinase C and presynaptic modulation of acetylcholine release in rabbit hippocampus. *Br. J. Pharmacol.,* 93: 525–534.

Anderson, N.G., Maller, J.L., Tonks, N.K. and Sturgill, T.W. (1990) Requirement for integration of signals from two distinct phosphorylation pathways for activation of MAP kinase. *Nature,* 343: 651–653.

Ase, K., Berry, N., Kikkawa, U., Kishimoto, A and Nishizuka, Y. (1988a) Differential down-regulation of protein kinase C subspecies in KM3 cells. *FEBS Lett.,* 236: 396–400.

Ase, K., Saito, N., Shearman, M.S., Kikkawa, U., Ono, Y., Igarashi, K., Tanaka, C. and Nishizuka, Y. (1988b) Distinct cellular expression of βI- and βII-subspecies of protein kinase C in the rat cerebellum. *J. Neurosci.,* 8; 3850–3856.

Benowitz, L.I. and Routtenberg, A. (1987) A membrane phosphoprotein associated with neural development, axonal regeneration, phospholipid metabolism, and synaptic plasticity. *Trends Neurosci.,* 10: 527–532.

Berridge, M.J., and Irvine, R.F. (1984) Inositol trisphosphate, a novel second messenger in cellular signal transduction. *Nature,* 312; 315–321.

Berry, N., Ase, K., Kikkawa, U., Kishimoto, A. and Nishizuka, Y. (1989) Human T cell activation by phorbol esters and diacylglycerol analogues. *J. Immunol.,* 143: 1407–1413.

Berry, N., Ase, K., Kishimoto, A. and Nishizuka, Y. (1990) Activation of resting human T cells requires prolonged stimulation of protein kinase C. *Proc. Natl. Acad. Sci. USA,* 87: 2294–2298.

Berry, N. and Nishizuka, Y. (1990) Protein kinase C and T-cell activation. *Eur. J. Biochem.,* 189: 205–214.

Blackshear, P.J., Witters, L.A., Girard, P.R., Kuo, J.F. and Quamo, S.N. (1985) Growth factor-stimulated protein phosphorylation in 3T3-L1 cells. *J. Biol. Chem.,* 260: 13304–13315.

Canonico, P.L., Favit, A., Catania, M.V. and Nicoletti, F. (1988) Phorbol esters attenuate glutamate-stimulated inositol phospholipid hydrolysis in neuronal cultures. *J. Neurochem.,* 51: 1049–1053.

Castagna, M., Takai, Y., Kaibuchi, K., Sano, K., Kikkawa, U. and Nishizuka, Y. (1982) Direct activation of calcium-activated, phospholipid-dependent protein kinase by tumor-promoting phorbol esters. *J. Biol. Chem.,* 257: 7847–7851.

Crepel. F. and Krupa, M. (1988) Activation of protein kinase C induces a long-term depression of glutamate sensitivity of cerebellar Purkinje cells. An in vitro study. *Brain Res.,* 458: 397–401.

Doerner, D., Pitler, T.A. and Alger, B.E. (1988) Protein kinase C activators block specific calcium and potassium current components in isolated hippocampal neurones. *J. Neurosci.,* 8: 4069–4078.

Dunkley, P.R., Heath, J.W., Harrison, S.M., Jarvie, P.E., Glenfield, P.J. and Rostas J.A.P. (1988) A rapid Percoll gradient procedure for isolation of synaptosomes directly from an S1 fraction: homogeneity and morphology of subcellular fractions. *Brain Res.,* 441: 59–71.

Exton, J.H. (1990) Signaling through phosphatidylcholine breakdown. *J. Biol. Chem.,* 265: 1–4.

Farago, A. and Nishizuka, Y. (1990) Protein kinase C in transmembrane signalling. *FEBS Lett.,* 268: 350–354.

Gispen, W.H., Leunissen, J.L.M., Oestreicher, A.B., Verkleij, A.J. and Zwiers, H. (1985) Presynaptic localization of B-50 phosphoprotein: the (ACTH)-sensitive protein kinase substrate involved in rat brain polyphosphoinositide metabolism. *Brain Res.,* 328: 381–385.

Hashimoto, K., Kishimoto, A., Aihara, H., Yasuda, I., Mikawa, K. and Nishizuka, Y. (1990) Protein kinase C during differentiation of human promyelocytic leukemia cell line, HL-60. *FEBS Lett.,* 263: 31–34.

Hashimoto, T., Ase, K., Sawamura, S., Kikkawa, U., Saito, N., Tanaka, C. and Nishizuka, Y. (1988) Postnatal development of a brain-specific subspecies of protein kinase C in rat. *J. Neurosci.,* 8: 1678–1683.

Hidaka, H., Tanaka, T., Onoda, K., Hagiwara, M., Watanabe, M., Ohta, H., Ito, U., Tsurudome, M. and Yoshida, T. (1988) Cell type-specific expression of protein kinase C isozymes in the rabbit cerebellum. *J. Biol. Chem.,* 263: 4523–4526.

Hoeffler, J.P., Deutsch, P.J., Lin, J. and Habener, J.F. (1989) Distinct adenosine $3',5'$-monophosphate and phorbol ester-responsive signal transduction pathways converge at the level of transcriptional activation by the interactions of DNA-binding proteins. *Mol. Endocrinol.,* 3: 868–880.

Hosoda, K., Saito, N., Kose, A., Ito, A., Tsujino, T., Ogita, K., Kikkawa, U., Ono, Y., Igarashi, K., Nishizuka, Y. and Tanaka, C. (1989) Immunocytochemical localization of the βI subspecies of protein kinase C in rat brain. *Proc. Natl. Acad. Sci. USA,* 86: 1393–1397.

Huang, K-P., Huang, F.L., Nakabayashi, H. and Yoshida, Y. (1988a) Biochemical characterization of rat brain protein kinase C isozymes. *J. Biol. Chem.,* 263: 14839–14845.

Huang, F.L., Yoshida, Y., Nakabayashi, H., Young, III, W.S. and Huang, K-P. (1988b) Immunocytochemical localization of protein kinase C isozymes in rat brain. *J. Neurosci.,* 8: 4734–4744.

Jacobson, R.D., Virág, I. and Skene, J.H.P. (1986) A protein associated with axon growth, GAP-43, is widely distributed and developmentally regulated in rat CNS. *J. Neurosci.,* 6: 1843–1855.

140

Kennedy, M.B. (1989) Regulation of synaptic transmission in the central nervous system: long-term potentiation. *Cell,* 59: 777–787.

Kikkawa, U., Ono, Y., Ogita, K., Fujii, T., Asaoka, Y., Sekiguchi, K., Kosaka, Y., Igarashi, K. and Nishizuka, Y. (1987) Identification of the structures of multiple subspecies of protein kinase C expressed in rat brain. *FEBS Lett.,* 217: 227–231.

Kishimoto, A., Kajikawa, N., Shiota, M. and Nishizuka, Y. (1983) Proteolytic activation of calcium-acvitated, phospholipid-dependent protein kinase by calcium-dependent neutral protease. *J. Biol. Chem.,* 258: 1156–1164.

Kose, A., Saito, N., Ito, H., Kikkawa, U., Nishizuka, Y. and Tanaka, C. (1988) Electron microscopic localization of type I protein kinase C in rat Purkinje cells. *J. Neurosci.,* 8: 4262–4268.

Kose, A., Ito, A., Saito, N. and Tanaka, C. (1990) Electron microscopic localization of γ- and βII-subspecies of protein kinase C in rat hippocampus. *Brain Res.,* 518: 209–217.

Matthies, H.J.G., Palfrey, H.C., Hirning, L.D. and Miller, R.J. (1987) Down regulation of protein kinase C in neuronal cells: effects on neurotransmitter release. *J. Neurosci.,* 7: 1198–1206.

Meiri, K.F., Pfenninger, K.H. and Willard, M.B. (1986) Growth-associated protein, GAP-43, a polypeptide that is induced when neurons extend axons, is a component of growth cones and corresponds to pp46, a major polypeptide of a subcellular fraction enriched in growth cones. *Proc. Natl. Acad. Sci. USA,* 83: 3537–3541.

Montz, H.P.M., Davis, G.E., Skaper, S.D., Manthorpe, M. and Varon, S. (1985) Tumor-promoting phorbol diester mimics two distinct neuronotrophic factors. *Dev. Brain Res.,* 23: 150–154.

Nichols, R.A., Haycock, J.W., Wang, J.K.T. and Greengard, P. (1987) Phorbol ester enhancement of neurotransmitter release from rat brain synaptosomes. *J. Neurochem.,* 48: 615–621.

Nishizuka, Y. (1984) The role of protein kinase C in cell surface signal transduction and tumour promotion. *Nature,* 308: 693–698.

Nishizuka, Y. (1986) Studies and perspectives of protein kinase C. *Science,* 233: 305–312.

Nishizuka, Y. (1988) The molecular heterogeneity of protein kinase C and its implications for cellular regulation. *Nature,* 334: 661–665.

Ono, Y., Kikkawa, U., Ogita, K., Fujii, T., Kurokawa, T., Asaoka, Y., Sekiguchi, K., Ase, K., Igarashi, K. and Nishizuka, Y. (1987) Expression and properties of two types of protein kinase C: alternative splicing from a single gene. *Science,* 236: 1116–1120.

Ono, Y., Fujii, T., Ogita, K., Kikkawa, U., Igarashi, K. and Nishizuka, Y. (1988) The structure, expression, and properties of additional members of the protein kinase C family. *J. Biol. Chem.,* 263: 6927–6932.

Oda, T., Shearman, M.S., and Nishizuka, Y. (1991) Synaptosomal protein kinase C subspecies: B. down-regulation promoted by phorbol ester and its effect on evoked norepinephrine release. *J. Neurochem.,* 56: 1263–1269.

Påhlman, S., Odelstad, L., Larsson, E., Grotte, G. and Nilsson, K. (1981) Phenotypic changes of human neuroblastoma cells in culture induced by 12-*O*-tetradecanoylphorbol-13-acetate. *Int. J. Cancer,* 28: 583–589.

Rodriguez-Pena, A. and Rozengurt, E. (1984) Disappearance of Ca^{2+}-sensitive, phospholipid-dependent protein kinase activity in phorbol ester-treated 3T3 cells. *Biochem. Biophys. Res. Commun.,* 120: 1053–1059.

Saito, N., Kikkawa, U., Nishizuka, Y. and Tanaka, C. (1988) Distribution of protein kinase C-like neurons in rat brain. *J. Neurosci.,* 8: 369–382.

Saito, N., Kose, A., Ito, A., Hosoda, K., Mori, M., Hirata, M., Ogita, K., Kikkawa, U., Ono, Y., Igarashi, K., Nishizuka, Y. and Tanaka, C. (1989) Immunocytochemical localization of βII subspecies of protein kinase C in rat brain. *Proc. Natl. Acad. Sci. USA,* 86: 3409–3413.

Sekiguchi, K., Tsukuda, M., Ase, K., Kikkawa, U. and Nishizuka, Y. (1988) Mode of activation and kinetic properties of three distinct forms of protein kinase C from rat brain. *J. Biochem.,* 103: 759–765.

Shearman, M.S., Sekiguchi, K. and Nishizuka, Y. (1989) Modulation of ion channel activity: a key function of the protein kinase C enzyme family. *Pharmacol. Rev.,* 41: 211–237.

Shearman, M.S., Shinomura, T., Oda, T. and Nishizuka, Y. (1991) Synaptosomal protein kinase C subspecies: A. dynamic changes in the hippocampus and cerebellar cortex concomitant with synaptogneses. *J. Neurochem.,* 56: 1255–1262.

Shuntoh, H., Taniyama, K., Fukuzaki, H., and Tanaka, C. (1988) Inhibition by cyclic AMP of phorbol ester-potentiated norepinephrine release from guinea pig brain cortical synaptosomes. *J. Neurochem.,* 51: 1565–1572.

Skene, J.H.P., Jacobson, R.D., Snipes, G.J., McGuire, C.B., Norden, J.J. and Freeman, J.A. (1986) A protein induced during nerve growth (GAP-43) is a major component of growth-cone membranes. *Science,* 233: 783–786.

Sposi, N.M., Bottero, L., Cossu, G., Russo, G., Testa, U. and Peschle, C. (1989) Expression of the protein kinase C genes during ontogenic development of the central nervous system. *Mol. Cell. Biol.,* 9: 2284–2288.

Stabel, S., Rodriguez-Pena, A., Young, S., Rozengurt, E. and Parker, P.J. (1987) Quantitation of protein kinase C by immunoblot – expression in different cell lines and response to phorbol esters. *J. Cell. Physiol.,* 130: 111–117.

Turner, R.S., Raynor, R.L., Mazzei, G.J., Girard, P.R. and Kuo, J.F. (1984) Developmental studies of phospholipid-sensitive Ca^{2+}-dependent protein kinase and its substrates and of phosphoprotein phosphatases in rat brain. *Proc. Natl. Acad. Sci. USA,* 81: 3143–3147.

Versteeg, D.H.G. and Ulenkate, H.J.L.M. (1987) Basal and electrically stimulated release of [³H]noradrenaline and [³H]dopamine from rat amygdala slices in vitro: effects of 4β-phorbol 12,13-dibutyrate, 4α-phorbol 12,13-didecanoate and polymyxin B. *Brain Res.,* 416: 343–348.

Vyas, S., Bishop, J.F., Gehlert, D.R. and Patel, J. (1990) Effects of protein kinase C down-regulation on secretory events and proopiomelanocortin gene expression in anterior pituitary tumor (AtT-20) cells. *J. Neurochem.,* 54: 248–255.

Weiel, J.E., Ahn, N.G., Seger, R. and Krebs, E.G. (1990) Communication between protein tyrosine and protein serine/threonine phosphorylation. In Y. Nishizuka, M. Endo and C. Tanaka (Eds.), *Advances in Second Messenger and Phosphoprotein Research,* Vol. 24, Raven Press, New York, pp 182–195.

Yoshida, Y., Huang, F.L., Nakabayashi, H. and Huang, K-P. (1988) Tissues distribution and developmental expression of protein kinase C isozymes. *J. Biol. Chem.,* 263: 9868–9873.

Young, S., Parker, P.J., Ullrich, A. and Stabel, S. (1987) Down-regulation of protein kinase C is due to an increased rate of degradation. *Biochem. J.,* 244: 775–779.

Zurgil, N. and Zisapel, N. (1985) Phorbol ester and calcium act synergistically to enhance neurotransmitter release by brain neurons in culture. *FEBS Lett.,* 185: 257–261.

W.H. Gispen and A. Routtenberg (Eds.)
Progress in Brain Research, Vol. 89
© 1991 Elsevier Science Publishers B.V.

CHAPTER 10

Protein kinase C subtypes and their respective roles

Kuo-Ping Huang, Freesia L. Huang, Charles W. Mahoney and Kuang-Hua Chen

Section on Metabolic Regulation, Endocrinology and Reproduction Research Branch, National Institute of Child Health and Human Development, National Institutes of Health, Bethesda, MD 20892, U.S.A.

Introduction

The protein kinase C (PKC) gene family consists of two groups of isozymes distinguishable by their requirements for Ca^{2+} (for review, see Nishizuka, 1988). The group A PKCs consist of type I, II, and III isozymes encoded, respectively, by the γ, βI and βII, and α genes (Huang et al., 1986, 1987a). These enzymes account for the majority of the Ca^{2+}/phosphatidylserine (PS)/diacylglycerol (DAG)-stimulated kinase activity in the CNS (Huang and Huang, 1986). The native PKCs encoded by the group B genes, δ, ϵ, and ζ, have not been well characterized. It is believed that PKC I, II, and III are stimulated at the membrane locations in response to the ligand-induced turnover of membrane phospholipids that causes increases in $[Ca^{2+}]_i$ and DAG. The increase in $[Ca^{2+}]_i$ is provided by the entry of Ca^{2+} from extracellular fluid via receptor-operated, voltage-gated, and second messenger-regulated Ca^{2+} channels (for review, see Jan and Jan, 1989) or by release of Ca^{2+} from intracellular stores following binding of inositol 1,4,5-trisphosphate (IP_3), generated from the phospholipase C-catalyzed hydrolysis of phosphatidylinositol 4,5-bisphosphate (PIP_2) (Berridge and Irvine, 1989), to its receptor (Ferris et al., 1989). DAG generated at the membrane is derived from the hydrolysis of phosphoinositides (Majerus et al., 1986), phosphatidylcholine (Pelech and Vance, 1989), and inositol-containing glycolipid (Low and Saltiel, 1988) by multiple forms of phospholipase C, or of phosphatidate by a phosphatase. Stimulation of PKCs in the nervous system has been implicated in the regulation of ion channels (Madison et al., 1986; Shearman et al., 1989b), enhancement of neurotransmitter release (Malenka et al., 1986; Nicoll et al., 1987), control of growth and differentiation (Burgess et al., 1986), and modification of neuronal plasticity (Routtenberg, 1985). Available evidence indicates that the various PKC isozymes have different biochemical characteristics and distinct regional,

---

Abbreviations: PKC, protein kinase C; DAG, diacylglycerol; MARCKS, myristoylated alanine-rich C kinase substrate; PS, phosphatidylserine; PA, phosphatidic acid; PC, phosphatidylcholine; PE, phosphatidylethanolamine; PG, phosphatidylglycerol; PI, phosphatidylinositol; PIP, phosphatidylinositol 4-phosphate; PIP_2, phosphatidylinositol 4,5-bisphosphate; IP_3, inositol 1,4,5-trisphosphate; NP-40, Nonidet P-40; C/EBP, CCAAT/enhancer binding protein.
Correspondence: Dr. Kuo-Ping Huang, Bldg. 10, Rm. B1-L400, NIH, Bethesda, MD 20892, U.S.A. Tel.: (301) 402-0253; Fax: (301) 480-8010.

cellular, and subcellular localizations which may contribute to the diverse responses to external stimuli (for reviews, see Huang, 1989, 1990).

Distinct expression of PKC isozymes in the CNS

All of the studies reported so far indicate that PKCs are heterogeneously distributed in the various brain regions (Huang et al., 1987b) and that neurons contain the highest level of PKC. Cellular distribution of the group B PKCs has not yet been studied. Immunochemical analysis with isozyme-specific antibodies reveal that PKC I is expressed in the CNS and spinal cord and PKC II and III in both the CNS and peripheral tissues (Huang et al., 1987a). In cerebellum, PKC I is localized largely in the cell bodies and dendrites of the Purkinje cells, but very little in the nuclei and the axons of these cells in the granule cell layer or the white matter. Since the dendrites of Purkinje cells receive input from the parallel fibers of granule cells, it is likely that PKC I functions in the postsynaptic signal transduction. Electron microscopic observations indicated that, in addition to cell bodies and dendrites, the axoplasm and synaptic vesicles of the Purkinje cell axons also contain PKC I (Kose et al., 1988), suggesting that this isozyme may also participate in presynaptic functions. Immunofluorescent staining with PKC II- and III-specific antibodies indicate that the former are localized mainly in cerebellar granule cells and the latter in both the granule cells and Purkinje cells. Since PKC II and III are poorly represented in the cerebellar molecular layer, it seems that the parallel fibers of the granule cells contain a relatively low level of these two enzymes. The dendritic localization of PKC II and III in granule cells has yet to be demonstrated. Even though PKC I and III colocalize in Purkinje cells, only PKC I, but not PKC III, is present in the dendrites of these cells. In cerebellum, PKC I appears to be localized only in the Purkinje cells, whereas PKC II and III are more widely distributed in the various neurons including basket and stellate cells.

Except for the Purkinje and granule cells in cerebellum and mitral cells (containing mainly PKC III) in the olfactory bulb, most neurons in the brains contain all three types of the group A PKCs (Huang et al., 1988). Hippocampal pyramidal cells and granule cells of the dentate gyrus are high in PKC I, II, and III, whereas cells of the strata oriens, radiatum, and lacunosum molecular, and molecular layer of the dentate gyrus contain much less PKC. Immunoreactivities of PKC I and II appear to uniformly cover the entire cell bodies of granule cells and pyramidal cells of the hippocampal formation, whereas PKC III is mainly cytoplasmic, indicating a possible difference in the subcellular localization of these enzymes. Distinct distribution of PKC isozymes is evident as well for the two alternatively spliced βI and βII (Hosoda et al., 1989).

PKC isozymes in the areas along the occipitotemporal cortex involved in the visual information processing pathway of monkey brain have been quantified by immunoblot analysis with isozyme-specific antibodies (Huang et al., 1989a). PKC I was found to have increasing concentrations rostral along the cerebral cortices of occipital and temporal lobes, and peaked at perirhinal and entorhinal areas. In comparison, PKC II and III appear to be uniformly present in the cortical areas of the visual information-processing pathway. Since neurobehavioral studies have demonstrated that the neocortical and limbic areas of the anterior and medial temporal regions participate more directly than the striate, prestriate, and posterior temporal regions in the storage of visual information, it seems likely that PKC I may participate in the plastic changes important for mnemonic function.

Distinct biochemical characteristics of PKC isozymes

PKC isozymes are structurally homologous, however, they exhibit distinctive differences in their sensitivities to stimulation by activators (Huang et al., 1988). Using phospholipid/detergent mixed micelles assay, these enzymes appear to have a similar requirement for Ca^{2+} ($A_{1/2} = 0.2$–0.6 μM) in the assays of kinase activity and phorbol ester binding. The phospholipid requirements of these PKC isozymes are distinguishable: PKC I is more sensitive to stimulation by cardiolipin in the absence of DAG than PKC II and III. In the presence of cardiolipin, the concentrations of DAG and phorbol 12,13-dibutyrate (PDBu) required for half-maximal activation of PKC I are nearly an order of magnitude lower than those for PKC II and III. In the presence of PS, binding of PDBu to PKC I evokes a corresponding stimulation of the kinase activity, whereas bindings of this phorbol ester to PKC II and III produce lesser degrees of kinase stimulation. These findings indicate that the binding of phorbol ester to the various PKC isozymes may evoke different degrees of activation of these enzymes. The various PKC isozymes also respond differently to stimulation by arachidonate (Naor et al., 1988) and its metabolites (Shearman et al., 1989a). In addition to their different responses to the various activators, the three PKC isozymes also have different K_m values for protein substrates. It is likely that these enzymes may be differentially stimulated by a variety of stimuli at different cellular locations.

Agonist-induced stimulation and translocation of PKC frequently results in an accelerated degradation of this enzyme. PKC I, II, and III exhibit differential sensitivity to proteolytic degradation in vitro by trypsin or calpain (Huang et al., 1989b; Kishimoto et al., 1989). PKC I is more sensitive to degradation by these proteases than PKC II and III. PKC III is most resistant to proteolysis. Degradation of PKC III can be facilitated by the addition of Ca^{2+}, PS, and DAG; none of these components alone is effective. While Ca^{2+} and PS together are sufficient to stimulate proteolytic degradation of PKC III, addition of DAG greatly reduces the Ca^{2+} requirement for such reaction. Conversion of PKC III from a protease-insensitive to -sensitive form also takes place in the presence of Ca^{2+} and PS/NP-40 mixed micelles, under which condition the kinase is not active in the phosphorylation assay. These findings indicate that binding of PKC to PS vesicles in the presence of Ca^{2+} is sufficient to promote proteolysis, whereas the activation of PKC requires additional activators such as DAG or phorbol ester. Thus, a persistent rise in $[Ca^{2+}]_i$ without a concomitant increase in DAG may lead to a degradation pathway predominantly. The effects of DAG and phorbol ester in promoting PKC degradation in the presence of Ca^{2+} are different with respect to the specificity for phospholipid. In the presence of DAG, only those phospholipids such as PS, cardiolipin, and PA, which are active in supporting kinase activation, are active. In contrast, the PMA-stimulated PKC degradation has a broader specificity for phospholipids; it is active with most acidic phospholipids. Neutral phospholipids, such as PC and PE, which are ineffective in the activation of PKC, are also ineffective in supporting PKC degradation either in the presence of DAG or PMA.

Differential sensitivity of PKC isozymes to proteolytic degradation was also found in the PMA-treated rat basophilic leukemia cells RBL-2H3 cells (Huang et al., 1989b). Treatment of these cells with PMA results in a translocation of PKC from the cytosolic to the particulate fractions. PKC II, which is mostly cytosolic in the untreated cells, is more rapidly degraded than PKC III, which is present in both the cytosolic and particulate fractions, following addition of PMA.

Acidic phospholipids are required for the stimulation of PKC in the protein phosphorylation assay. However, incubation of PKCs with acidic phospholipids, PS, PA, or PG, in the absence of divalent metal ion results in a differential inactivation of PKC isozymes (Huang and Huang, 1990). The phospholipid-induced inactivation of PKC is concentration- and time-dependent and only affects the kinase activity without influencing phorbol ester binding. PKC I is most susceptible to the phospholipid-induced inactivation and PKC III the least. Addition of divalent cation such as Ca^{2+} or Mg^{2+} suppresses the phospholipid-induced inactivation of PKC. The phospholipid-induced inactivation of PKC is a result of direct interaction of PKC with acidic phospholipid in the absence of divalent cation. PKC I interacts more readily than PKC II and PKC III is the least favorable to interact with the phospholipid. The phospholipid-induced inactivation of PKC is a result of direct interaction of the acidic phospholipid with the catalytic domain in the absence of Ca^{2+}. In the presence of Ca^{2+}, these phospholipids mainly interact with the regulatory domain of PKC resulting in the activation of the kinase.

Association of PKC isozymes with phosphoinositides

The inositol-containing phospholipids, phosphatidylinositol (PI), phosphatidylinositol 4-phosphate (PIP), and phosphatidylinositol 4,5-bisphosphate (PIP$_2$), account for nearly 5–10% of the total cellular phospholipid in the mammalian brain (Hawthorne and Kemp, 1964). Among them, the polyphosphoinositides, PIP and PIP$_2$, represent a minor part of this class of phospholipid. The phosphoinositides as a whole are not as effective as PS in supporting PKC activity in the presence of DAG and Ca^{2+} nor the phorbol ester binding. However, they interact more avidly with PKC, especially PKC I, than PS. The rates of

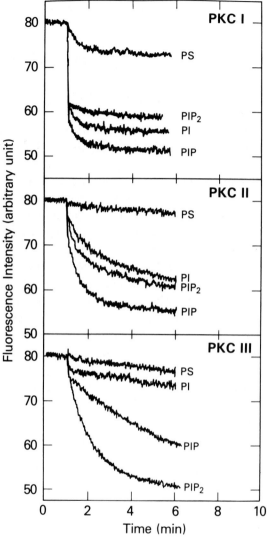

Fig. 1. Interaction of PKC isozymes with the various phospholipids. PKC I, II and III in 1.6 ml of 10 mM Tris-Cl buffer, pH 7.5, containing 0.5 mM DTT, 0.25 mM EDTA, 0.25 mM EGTA and 5% glycerol were excited at 280 nm under constant mixing at 22–25°C. Emission fluorescence at 340 nm was monitored following the addition of 6 μM of PS, PI, PIP and PIP$_2$ suspended in 20 mM Tris-Cl buffer, pH 7.5.

interaction of PKC I with PI, PIP, and PIP$_2$ are much faster than PKC II and III (Fig. 1), suggesting that a preferential binding of PKC I with these phospholipids may take place in vivo. Overall, the polyphosphoinositides have a greater affinity for all PKCs than PI. Interaction of PKC with polyphosphoinositides results in the formation of stable complexes consisting of one PKC molecule/micelle as determined by gel filtration chromatography. These complexes can further interact with PS to form larger complexes capable of binding phorbol ester in the presence of Ca^{2+} and express kinase activity in the presence of Ca^{2+} and DAG. Because of the high affinity binding between PKCs and polyphosphoinositides, we speculate that these phospholipids may serve as anchoring sites for PKC at the membrane locations (Fig. 2). The attachment of PKCs to membrane phosphoinositides places the kinase in proximity to DAG generated from the hydrolysis of these phospholipids by PI-specific phospholipase C. In addition, the binding of PKCs to PIP_2 could also confer moderate stimulation of the kinase at increased $[Ca^{2+}]_i$, which promotes the interaction of PKC/PIP_2 complexes with PS to form a partially active kinase (Chauhan and Brockerhoff, 1988). Whether the differential affinity among the various PKC isoforms for the inositol-containing phospholipids could also con-

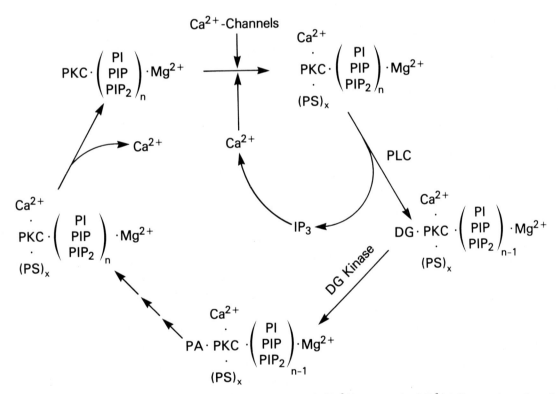

Fig. 2. Stimulation of phosphoinositides-associated PKC. Increase in $[Ca^{2+}]_i$ as a result of Ca^{2+} influx or release from the internal stores promotes the interaction of phosphoinositides-associated PKC with PS. PKC becomes activated when the phosphoinositides are hydrolyzed by phospholipase C generating DG. The kinase activity subsides when DG is converted to phosphatidate and other phospholipids. When $[Ca^{2+}]_i$ drops, PKC and PS interaction weaken and the kinase returns to the basal state. By positioning the kinase in proximity to the source of DG, the kinase becomes more responsive to the external signal.

fer the distinct subcellular localization of these enzymes requires further investigation.

Developmental expression of PKC isozymes in rat brain

In fetal rat brain, the levels of all three PKCs are very low. Both PKC II and III increase progressively from fetal stage to 2–3 weeks of age and remain constant thereafter up to 6 weeks (Yoshida et al., 1988). The amount of immunoreactive PKC I appears to be very low in the fetus and within 1 week of age. A rapid increase occurs between 2 and 3 weeks of age. This delayed pattern of PKC I expression was also observed during the development of the cerebellum.

Throughout postnatal development, Purkinje cells are the most prominent constituents in the cerebellar cortex that expresses PKC I (Huang et al., 1990). At one week postnatal, the monolayered Purkinje cells are clearly visible following staining with antibody specific for PKC I (Fig. 3A); occasionally some have newly sprouted primary dendrites but rarely secondary processes. At 2 weeks of age (Fig. 3B), the brightly stained Purkinje cell somata with apically oriented primary dendrites and fine branches at their tips are all visible, however, there is a limited spread of the arborizing dendrites. At three weeks of age (Fig. 3C), the flattened arborization of Purkinje cells appears to be fully developed. Staining of these cells and their dendrites by PKC I-antibody is similar to that of adult rat cerebellum. Thus, at various stages of postnatal development of cerebellar cortex, PKC I is present mainly in the cell bodies and the developing dendrites of the Purkinje cells, and no other component was found to contain significant quantity of PKC I.

In adult rat cerebellum, PKC II is located mainly in the neurons of the granular layer. At one week of age, PKC II is present mainly in the outer germinal layer of cerebellar cortex (Fig.

3D). There was some diffuse staining in the internal granular layer, indicating a few granule cells have reached their destination at this stage of development. The medullary layer of the future white matter is devoid of this enzyme. PKC II in the two week-old rat cerebellum (Fig. 3E) is localized in the well-formed internal granular layer and the remaining thin external germinal layer. By three weeks of age, there is a nearly complete absence of the external germinal layer and PKC II is present largely in the internal granular layer as in the adult cerebellum (Fig. 3F).

In the adult rat cerebellum, PKC III is associated with both Purkinje and granule cells. Staining of cerebellar cortex of 1-week-old animals with PKC III-specific antibody showed a weak staining of both the Purkinje and granule cells (Fig. 3G) in the external and those early arriving ones in the internal layers. At 2 weeks of age (Fig. 3H), the external germinal layer was stained more brightly than the Purkinje cells and the internal granule cells. The staining of Purkinje cells was mostly in the cytoplasm of the neurons and not in their dendrites. At 3 weeks of age, with the disappearing external germinal layer, the Purkinje and granule cells remained strongly stained by PKC III antibody (Fig. 3I). This staining pattern remained unchanged in adult rat cerebellum. Thus, throughout all stages of postnatal development, PKC III is present in both Purkinje and granule cells but apparently absent from the processes of these cells.

Both PKC II and III are present in the granule cells, however, PKC II appears mainly in the center of the somata while PKC III has mostly cytosolic appearance. Unlike PKC I, which is present in cell bodies and dendrites of Purkinje cells, PKC III appears to be only in the Purkinje cell bodies. These results demonstrate that the developmental expression of PKC isozymes is under separate control, and that their distinct cellular and subcellular localizations may dictate their unique functions in cerebellum. The observation that a sudden increase of PKC I between 2–3 weeks postnatal parallels the synapse formation between parallel fiber of granule cells and dendritic spine of the Purkinje cells suggests the

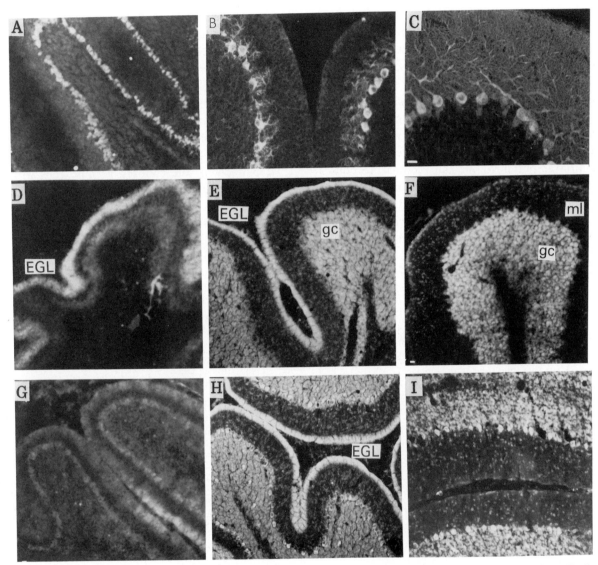

Fig. 3. Developmental expression of PKC isozymes in rat cerebellum. Immunofluorescent staining of cerebellar cortex by antibodies specific for PKC I (A, B and C), II (D, E and F) and III (G, H and I). Sagittal sections of cerebellum of 1-week-old (A, D and G), 2-week-old (B, E and H) and 3-week-old (C, F and I) rats were used. PKC I antibody stained mainly the Purkinje cell bodies and dendrites throughout the development. PKC II antibody stained the cerebellar granule cells in the external germinal layer (EGL) of the 1- and 2-week-old rats and mainly the granular layer of the 3-week-old rats. PKC III antibody stained both granule cells and Purkinje cells throughout the development.

involvement of PKC I in such synaptic transmission.

Expression and translocation of PKC in the primary culture of rat cerebellar granule cells

Granule cells in culture have been used frequently in pharmacological characterizations of cell surface receptors and their specific ligand-mediated responses. These cells are rich in L-glutamate-synthesizing enzyme glutaminase (Patel and Balazs, 1975) and they release glutamate in response to depolarizing stimuli (Gallo et al., 1982, 1987). These cells also utilize glutamate as neurotransmitter to mediate cellular responses (Garthwaite and Brodbelt, 1990). Granule cells in culture also express PKC II and III, but not PKC I. Early in the culturing, PKC II was detected in both the cytosol and membrane fractions; as the culture progressed and the neurons differentiated, most of the PKC II becomes membrane-associated and resists extraction by buffer containing metal chelators. The amount of immunoreactive PKC III was very low early in the culture; a progressive increase was observed during granule cell differentiation. The increase in PKC III level toward the later stage of culturing is reflected by an initial increase in the membrane followed by the cytosolic fractions. Staining of granule cells with PKC II- and III-specific antibodies revealed that these enzymes are mostly located in the cell bodies during early stage of culturing and partial staining at the processes during the later stage.

Treatment of the cultured granule cells with glutamate results in a translocation of PKC to the particulate fractions. Two populations of PKC II were identified: one was extractable by 0.5% NP-50 having a M_r of 82 000 on SDS-PAGE and the

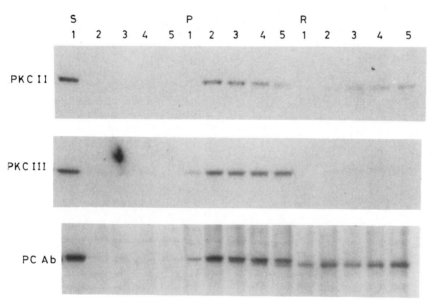

Fig. 4. Phorbol ester-mediated translocation of PKC in cultured granule cells. Granule cells after 8 days in culture were treated with 50 nM PMA for 1, 0 min; 2, 10 min; 3, 30 min; 4, 60 min; and 5, 180 min. Cells were homogenized with 20 mM Tris-Cl, pH 7.5, containing 1 mM DTT, 2 mM EDTA, 5 mM EGTA, 10% glycerol, 2 μg/ml leupeptin and 0.5 mM PMSF and fractionated into soluble (S), 0.5% NP-40 extractable membrane (P) and the residual NP-40 nonextractable membrane (R) fractions. Proteins of 50 μg each were analyzed by immunoblot with PKC II- and III-specific antibodies and polyclonal antibodies that recognize both isozymes.

other not extractable by this detergent exhibiting a M_r of 80 000. This latter species appears to associate tightly with the membrane skeleton. Conversion of the 82-kDa to the 80-kDa PKC following addition of glutamate could be due to proteolysis or dephosphorylation. Accumulation of the membrane skeleton-associated PKC II was also observed for the phorbol ester-treated granule cells (Fig. 4). In contrast, translocation of PKC III to the membrane fraction does not result in an accumulation of the modified form of PKC in the membrane skeleton. It appears that selective translocation of PKC II to the membrane skeleton may enable this isozyme to phosphorylate its membrane-associated substrates such as neuromodulin (Cimler et al., 1985; Meiri and Gordon-Weeks, 1990) and MARCKS (Stumpo et al., 1989).

Characterization of the 5'-flanking region of the rat PKC γ gene

PKC I, encoded by the γ gene, is specifically expressed in the CNS and its synthesis is controlled during development. One of the important steps involved in the regulation of gene expression is the interaction of regulatory proteins with specific DNA sequences upstream of the transcriptional initiation site. A 3.6-kb rat genomic segment containing the 5'-flanking region, the first exon, and first intron has been identified. Promoter activity of a DNA segment spanning the 5'-flanking region has been demonstrated by both in vitro transcription and chloramphenicol acetyltransferase assay. This promoter-active fragment contains several specific nuclear protein binding regions as determined by foot-print analysis. Some of them contain sequence elements similar to other defined transcriptional factor binding sites such as those for SP1, AP2, c-*myc*, AP1, CREB, and EnhC (Fig. 5). These transcriptional factors may be working in conjunction with each other to modulate the level of transcription. Of particular interest is the presence of both AP 1 and AP2 binding sites in the promoter region. Since the functional expression of these two binding proteins is stimulated by phorbol ester (Angel et al., 1987; Lee et al., 1987; Mitchell et al., 1987), it seems possible that the expression of PKC γ gene may be under a positive control by the activation of PKC. In the fetal brain, the level of PKC I is very low in comparison with those of PKC II and III. Thus, activation of the latter two kinases, which usually coexist with PKC I in the

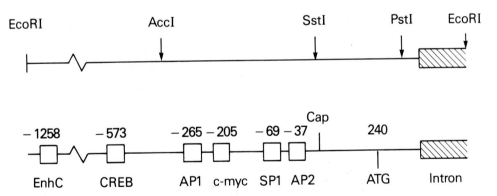

Fig. 5. Schematic diagram of the 5'-flanking region of rat PKC γ gene. The restriction sites and potential transcription factor-binding sites of the PKC γ gene spanning the 5'-flanking region, first exon and first intron are as indicated. The abbreviations of transcription factors are: AP2, activator protein 2; SP1, stimulatory protein 1; APl, activator protein 1; CREB, cAMP regulatory element-binding protein; EnhC, enhancer core.

152

same neurons, may provide a positive signal for the expression of PKC γ gene. This putative control mechanism may contribute to the delayed increase of PKC I as compared to PKC II and III during brain development.

Phosphorylation of CCAAT/enhancer binding protein by PKC

PKC has been implicated in the regulation of gene expression by phosphorylation of transcriptional regulatory proteins. CCAAT/enhancer binding protein (C/EBP) (Vinson et al., 1989) was chosen as a model to define the functional effect of phosphorylation of this protein. C/EBP is characterized by a leucine repeat-dimerization interface and an upstream basic region containing a DNA-binding domain. Several truncated forms of the C/EBP containing the leucine zipper and basic DNA-binding region were tested as substrates of PKC isozymes. The rates and extents of phosphorylation of these proteins are dependent on PKC isozymes; among them, PKC II and III are more active than PKC I. Amino acid sequence analysis of the major ^{32}P-labeled peptides derived from an 87-amino acid truncated C/EBP phosphorylated by PKC II revealed the phosphorylation of Ser^{277} and Ser^{299} located within the basic DNA-binding region (Fig. 6). Phosphorylation of Ser^{299} results in an inhibition of binding of the truncated C/EBP with a ^{32}P-labeled CCAAT probe as demonstrated by gel retardation experiments. Binding of the DNA probe with C/EBP prevents the phosphorylation by PKC, lending further support to the contention that the phosphorylation site (Ser^{299}) is in close proximity to the DNA binding region (Vinson et al., 1989). Phosphorylation of this site introduces a negatively charged group so that the interaction of C/EBP with DNA is attenuated. Since PKC II has been identified in rat liver nuclei (Rogue et al., 1990), it seems likely that

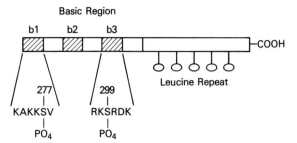

Fig. 6. Phosphorylation of the basic region of a truncated form of C/EBP by PKC. An 87-amino acid truncated form of C/EBP containing a leucine repeat-dimerization interface and an upstream basic region for DNA binding was phosphorylated by PKC at Ser^{277} and Ser^{299} located at the basic region b1 and b3, respectively. Phosphorylation of this protein resulted in an attenuation of binding of the C/EBP to a CCAAT probe.

this isozyme may be involved in the regulation of gene expression.

Conclusions

The Ca^{2+}/PS/DAG-stimulated PKC subtypes are a family of closely related enzymes highly abundant in CNS. These enzymes exhibit distinct biochemical characteristics, cellular and subcellular localizations, and developmental expression; however, direct evidence linking the activation of each of these enzymes with a particular response is not yet available. The difficulty encountered in defining the functional roles of these enzymes is, in part, due to a lack of a specific activator or inhibitor, as well as specific physiological substrates for each of these enzymes. Selective subcellular localizations of these enzymes have thus far provided the best indication of possible functional distinction among these enzymes. For example, the presence of PKC I in the cell bodies and dendrites of Purkinje cells as compared to the mostly cell body localization of PKC III is indicative of different functional roles of these two enzymes. In addition, the localization of PKC II in the nuclei has implicated its role in the regulation of nuclear events. Because of the un-usual association of PKC isozymes with distinct cellular compartments, we have investigated the membrane lipid components capable of confer-ring such a specificity. Among those tested, the inositol-containing phospholipids appear to pro-vide the best selectivity for PKC I. Thus, associa-tion of PKC I with these phospholipids not only places this enzyme near the site of its activator, DAG, generated by the activation of phospholi-pase C, but also its substrates which share the same binding affinity for these phospholipids as PKC I. Several actin-binding proteins, such as profilin (Lassing and Lindberg, 1985) and gelsolin (Janmey and Stossel, 1987), and perhaps others may be potential targets of this kinase. Stimula-tion of cultured cerebellar granule cells with glu-tamate or phorbol ester leads to the selective association of a modified form of PKC II with membrane skeleton, whereas no such association was observed for PKC III. It appears that some of the membrane-skeleton associated proteins, such as neuromodulin and MARCKS, could be potential targets of PKC II in cerebellar granule cells. PKC II appears to be a prominent nuclear protein kinase, which phosphorylates the trun-cated form of C/EBP at the basic region, and thus may regulate the binding to DNA and con-trol gene expression. Several other transcriptional factors having similar structural organization and sequence homology as C/EBP may also be the targets of PKC II. Functional manifestation of the various PKC isozymes depend not only on the strength and duration of the various activators but also on the levels of these enzymes. Thus, modification of the transcriptional control of PKC gene expression will be a useful approach to define the functional roles of these enzymes. Ge-nomic structure of PKC I has been analyzed and its potential regulatory sites identified. This infor-mation could be applicable to the in vitro manip-ulation of PKC I gene expression in neurons.

Further study to elucidate the genomic structures of the other PKC subtypes is essential to make use of genetic approaches to define the func-tional roles of these kinases.

References

Angel, P., Imagawa, M., Chiu, R., Stein, B., Imbra, R.J., Rahmsdorf, H.J., Jonat, C., Herrlich, P, and Karin, M. (1987) Phorbol ester-inducible genes contain a common cis element recognized by a TPA-modulated trans-acting fac-tor. *Cell,* 49: 729–739.

Berridge, M.J. and Irvine, R.F. (1989) Inositol phosphates and cell signalling. *Nature,* 341: 197–205.

Burgess, S.K., Sahyoun, N., Blanchard, S.G., LeVine, H., III, Chang, K.-J. and Cuatrecasas, P. (1986) *J. Cell Biol.,* 102: 312–319.

Chauhan, V.P.S. and Brockerhoff, H. (1988) Phosphatidyli-nositol-4,5-bisphosphate may antecede diacylglycerol as activator of protein kinase C. *Biochem. Biophys. Res. Com-mun.,* 155: 18–23.

Cimler, B.M., Andreasen, T.J., Andreasen, K.I. and Storm, D.R. (1985) P-57 is a neural specific calmodulin-binding protein. *J. Biol. Chem.,* 260: 10784–10788.

Ferris, C.D., Huganir, R.L., Supattapone, S. and Snyder, S.H. (1989) Purified inositol 1,4,5-trisphosphate receptor medi-ates calcium flux in reconstituted lipid vesicles. *Nature,* 342: 87–89.

Gallo, V., Ciotti, M.T., Coletti, A., Aloisi, F. and Levi, G. (1982) Selective release of glutamate from cerebellar gran-ule cells differentiating in culture. *Proc. Natl. Acad. Sci. USA,* 79: 7919–7923.

Gallo, V., Kingsberg, A., Balazs, R. and Jorgensen, O.S.. (1987) The role of depolarization in the survival and differentiation of cerebellar granule cells in culture. *J. Neurosci.,* 7: 2203–2213.

Garthwaite, J. and Brodbelt, A.R. (1990) Glutamate as the principal mossy fiber transmitter in rat cerebellum: phar-macological evidence. *Eur. J. Neurosci.,* 2: 177–180.

Hawthorne, J.N. and Kemp, P. (1964) The brain phospho-inositides. *Adv. Lipid Res.,* 2: 127–166.

Hosoda, K., Saito, N., Kose, A., Ito, A., Tsujino, T., Ogita, K., Kikkawa, U., Ono, Y., Igarashi, K., Nishizuka, Y. and Tanaka, C. (1989) Immunocytochemical localization of the βI subspecies of protein kinase C in rat brain. *Proc. Natl. Acad. Sci. USA,* 86: 1393–1397.

Huang, F.L., Yoshida, Y, Nakabayashi, H., Knopf, J.L., Young, W.S., III and Huang K.-P. (1987a) Immunochemical iden-tification of protein kinase C isozymes as products of discrete genes. *Biochem. Biophys. Res. Commun.,* 149: 946–952.

Huang, F.L., Yoshida, Y., Nakabayashi, H. and Huang, K.-P.

154

(1987b) Differential distribution of protein kinase C isozymes in the various regions of brain. *J. Biol. Chem.*, 262: 15714–15720.

Huang, F.L., Yoshida, Y., Nakabayashi, H., Young, W.S., III and Huang, K.-P. (1988) Immunocytochemical localization of protein kinase C isozymes in rat brain. *J. Neurosci.*, 8: 4734–4744.

Huang, F.L., Yoshida, Y., Nakabayashi, H., Friedman, D.P., Ungerleider, IL.G., Young, W.S. III and Huang, K.-P. (1989a) Type I protein kinase C isozyme in the visual information processing pathway of monkey brain. *J. Cell. Biochem.*, 39: 401–410.

Huang, F.L., Yoshida, Y., Cunha-Melo, J.R., Beaven, M. and Huang, K.-P. (1989b) Differential down-regulation of protein kinase C isozymes. *J. Biol. Chem.*, 264: 4238–4243.

Huang, F.L., Young, W.S., III, Yoshida, Y. and Huang, K.-P. (1990) Developmental expression of protein kinase C isozymes in rat cerebellum. *Dev. Brain Res.*, 52: 121–130.

Huang, K.-P. (1989) The mechanism of protein kinase C activation. *Trends Neurosci.*, 12: 425–432.

Huang, K.-P. (1990) Role of protein kinase C in cellular regulation. *BioFactors*, 2: 171–178.

Huang, K.-P. and Huang, F.L. (1986) Immunochemical characterization of rat brain protein kinase C. *J. Biol. Chem.*, 261: 14781–14787.

Huang, K.-P. and Huang, F.L. (1990) Differential sensitivity of protein kinase C isozymes to phospholipid-induced inactivation. *J. Biol. Chem.*, 265: 738–744.

Huang, K.-P., Nakabayashi, H. and Huang, F.L. (1986) Isozymic forms of rat brain Ca^{2+}-activated and phospholipid-dependent protein kinase. *Proc. Natl. Acad. Sci. USA*, 83: 8535–8539.

Huang, K.-P., Huang, F.L., Nakabayashi, H. and Yoshida, Y. (1988) Biochemical characterization of rat brain protein kinase C isozymes. *J. Biol. Chem.*, 263: 14839–14845.

Jan, L.Y. and Jan, Y.N. (1989) Voltage-sensitive ion channels. *Cell*, 56: 13–25.

Janmey, P.A. and Stossel, T.P. (1987) Modulation of gelsolin function by phosphatidylinositol 4,5-bisphosphate. *Nature*, 325: 362–364.

Kishimoto, A., Mikawa, K., Hashimoto, K., Yasuda, I., Tanaka, S., Tominaga, M., Kuroda, T. and Nishizuka, Y. (1989) Limited proteolysis of protein kinase C subspecies by calcium-dependent neutral protease (calpain). *J. Biol. Chem.*, 264: 4088–4092.

Kose, A., Saito, N., Ito, H., Kikkawa, U., Nishizuka, Y. and Tanaka, C. (1988) Electron microscopic localization of type I protein kinase C in rat Purkinje cells. *J. Neurosci.*, 8: 4262–4268.

Lassing, I. and Lindberg, U. (1985) Specific interaction between phosphatidylinositol 4,5-bisphosphate and profilactin. *Nature*, 314: 472–474.

Lee, W., Mitchell, P. and Tjian, R. (1987) Purified transcrip-

tion factor AP-1 interacts with TPA-inducible enhancer elements. *Cell*, 49: 741–752.

Low, M.G. and Saltiel, A.R. (1988) Structural and functional roles of glycosyl-phosphatidylinositol in membranes. *Science*, 239: 268–275.

Madison, D.V., Malenka, R.C. and Nicoll, R.A. (1986) Phorbol esters block a voltage sensitive chloride current in hippocampal pyramidal cells. *Nature*, 321: 695–697.

Majerus, P.W., Connolly, T.M., Deckmyn, H., Ross, T.S., Bross, T.E., Ishii, H., Bansal, V.S., Wilson, D.B. (1986) The metabolism of phosphoinositide-derived messenger molecules. *Science*, 234: 1519–1526.

Malenka, R.C., Madison, D.V. and Nicoll, R.A. (1986) Potentiation of synaptic transmission in the hippocampus by phorbol ester. *Nature*, 321: 175–177.

Malenka, R.C., Ayoub, G.S. and Nicoll, R.A. (1987) Phorbol esters enhance transmitter release in rat hippocampal slices. *Brain Res.*, 403: 198–203.

Meiri, K.F. and Gordon-Weeks, P.R. (1990) GAP-43 in growth cones is associated with areas of membrane that are tightly bound to substrate and is a component of a membrane skeleton subcellular fraction. *J. Neurosci.*, 10: 256–266.

Mitchell, P.J., Wang, C. and Tjian, R. (1987) Positive and negative regulation of transcription in vitro: enhancer-binding protein AP-2 is inhibited by SV40 T antigen. *Cell*, 50: 847–861.

Naor, Z., Shearman, M.S., Kishimoto, A. and Nishizuka, Y. (1988) Calcium-independent activation of hypothalamic type I protein kinase C by unsaturated fatty acids. *Mol. Endocrinol.*, 2: 1043–1048.

Nicoll, R.A., Haycock, J.W., Wang, J.K.T. and Greengard, P. (1987) Phorbol ester enhancement of neurotransmitter release from rat brain synaptosomes. *J. Neurochem.*, 48: 615–621.

Nishizuka, Y. (1988) The molecular heterogeneity of protein kinase C and its implication for cellular regulation. *Nature*, 334: 661–665.

Patel, A.J. and Balazs, R. (1975) Effect of x-irradiation on the biochemical maturation of rat cerebellum: metabolism of [^{14}C]glucose and [*14C*]acetate. *Rad. Res.*, 62: 456–469.

Pelech, S.L. and Vance, D.E. (1989) Signal transduction via phosphatidylcholine cycles. *Trends Biochem. Sci.*, 14: 28–30.

Rogue, P., Labourdette, G., Masmoudi, A., Yoshida, Y., Huang, F.L., Huang, K.-P., Zwiller, J., Vincendon, G. and Malviya, A.N. (1990) Rat liver nuclear protein kinase C is the isozyme type II. *J. Biol. Chem.*, 265: 4161–4165.

Routtenberg, A. (1985) Protein kinase C activation leading to protein F1 phosphorylation may regulate synaptic plasticity by presynaptic terminal growth. *Behav. Neural. Biol.*, 44: 186–200.

Shearman, M.S., Naor, Z., Sekiguchi, K., Kishimoto, A. and Nishizuka, Y. (1989a) Selective activation of the γ-subspecies of protein kinase C from bovine cerebellum by arachidonic acid and its lipoxygenase metabolites. *FEBS Lett.*, 243: 177–182.

Shearman, M.S., Sekiguchi, K. and Nishizuka, Y. (1989b) Modulation of ion channel activity: a key function of the protein kinase C enzyme family. *Pharmacol. Rev.*, 41: 211–237.

Stumpo, D.J., Graff, J.M., Albert, K.A., Greengard, P. and Blackshear, P.J. (1989) Molecular cloning, characterization and expression of a cDNA encoding the "80- to 87-kDa"

myristoylated alanine-rich C kinase substrate: a major cellular substrate for protein kinase C. *Proc. Natl. Acad. Sci. USA*, 86: 4012–4016.

Vinson, C.R., Sigler, P.B. and McKnight, S.L. (1989) Scissors-grip model for DNA recognition by a family of leucine zipper proteins. *Science*, 246: 911–916.

Yoshida, Y., Huang, F.L., Nakabayashi, H. and Huang, K.-P. (1988) Tissue distribution and developmental expression of protein kinase C isozymes. *J. Biol. Chem.*, 263: 9869–9873.

W.H. Gispen and A. Routtenberg (Eds.)
Progress in Brain Research, Vol. 89
© 1991 Elsevier Science Publishers B.V.

CHAPTER 11

Regional distribution and properties of an enzyme system in rat brain that phosphorylates ppH-47, an insoluble protein highly labelled in tissue slices from the hippocampus

Richard Rodnight, Carlos A. Gonçalves, Rodrigo Leal, Elisabete Rocha, Christianne G. Salbego and Susana T. Wofchuk

Departamento de Bioquimica, Instituto de Biociencias, Universidade Federal do Rio Grande do Sul, 90.050 Porto Alegre, Brazil

Introduction

Substrates of the protein kinases are remarkably numerous in the mammalian brain. Yet despite compelling evidence for the involvement of protein phosphorylation in many aspects of neuronal signal transduction (for recent reviews, see Greengard, 1987; Hemmings et al., 1989), relatively few molecular roles for specific phosphoproteins have been discovered. An important approach to this problem attempts to correlate knowledge of the morphological and chemical anatomy of the brain with regional differences in the activity of protein phosphorylating systems. A striking example of this strategy led to the discovery of two phosphoproteins whose occurrence is restricted to neurones possessing dopamine D_1 receptors, namely DARPP-32 and ARPP-21 (Hemmings and Greengard, 1986; Meister et al., 1988; Ouimet et al., 1989). In its phosphorylated form DARPP-32 functions as a potent inhibitor of protein phosphatase-1 (Hemmings et al., 1987)

and activation of NMDA receptors induces its dephosphorylation and abolishes its inhibitory potential (Halpain et al., 1990). The molecule has recently been cloned and sequenced (Erhlich et al., 1990).

The technical procedures which resulted in the discovery of DARPP-32 and ARPP-21 consisted of labelling subcellular fractions with γ-[^{32}P]ATP and analysing the tissue by unidimensional electrophoresis (Walaas et al., 1983a,b). This subcellular approach is very powerful, but as has often been pointed out (e.g., Rodnight et al., 1986), it provides no information on the activity of protein phosphorylating systems in the intact tissue under more physiological conditions. For this purpose, a complementary approach is required: that of labelling the phosphoproteins in the intact tissue with [^{32}P]phosphate as primary donor, either in vivo (e.g., Rodnight et al., 1985) or in slices or synaptosomes in vitro (e.g., Rodnight et al., 1988; Wang et al., 1988). However, because of the enormous complexity of protein phosphorylation in the brain, only very limited information can be obtained by the analysis of unfractionated cell-containing preparations by unidimensional electrophoresis. Moreover, subcellular fractionation of whole labelled tissue is excluded by the difficulty of controlling for dephosphorylation during centrifugation. Analysis by high resolution two-dimensional electrophoresis provides an answer to this problem, with the limitation that some phosphoproteins are poorly focussed in the first dimension.

The work described in this chapter orginated from a preliminary survey of protein phosphorylating patterns in micro-slices prepared from various regions of the rat brain. A phosphopeptide of M_r approximately 47 kDa (later designated ppH-47) was found to be highly labelled in slices of hippocampus compared to cerebral cortex and caudate nucleus. A more detailed study (Rodnight and Leal, 1990) then showed that the activity of the system that phosphorylates this protein substrate has a characteristic distribution in the rat brain. Further research showed that ppH-47 is derived from a highly insoluble protein, which is probably located in a postsynaptic structure. The phosphorylation of this substrate is stimulated by glutamate in immature animals and the expression of the phosphorylating system is developmentally regulated (Gonçalves et al., 1990; Wofchuk and Rodnight, 1990).

Methodology

The methods used in these studies were based on Rodnight et al., (1988) with modifications introduced in later publications (Wofchuk and Rodnight, 1990; Gonçalves et al., 1990; Rodnight and Leal, 1990). Briefly, microslices from different regions of the brain were prepared by punching out 1 mm discs from macro-slices cut in a McIlwain chopper. The microslices were pre-incubated in Krebs-Ringer medium, transferred to fresh medium containing [^{32}P]phosphate and incubated for 0.5–1 h. The labelled slices were dissolved in isoelectric focussing buffer and analysed by non-equilibrium two-dimensional electrophoresis (NEPHGE; O'Farrell et al., 1977), using 8% polyacrylamide gels for the second dimension. Phosphopeptides were detected by autoradiography of the dried gels.

The use of NEPHGE greatly facilitated the analysis since excellent separations of the phosphopeptides of interest could be obtained in the first dimension with runs of 3–4 h. We used 8% gels for the second dimension to improve the separation of the major protein kinase C substrate B-50/GAP43/F1 (4 in Fig. 1) from a highly labelled complex of two phosphopeptides of 40 kDa (6 in Fig. 1).

Properties of the ppH-47 protein phosphorylating system

Phosphopeptide ppH-47 was first observed on autoradiographs prepared from microslices of rat hippocampus incubated with [^{32}P]phosphate (Rodnight et al., 1988). Subsequently, much lower levels of phosphorylation of this substrate were found in microslices from most areas of the cerebral cortex. Typical autoradiographs are shown in Fig. 1 which compares the phosphorylation pattern in the frontal cortex (A) with that in the hippocampus (B); arrows mark the position of ppH-47. The isoelectric point of ppH-47, determined by equilibrium focussing, was approximately 6.3; phosphoamino acid analysis showed phosphorylserine to be the only ^{32}P-labeled amino acid present.

Silver staining of gels prepared from slices of hippocampus, globus pallidus and cerebellar cortex revealed a faint spot which corresponded exactly with the position of ppH-47 on the autoradiographs. This spot could not be detected on gels from areas with low activity of the phosphorylating system (cerebral cortex and caudate nucleus).

Assuming that the silver-stained spot is indeed ppH-47 we estimate that the protein constitutes less than 0.05% of total protein in the hippocampus, and less than 0.01% in the other areas.

Solubility

Extraction of slice protein with a salt solution (100 mM NaF, 50 mM NaCl, 10 mM EDTA) containing 0.5% Triton X-100, solubilized 90% of the labelled phosphopeptides, but left over 95% of ppH-47 in the insoluble fraction (Gonçalves et al., 1990). A second extraction with Triton, or several other detergents including CHAPS, cholate and deoxycholate failed to solubilize more than a very small fraction of the phosphopeptide. However, the phosphopeptide was readily solubilized by 6 M urea. Similar results were obtained with adult and immature animals.

Triton X-100 is known to solubilize extra-junctional membranes leaving synaptic junctional structures intact (Matus et al., 1980). It is probable therefore that ppH-47 is associated with a component of the synaptic junction, rather than with mitochondria or nuclear membranes, for ex-

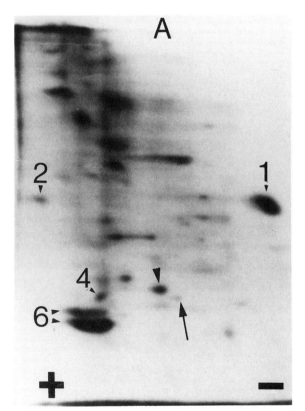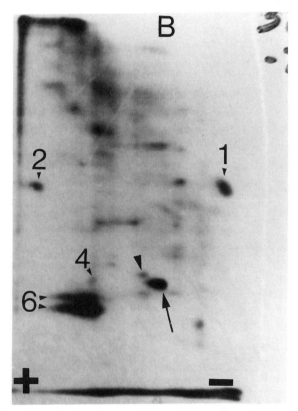

Fig. 1. Autoradiographs prepared from 1 mm diameter microslices from the frontal cortex (A) and hippocampus (B). The microslices were incubated for 60 min in medium containing 2.5 mCi/ml of ^{32}P and analysed by non-equilibrium IEF followed by SDS polyacrylamide electrophoresis in 8% gels. Presumed identity of the numbered phosphopeptides: 1: synapsin 1; 2: MARCKS (82–87 kDa protein kinase C substrate); 4: B-50/GAP43/F1; 6: unknown complex of two phosphopeptides of 40–42 kDa. The arrow head points to ppC-50 and the arrow to ppH-47.

ample. This conclusion is supported by the ontogeny of the system which corresponds to the period of rapid synaptogenesis in the rat and by the fact that phosphorylation of ppH-47 in immature animals is stimulated by glutamate, suggesting that the substrate may be associated with a glutamate receptor. Further, a postsynaptic site is suggested by the observation that phosphorylation of ppH-47 was not detected in hippocampal synaptosomes analysed by identical procedures.

Ontogeny

Labelling of ppH-47 was first detectable in hippocampal slices around day 8 after birth (Fig. 2) and a day or two later in cerebral cortical slices (data not shown). As mentioned above this delayed ontogenesis matches the onset of massive synaptogenesis, which in the rat occurs during the second postnatal week (Aghajanian and Bloom, 1967). Adult levels of the system were not reached until 50–60 days postnatal. It is of interest to note

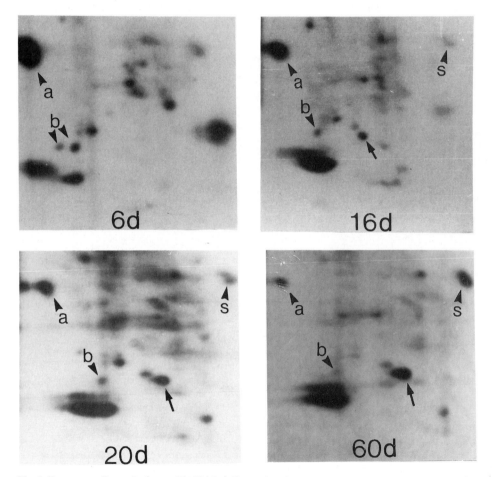

Fig. 2. Part autoradiographs (range 35–90 kDa) illustrating the ontogeny of the labelling of ppH-47 (arrows) in hippocampal slices. Orientating phosphopeptides are indicated as follows: a: MARCKS; b: B-50/GAP43/F-1; s: synapsin I. The labelling of synapsin in the 20 day panel is atypically weak.

that in some experiments the major increase in ppH-47 phosphorylation occurred during the third postnatal week (see Fig. 2) suggesting that the development of the system is associated with the process of synaptic maturation, rather than with the initial stages of synapse formation. The ontogenetic profile therefore appears to resemble that of the α-subunit of the Ca^{2+}/calmodulin protein kinase II in the postsynaptic densities (Rostas et al., 1988), rather than the presynaptic protein synapsin I, the phosphorylation of which increases linearly from day 10 to day 30 (Lohmann et al., 1978).

It is important to remember that these observations provide information on the postnatal development of the ppH-47 phosphorylation system rather than the ontogeny of the substrate protein. Thus the delay in the appearance of ppH-47 phosphorylation could be due to the late synthesis of a protein kinase or phosphatase specific for the protein.

Dependency of the system on Ca^{2+}

In previous work (Rodnight et al., 1988) we found that when microslices from adult animals were prepared, pre-incubated and labelled in Ca^{2+}-free medium containing 1 mM EGTA, no phosphorylation of ppH-47 was observed. Slices prepared in Ca^{2+}-free media and then labelled in normal medium containing 1 mM Ca^{2+}, also failed to phosphorylate the substrate. Therefore it appears that the absence of Ca^{2+} in the preparation medium irreversibly damages the phosphorylation system. Inclusion of as little as 0.1 mM Ca^{2+} in the preparation medium prevented this damage and permitted normal levels of phosphorylation in the presence of 1 mM Ca^{2+} in the labelling medium. Further, slices prepared in medium containing 1 mM Ca^{2+} and subsequently incubated in medium containing EGTA instead of Ca^{2+}, could phosphorylate ppH-47 to some 75% of the normal level. Maximal phosphorylation required 1 mM Ca^{2+}, half-maximal 0.1–0.2 mM Ca^{2+}.

Surprisingly, in immature animals (< 20 days postnatal) Ca^{2+}-dependency was not observed. In fact the phosphorylation of ppH-47 was enhanced in the presence of EGTA (Fig. 3). Dependency of the system on Ca^{2+} began to appear during the fourth week postnatal and was complete by the ninth week.

Dependency of the phosphorylation reaction on Ca^{2+} in the adult tissue is almost certainly mainly located at the protein kinase rather than the phosphatase step: if the overall reaction in-

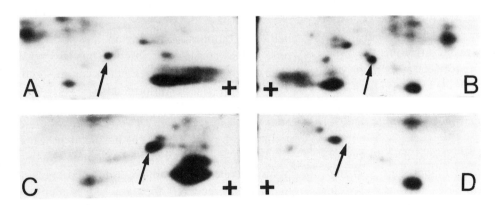

Fig. 3. Autoradiographs from hippocampal slices illustrating the effect of substituting Ca^{2+} in the incubation medium by EGTA. A and B: immature animal (14 days); C and D: adult animal (60 days). A and C: normal medium containing 1 mM Ca^{2+}; B and D: with 1 mM EGTA and zero Ca^{2+}. The position of ppH-47 is marked by the arrow.

162

volved a Ca^{2+}-sensitive phosphatase, labelling of vacant sites would occur in the initial stages of incubation. However we do not know at present whether ppH-47 is directly phosphorylated by a Ca^{2+}-dependent kinase, or via an enzyme cascade involving a Ca^{2+}-dependent kinase activating by phosphorylation a Ca^{2+}-independent ppH-47 kinase. Since the phosphorylation reaction in slices is not stimulated by active phorbol esters or inhibited by calmodulin antagonists (Salbego and Rodnight, 1989, and unpublished) it would appear that neither of the two principal Ca^{2+}-requiring kinases in brain – protein kinase C and the Ca^{2+}/calmodulin kinase II – are involved. However, further work at the subcellular level is needed to confirm this conclusion.

The irreversible inactivation of the reaction which occurs when slices are prepared in Ca^{2+}-free media is probably related to a loss of internal Ca^{2+} consequent on the massive postmortem increase in internal Na^+, which presumably stimulates Na^+/Ca^{2+} exchange (McBurney and Neering, 1987). However, whether the inactivation is due to proteolysis or dissociation of the complex is unknown.

We have no explanation at present for the remarkable late development of the Ca^{2+}-dependency. This encompassses a period when synaptogenesis and the structural architecture of the rat brain is largely complete. Presumably intracellular changes, possibly affecting the availability of internal Ca^{2+} stores, are occurring during this period.

Apparent absence of the ppH-47 system in synaptosomes

When synaptosomes prepared from hippocampus were labelled with [^{32}P]phosphate and analysed by identical procedures to those used for slice tissue, we obtained the phosphorylation pattern shown in Fig. 4. Although this pattern has many similarities with that given by slices, la-

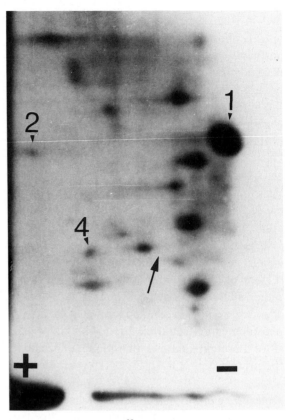

Fig. 4. Autoradiograph of ^{32}P-labelled synaptosomes analysed by the standard procedure and showing the apparent absence of ppH-47. Arrow points to the position occupied by ppH-47 on an autoradiograph of a mixture of labelled slice protein and labelled synaptosomes. For the presumed identity of the numbered phosphopeptides see Fig. 1.

belling of a phosphopeptide in the position of ppH-47 was consistently absent. Thus it is possible that the substrate does not occur in significant quantities in nerve terminals.

Stimulation of phosphorylation by glutamate

In immature animals in the postnatal age range 10–16 days, phosphorylation of ppH-47 in hippocampal slices was strongly stimulated by glutamate in a reaction apparently dependent on Ca^{2+} (Wofchuk and Rodnight, 1990; Table I). In a smaller series of experiments glutamate also stimulated the phosphorylation of ppH-47 3-fold in slices of cerebellar cortex (mean normalised peak height in the presence of glutamate: 332 ± 68, $P < 0.03$). No effect of glutamate was found in adult animals. The stimulatory action of glutamate in young animals is probably receptor-mediated, although so far we have been unable to demonstrate a similar strong stimulation of phosphorylation using the glutamate receptor agonists NMDA, quisqualate, kainate or ibotenate. Possibly more than one receptor is involved, in which case a mixture of agonists would be necessary to obtain the same effect as glutamate. Further work using an improved assay system is needed to elucidate this problem.

In these experiments we also observed a weak stimulation of the phosphorylation of the protein kinase substrates MARCKS and B-50/GAP43/F1 (2 and 4 respectively in Fig. 1) and this effect appeared to be dependent on Ca^{2+} (Table I). Excitatory amino acids have been shown to stimulate inositol phospholipid hydrolysis in cultured striatal neurones (Sladeczek et al., 1985), hippocampal slices from neonatal animals (Nicoletti et al., 1986a) and in cultured cerebellar granule cells (Nicoletti et al., 1986b). Since the receptor-activated hydrolysis of phosphatidylinositol 4,5-bisphosphate yields, in addition to inositol phos-

phates, the protein kinase C activator, diacylglycerol (Nishizuka, 1984), the small stimulation of the two main substrates of protein kinase C in brain is probably a consequence of the activation of the phosphatidylinositol cycle by glutamate. However, further work using the appropriate receptor antagonists is required to confirm this conclusion.

It is of interest to note that the 10-fold stimulation of inositol phospholipid hydrolysis by glutamate agonists reported by Nicoletti et al., (1986a) was only observed in hippocampal slices from 6–8-day-old rats, the response to the amino acid declining to zero after 35 days. This result is another example of the way in which the developing nervous system in the rat responds to extracellular glutamate differently to the adult tissue.

Detailed regional distribution of the system

A semi-quantitative study of the variation in labelling potential in 19 areas of the rat brain revealed a striking regional profile for the ppH-47 phosphorylation system (Rodnight and Leal, 1990). These results, together with those from another 5 areas, are presented in Table II. The data were derived from a series of experiments in

TABLE I

Effect of glutamate on protein phosphorylation in hippocampal slices from immature rats

	Ca^{2+}	Mg^{2+}	Mean normalised peak heights (control = 100)	Mean differences (percent)	P value
ppH-47 (10, 16)	+	+	280 ± 37	+180	< 0.001
ppH-47 (3, 6)	−	+	132 ± 27	+32	NS
ppH-47 (3, 6)	+	−	238 ± 27	+138	< 0.04
B-50 (10, 21)	+	+	142 ± 14	+42	< 0.007
B-50 (3, 6)	−	+	91 ± 3.5	−9	< 0.05
MARCKS (10, 22)	+	+	136 ± 15	+36	< 0.023
MARCKS (3, 6)	−	+	95 ± 2.4	−5	NS

Glutamate concentration was 1 mM. Ca^{2+} and Mg^{2+} concentrations were 1 mM and 1.23 mM, respectively; in incubations without Ca^{2+}, 1 mM EGTA was present. The number of experiments followed by the number of slices analysed is given in parentheses. Significance was determined by a paired t-test. From Wofchuk and Rodnight (1990).

which phosphorylation activity in hippocampal slices was compared with other areas by densitometric scanning of the autoradiograms. In each experiment the peak height of the spot corresponding to ppH-47 from the hippocampal slice was adjusted to 100 and those from other areas expressed as percentages of this value. Besides the hippocampus particularly high rates of phosphorylation were found in the interpeduncular nucleus, habenula and globus pallidus. Other areas that exhibited a relatively high rate of labelling ($> 29\%$ of the hippocampal level) were the periventricular gray matter of the aqueduct, lateral septum, cerebellar cortex, entorhinal cortex, hypothalamus, mammillary nuclei, amygdala and substantia nigra. By contrast labelling of ppH-47 in the caudate nucleus, thalamic nuclei, colliculi and cortical areas other than the entorhinal cortex was consistently low or absent. It is worth noting that the range of phosphorylating activity is remarkably large: over 100-fold between the hippocampus and the two colliculi.

TABLE II

Semi-quantitative densitometric analysis of regional variation in the labelling of ppH-47, ppC-50 and MARCKS

Region	Peak heights as percentage of area most highly labelled			Ratio of Peak height ppH-47 / Peak height ppC-50
	ppH-47	ppC-50	MARCKS	
Hippocampus	100	12	18	12.9
Interpeduncular nucleus	90	9	20	8.5
Habenula	68	12	70	3.3
Globus pallidus	70	27	44	4.0
Periventricular gray [b]	54	47	78	1.8
L. septum [a]	53	13	7	6.8
Cerebellar cortex	42	15	100	4.3
Entorhinal cortex	39	27	18	2.5
Hypothalamus [c]	32	6	92	8.3
Mammillary nuclei	32	25	88	2.3
Amygdala [d]	30	42	24	1.2
Substantia nigra	29	23	10	2.0
Cingulate cortex	10	50	15	0.3
Retrosplenial cortex	8	39	10	0.3
Parietal cortex	6	52	5	0.2
Temporal cortex	6	75	4	0.1
Frontal cortex	5	73	11	0.1
Nucleus accumbens	4	98	30	< 0.1
Posterior caudate	4	100	18	0.2
Anterior caudate	2	37	9	< 0.1
Thalamus, ventromedial nucleus	1	5	3	0.6
Thalamus, ventrolateral nucleus	1	1	3	0.6
Inferior colliculus	< 1	13	5	< 0.1
Superior colliculus	< 1	18	71	< 0.1

Modified from Rodnight and Leal (1990).

[a] Level of fornix.

[b] Level of section C in Fig. 3 of Rodnight and Leal, 1990.

[c] Ventromedial and arcuate nuclei.

[d] Anterior nuclei.

Within the hippocampus the rate of labelling of ppH-47 was approximately equal in microslices taken from the dorsal and ventral hippocampus and from the dentate gyrus, CA1 and CA3 areas and the subiculum; in the pre-subiculum rates equal to those found in the entorhinal cortex were observed.

To highlight the characteristic distribution of the ppH-47 system, Table II also includes data on the regional distribution of two other protein phosphorylating systems, the substrates of which are an unknown phosphopeptide designated ppC-50 and the MARCKS phosphopeptide. Phosphopeptide ppC-50 (M_r approximately 50 kDa; arrowhead in Fig. 1) is slightly more acid than ppH-47. The highest rates of labelling of this molecule were observed in the caudate nucleus and in most areas of the cerebral cortex. The MARCKS phosphopeptide is the well known 82 to 87 kDa substrate of protein kinase C (Stumpo et al., 1989; 2 in Fig. 1). In this case microslices of the cerebellar cortex exhibited the highest rates of labelling, followed by the hypothalamus, mammillary nuclei, periventricular gray matter of the aqueduct and superior colliculus. It is of interest to note that this regional distribution of the MARCKS phosphorylating system only partly agrees with a study of the distribution of the substrate determined by immunocytochemistry (Ouimet et al., 1990), where the amygdala and the striatum exhibited the highest content of the protein. This suggests that factors other than substrate level determine the characteristic distribution of the MARCKS phosphorylating system in incubated brain slices.

In some areas phosphorylation rates for ppH-47 and MARCKS were roughly correlated, but there were several striking exceptions: in particular in the hippocampus, lateral septum and superior colliculus. A more consistent inverse correlation was observed between the labelling of ppH-47 and ppC-50, as shown by the last column of Table II. However, regional variations in the labelling of ppC-50 were less pronounced than in the case of the other two phosphopeptides.

Precise elucidation of the nature of the neuronal pathways reflected by the very characteristic distribution of the ppH-47 system must await more detailed anatomical studies, including cytochemical localization of the substrate. Nevertheless some preliminary comments are justified. First, we note that although the system is stimulated by glutamate, it cannot be concluded that the substrate is uniquely associated with a glutamate receptor, since several areas exhibiting low activity, such as the caudate nucleus are massively innervated by glutamergic fibres. Secondly, it is striking how many structures that form part of the limbic or "olfactory" forebrain and midbrain exhibit a high rate of labelling of ppH-47. Thus, with the exception of the globus pallidus, cerebellar cortex and substantia nigra, the 12 areas in Table II with 29% or more of the hippocampal activity level, are all considered part of the limbic system (Zeman and Innes, 1963). The very high activity of the system in the globus pallidus was unexpected in view of the close association of this structure with the neostriatum where the activity is amongst the lowest. In primates the globus pallidus receives a glutamergic input from the subthalamic nucleus (Graybiel, 1990), but we are not aware whether this pathway exists in the rat. With regard to the interpeduncular nucleus, where activity was almost equal to that in the hippocampus, there are intriguing pointers to the existence of a glutamergic pathway from the habenula to this structure via the fasciculus retroflexus (Brown et al., 1984; see also Ottersen and Storm-Mathisen, 1984) which await confirmation. The existence of pathways from the septum and hippocampus to the habenula has been known for decades (Zeman and Innes, 1963), but the precise nature of the transmitter(s) in these tracts appears unknown at present.

Conclusions

Using a micro-slice technique followed by analysis by two-dimensional electrophoresis we have demonstrated the presence in rat brain of an enzyme system that phosphorylates a protein substrate designated ppH-47. The highly insoluble nature of the phosphopeptide, the fact that its developmental profile temporally coincides with synaptogenesis and its apparent absence in ^{32}P-labelled synaptosomes point to a postsynaptic location, possibly in the postsynaptic densities. With regard to the regional distribution of the system, the characteristic profile could reflect regional variations in the level of the ppH-47 substrate or in the activity of the enzyme complex. The fact that we could only detect a silver-stained spot exactly corresponding to the labelled phosphopeptide in gels from areas high in the system (hippocampus, globus pallidus and cerebellar cortex) suggests that substrate variation is the main factor determining the profile, but until an antibody to the protein is available the question must remain open. The remarkable concentration of the system in limbic areas of the brain encourages the tentative speculation that it may be involved in the modulation of neuronal pathways concerned with emotion and behaviour.

Stimulation of the system by glutamate in immature animals is an enigma, since we have so far been unable to identify a receptor and stimulation by the amino acid could not be demonstrated in adult animals. It is possible that the latter inconsistency is related to the other curious developmental paradox shown by the system: the remarkable change in sensivity to Ca^{2+} with age. Unfortunately progress in investigating the precise mechanisms involved in the actions of glutamate and Ca^{2+} is presently hindered by the lack of a satisfactory quantitative assay for the phosphorylation reaction. Probably we will have to wait for a specific antibody and the development of an assay based on immunoprecipitation before substantial progress in these aspects can be expected.

Acknowledgements

This work was supported by the Brazilian research funding agencies FINEP, CNPq and FARPERGS.

References

Aghajanian, G.K. and Bloom, F.E. (1967) The formation of synaptic junctions in the developing rat brain: a quantitative electron microscopic study. *Brain Res.*, 6: 716–727.

Brown, D.A., Docherty, R.J. and Halliwell, J.V. (1984) Chemical transmission in the rat interpedunduclar nucleus in vitro. *J. Physiol.*, 341: 655–670.

Ehrlich, M.E., Kurihara, T. and Greeengard, P. (1990) Rat DARPP-32: cloning, sequencing and characterization of cDNA. *J. Mol. Neurosci.*, 2: 1–10.

Gonçalves, C.A., Salbego, C.G., Wofchuk, S.T., Rocha, E. and Rodnight, R. (1990) Properties of a protein phosphorylating system that labels a 47 kDa phosphoprotein (ppH-47) in slices of rat hippocampus. *Neurosci. Res. Commun.*, 6: 129–134.

Graybiel, A.M. (1990) Neurotransmitters and neuromodulators in the basal ganglia. *Trends Neurosci.*, 13: 244–254.

Greengard. P. (1987) Neuronal phosphoproteins. Mediators of signal transduction. *Mol. Neurobiol.*, 1: 81–119.

Halpain, S., Girault, J.A. and Greengard, P. (1990) Activation of NMDA receptors induces dephosphorylation of DARPP32 in rat striatal slices. *Nature*, 343: 369–372.

Hemmings, H.C., Jr. and Greengard, P. (1986) DARPP-32, a dopamine and adenosine 3':5'-monophosphate regulated phosphoprotein: regional, tissue and phylogenetic distribution. *J. Neurosci.*, 6: 1469–1481.

Hemmings, H.C., Nestler, E.J., Walaas, S.I., Ouimet, C.C. and Greengard, P. (1987) Protein phosphorylation and neuronal function: DARPP-32, an illustrative example. In G.E. Edelman, W.E. Gall and W.M. Cowan (Eds.), *Synaptic Function*, John Wiley, New York, pp.213–240.

Hemmings, H.C., Jr., Nairn, A.C., McGuiness, T.L., Huganir, R.L. and Greengard, P. (1989) Role of protein phosphorylation in neuronal signal transduction. *FASEB J.*, 3: 1583–1592.

Lohmann, S.M., Ueda. T. and Greengard, P. (1978) Ontogeny of synaptic phosphoproteins in brain. *Proc. Natl. Acad. Sci. USA*, 75: 4037–4041.

McBurney, E.M. and Neering, I.R. (1987) Neuronal calcium homeostasis. *Trends Neurosci.*, 10: 164–169.

Matus, A., Pehling, G., Ackermann, M. and Maeder, J. (1980) Brain postsynaptic densities: their relationship to glial and neuronal filaments. *J. Cell Biol.*, 87: 346–359.

Meister, B., Hokfelt, T., Tsuro, Y., Hemmings, H., Ouimet, C., Greengard, P. and Goldstein, M. (1988) DARPP-32, a dopamine- and cyclic AMP-regulated phosphoprotein in tanycytes of the mediobasal hypothalamus: distribution and relation to dopamine and lutenizing hormone-releasing hormone neurons and other glial elements. *Neuroscience*, 27: 607–622.

Nishizuka, Y. (1984) The role of protein kinase C in cell surface signal transduction and tumour promotion. *Nature*, 308: 693–698.

O'Farrell, P.Z., Goodman, H.M., and O'Farrell, P.H. (1977) High resolution two-dimensional electrophoresis of basic as well as acidic proteins. *Cell*, 12: 1133–1142.

Ottersen, O.P. and Storm-Mathisen, J. (1984) Neurons containing or accumulating transmitter amino acids. In A. Björklund, T. Hökfelt and M.H. Kuhar (Eds.), *Handbook of Chemical Anatomy, Vol.3*, Elsevier, Amsterdam, p. 148.

Ouimet, C.C., Hemmings, H.C., Jr. and Greengard, P. (1989) ARPP-21, a cyclic AMP-regulated phosphoprotein enriched in dopamine-innervated brain regions. II. Immunocytochemical localization in rat brain. *J. Neurosci.*, 9: 865–875.

Ouimet, C.C., Wang, J.K.T., Walaas, S.I., Albert, K.A. and Greengard, P. (1990) Localisation of the MARCKS (87 kDa) protein, a major specific substrate for protein kinase C in rat brain. *J. Neurosci.*, 10: 1683–1698.

Rodnight, R., Trotta, E.E. and Perrett, C. (1985) A simple and economical method for studying protein phosphorylation in vivo in the rat brain. *J. Neurosci. Methods*, 13: 87–95.

Rodnight, R. and Leal, R. (1990) Regional variations in protein phosphorylating activity in rat brain studied in microslices labelled with [^{32}P]phosphate. *J. Mol. Neurosci.*, 2: 115–122.

Rodnight, R., Perrett, C. and Soteriou, S. (1986) Two-dimensional patterns of neural phosphoproteins from the rat labeled in vivo under anaesthesia and in vitro in slices and synaptosomes. In W.H. Gispen and A. Routtenberg (Eds.), *Phosphoproteins in Neuronal Function, Progress in Brain Research, Vol. 69*, Elsevier, Amsterdam, pp. 373–381.

Rodnight, R., Zamani, R. and Tweedale, A. (1988) An investigation of experimental conditions for studying protein phosphorylation in microslices of rat brain by two dimensional electrophoresis. *J. Neurosci. Methods*, 24: 27–38.

Rostas, J.A.P., Seccombe, M. and Weinberger, R.P. (1988) Two developmentally regulated isoenzymes of calmodulin-stimulated protein kinase II in rat forebrain. *J. Neurochem.*, 50: 945–953.

Salbego, C. and Rodnight, R. (1989) Ontogenetic development of protein kinase C activity towards endogenous substrates in rat brain studied in microslices and by two-dimensional electrophoresis. *J. Neurochem.*, 52 (Suppl.): 167C.

Stumpo. D.J., Graff, J.M., Albert, K.A., Greengard, P. and Blackshear, P.J. (1989) Molecular cloning, characterization, and expression of a cDNA encoding the "80- to 87-kDa" myristoylated alanine-rich C kinase substrate: a major cellular substrate for protein kinase C. *Proc. Natl. Acad. Sci. USA*, 86: 4012–4016.

Walaas. S.I., Nairn, A.C. and Greengard, P. (1983a) Regional distribution of calcium- and cyclic adenosine 3′:5′-monophosphate-regulated protein phosphorylation systems in mammalian brain. I. Particulate systems. *J. Neurosci.*, 3: 291–301.

Walaas, S.I., Nairn, A.C. and Greengard, P. (1983b) Regional distribution of calcium- and cyclic adenosine 3′:5′-monophosphate-regulated protein phosphorylation systems in mammalian brain. II. Soluble systems. *J. Neurosci.*, 3: 302–311.

Wang, J.K.T., Walaas, S.I. and Greengard, P. (1988) Protein phosphorylation in nerve terminals: comparison of calcium/calmodulin dependent and calcium/diacylglycerol-dependent systems. *J. Neurosci.*, 8: 281–288.

Wofchuk, S.T.,and Rodnight, R. (1990) Stimulation by glutamate of the phosphorylation of two substrates of protein kinase C, B-50/GAP and MARCKS, and of ppH-47, a protein highly labelled in incubated slices from the hippocampus. *Neurosci. Res. Commun.*, 6: 135–140.

Zeman, W. and Innes, J.R.M. (1963) *Craigie's Neuroanatomy of the Rat*, Academic Press, New York, pp. 144–155.

W.H. Gispen and A. Routtenberg (Eds.)
Progress in Brain Research, Vol. 89
© 1991 Elsevier Science Publishers B.V.

CHAPTER 12

Molecular and cellular studies on brain calcium/calmodulin-dependent protein kinase II

T.R. Soderling, K. Fukunaga, D.A. Brickey, Y.L. Fong, D.P. Rich, K. Smith
and R.J. Colbran

Department of Molecular Physiology and Biophysics, Vanderbilt University, Nashville, TN 37232-0615, U.S.A.

Introduction

Elevation of intracellular free Ca^{2+} ($[Ca_i^{2+}]$) is a common cellular response in signal transduction pathways of many hormones and neurotransmitters. Of the multiple physiological responses triggered by alterations in $[Ca_i^{2+}]$, phosphorylation/dephosphorylation catalyzed by Ca^{2+}-dependent protein kinases and phosphatase(s) is a frequently utilized mechanism. Several Ca^{2+}-dependent protein kinases from various tissues have been purified and characterized including phosphorylase kinase, myosin light chain kinase, protein kinase C, and Ca^{2+}/calmodulin (CaM)-dependent protein kinases I, II, and III. Most of these Ca^{2+}-dependent protein kinases phosphorylate a very restricted number of protein substrates. However, protein kinase C and Ca^{2+}/CaM-dependent protein kinase II (CaM-kinase II) phosphorylate numerous proteins, and these two multifunctional protein kinases are analogous in this respect to the well-studied cAMP-dependent protein kinase (cAMP-kinase). This article discusses brain CaM-kinase II with particular focus on its unqiue regulatory properties, its regulation in cultured brain cells, and its physiological functions.

CaM-kinase II has widespread tissue distribution as oligomeric isozyme forms and is particularly abundant in brain (reviewed in Colbran and Soderling, 1990a; Schulman, 1988). In certain regions of the brain, such as hippocampus, it constitutes up to 2% of total protein (Erondu and Kennedy, 1985) which probably makes it the most abundant enzyme in these tissues. CaM-kinase II is localized presynaptically where it is involved in Ca^{2+}-dependent regulation of neurotransmitter biosynthesis and exocytosis. At excitatory synapses in forebrain there is a thickening of the postsynaptic membrane called the postsynaptic density (PSD), and CaM-kinase II constitutes about 30–50% of the protein in the PSD (Kennedy et al., 1983). These excitatory synapses are subject to a usage-dependent enhancement of synaptic transmission called long-term potentiation (LTP), a popular model for learning and memory. Recent

Correspondence: Dr. T.R. Soderling, Vollum Institute L-474, Oregon Health Sciences University, 3181 S.W. Sam Jackson Park Rd., Portland, OR 97201, U.S.A.

results implicate CaM-kinase II in the mechanisms of LTP (Malenka et al., 1989; Malinow et al., 1989). Other neuronal functions of CaM-kinase II will undoubtedly materialize as specific probes of its activity are developed.

Substrate specificity determinants

The primary sequence determinant in protein substrates of CaM-kinase II is -Arg-X_1-X_2-Ser/Thr- (X_n is any amino acid). A second basic residue at position X_1 is a negative determinant for CaM-kinase II (Soderling et al., 1986) whereas it is a strong positive determinant for cAMP-kinase (Kemp et al., 1977). Although CaM-kinase II and cAMP-kinase often phosphorylate the same proteins, their primary sites of phosphorylation are often different (reviewed in Colbran et al., 1989a). The kinetic consequences of these different phosphorylation sites can either be the same (e.g., cardiac phospholamban, see Simmerman et al., 1986) or different (e.g., tyrosine hydroxylase; Atkinson et al., 1987). In addition to the primary sequence determinants, there are indications of higher order structural determinants. For example, most synthetic peptide substrates are not such good substrates as the protein from which they are derived. Direct evidence for higher order determinants comes from studies on phenylalanine hydroxylase, a protein which is phosphorylated on the same site by both protein kinases (Doskeland et al., 1984). Binding of phenylalanine at an allosteric site enhances the phosphorylation rate by cAMP-kinase but inhibits the rate of phosphorylation by CaM-kinase II. This result raises the interesting theoretical possibility that the rates of phosphorylation of the same site by these two protein kinases could be differentially altered by changes in the intracellular concentration of allosteric modulators.

Regulatory properties of CaM-kinase II

A major effort of our laboratory over the past five years has been the elucidation of the regulatory properties of purified rat brain CaM-kinase II. CaM-kinase II, like most other serine/threonine protein kinases, undergoes intramolecular autophosphorylation at multiple serine and threonine residues. This autophosphorylation converts the kinase from an initial totally Ca^{2+}/CaM-dependent enzyme to a form that expresses about 50–80% of its total activity (assayed with Ca^{2+}/CaM) in the presence of EGTA (reviewed in Colbran and Soderling, 1990a). Biochemical analyses indicate this is due to autophosphorylation of Thr^{286}. Our investigations of the regulation of CaM-kinase II by binding of Ca^{2+}/CaM and by autophosphorylation were dependent on elucidation of the amino acid sequences, deduced from the cDNA sequences (Lin et al., 1987; Bulliet et al., 1988), of the three types of subunits (each of 50–60 kDa) of brain CaM-kinase II holoenzyme (650–700 kDa). These three polypeptides have 95% identical amino acid sequences over the NH_2-terminal 2/3 of the subunits. They differ primarily by insertions/deletions within the COOH-terminal 1/3 of the subunits (Fig. 1). Homology comparisons to other protein kinases indicate that the catalytic domain is close to the NH_2-terminus, and a putative CaM-binding sequence is situated just NH_2-terminal to the region of insertions/deletions. The COOH-terminus is thought to be involved in subunit assembly and perhaps in subcellular localization of the kinase.

Our approach to understanding the regulatory properties of CaM-kinase II has been to reconstitute the regulatory properties of the oligomeric holoenzyme using a monomeric 30–32 kDa fragment of the kinase in which the C-terminal 1/3 of the subunit, including the CaM-binding domain and the autophosphorylation site (i.e., residue 286; residue numbers will refer to the α subunit sequence unless stated otherwise), has been proteolytically removed. This constitutively-active NH$_2$-terminal fragment of CaM-kinase II is potently inhibited ($K_i = 0.2$ μM) by a synthetic peptide containing the sequence of the α subunit from residues 281–309 (Colbran et al., 1988). Kinetic analyses of the inhibition showed that peptide 290–309 inhibits competitively with peptide substrates (Payne et al., 1988) whereas peptide 281–309 inhibits competitively with Mg^{2+}/ATP (Colbran et al., 1989b). This suggests that residues 281–289 interact with the ATP-binding motif and residues 290–309 interact with the protein substrate binding elements of the catalytic domain. This conclusion was strengthened by the observation that peptide 290–309 did not protect the ATP-binding site from phenylglyoxal inactivation whereas peptide 281–309 afforded complete protection (Colbran et al., 1989b). Peptide 281–309 also contains the CaM-binding do-

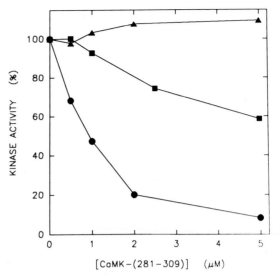

Fig. 2. Effects of Ca^{2+}/CaM and phosphorylation on the inhibition by peptide 281–309. Proteolyzed CaM-kinase II was assayed in the presence of EGTA ($\bullet$, $\blacksquare$), or Ca^{2+}/CaM ($\blacktriangle$) together with the indicated concentrations of Thr^{286}-phosphorylated ($\blacksquare$) or non-phosphorylated ($\bullet$, $\blacktriangle$) peptide 281–309. Reproduced by permission from the data of Colbran et al. (1988, 1989).

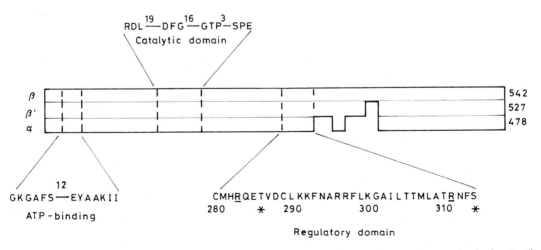

Fig. 1. Schematic representation of the subunits of brain CaM-kinase II. The three highly homologous subunits (α, β, β') of rat brain CaM-kinase II are depicted with the numbers on the right indicating the amino acids per subunit. Progressing from the NH$_2$-terminus to the COOH-terminus are the ATP-binding motif, the conserved amino acid triads of the putative catalytic domain, the multiregulatory domain including the CaM-binding and autoinhibitory elements and the positions of deletions in the α and β' subunits.

172

main (residues 296–309), and as shown in Fig. 2 binding of Ca^{2+}/CaM to peptide 281–309 completely abolishes its inhibitory properties (Colbran et al., 1988). If Thr^{286} is phosphorylated in peptide 281–309, its inhibitory potency is decreased by at least 10-fold (Fig. 2; Colbran et al., 1989b).

In addition to the extremely rapid autophosphorylation of Thr^{286}, which precedes phosphorylation of exogenous substrates (Kwiatkowski et al., 1988), there are at least 6–7 other sites of autophosphorylation. Some of these sites are only autophosphorylated by the Ca^{2+}-independent form of the kinase in the absence of Ca^{2+}/CaM (Hashimoto et al., 1987). This suggests that some of these sites may be within the CaM-binding domain, and this hypothesis is strengthened by the fact that this autophosphorylation results in a loss of CaM-binding. Recent studies have identified these sites of autophosphorylation as $Thr^{305/306}$ (Patton et al., 1990). These sites are located within the hydrophobic pocket of the

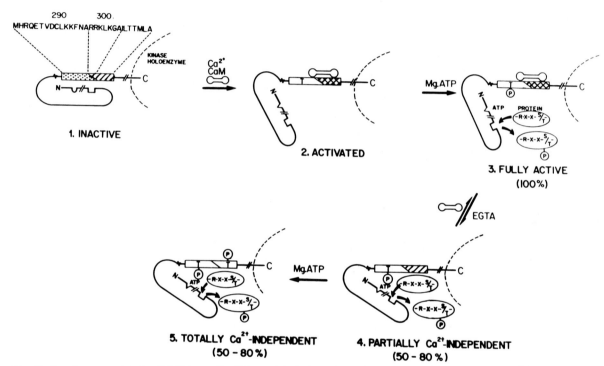

Fig. 3. Regulatory model for CaM-kinase II. This model depicts the activation of CaM-kinase II by Ca^{2+}/CaM and by autophosphorylation. The amino acid backbone of a single subunit of the oligomeric enzyme is depicted by the solid line. The semi-circle and square blocks indicate the ATP- and protein-substrate binding elements of the catalytic domain and the ➤ denotes a putative flexible hinge region (residues 270–280). The COOH-terminus is thought to be involved in subunit assembly. The crosshatched and stippled bar represents the CaM-binding and autoinhibitory domains. As isolated the kinase is inactive due to the occupation of the catalytic domain by the autoinhibitory domain (species 1). Initial activation requires binding of Ca^{2+}/CaM which induces a conformational change in the overlapping autoinhibitory domain which relieves its inhibitory interaction with the catalytic domain (species 2). In the presence of Mg^{2+}/ATP the fully activated kinase catalyzes rapid autophosphorylation of Thr^{286} as well as phosphorylation of exogenous substrates containing the consensus recognition sequence -R-X-X-S/T- (species 3). Upon removal of Ca^{2+}/CaM the kinase remains partially active (Ca^{2+}-independent) due to the presence of the negative charge of the phosphorylated Thr^{286} (species 4). The Ca^{2+}-independent form of the kinase can autophosphorylate at $Thr^{305/306}$ which then prevents subsequent binding of Ca^{2+}/CaM (species 5). Reproduced by permission from Soderling (1990).

CaM-binding domain, and their phosphorylation in peptide 290–309 blocks CaM-binding (Colbran and Soderling, 1990b). It is somewhat surprising the these two Thr residues are autophosphorylated in the kinase since there is no Arg three residues NH_2-terminal, but they are probably held close to the catalytic domain by other structural restraints. This is consistent with the result obtained when the recognition determinant (Arg^{283}) for Thr^{286} is mutated to Gln or Glu (see next section). Ser^{314}, which conforms to a consensus recognition sequence, is also autophosphorylated by the Ca^{2+}-independent form of the kinase (Patton et al., 1990), but its phosphorylation has little or no effect on CaM-binding since it lies outside the hydrophobic pocket of the CaM-binding domain (Colbran and Soderling, 1990b).

The above results have been formulated into a regulatory model (Fig. 3) which is consistent with the existence of autoinhibitory domains in other protein kinases (Soderling, 1990). In the holoenzyme residues 281–302 constitute an autoinhibitory domain that interacts with and inhibits the NH_2-terminal catalytic site elements (species 1, Fig. 3). Binding of Ca^{2+}/CaM to residues 296–309 (species 2) disrupts the autoinhibitory domain, perhaps by inducing α-helical structure, and frees the catalytic domain which can then bind Mg^{2+}/ATP and protein substrate in an active conformation. The activated kinase catalyzes phosphorylation of substrates at consensus recognition determinants as well as autophosphorylation at the consensus sequence Arg-Gln-Glu-Thr^{286} (species 3). Following autophosphorylation of Thr^{286}, the kinase remains partially active when Ca^{2+}/CaM is removed (i.e., partially Ca^{2+}/CaM-independent) since the negative charge prevents effective interaction with the catalytic domain (species 4). If Ca^{2+}/CaM is removed from the kinase, additional sites of autophosphorylation within the CaM-binding domain are exposed. Ca^{2+}-independent autophosphorylation of $Thr^{305/306}$ prevents subsequent binding of Ca^{2+}/CaM (Colbran and Soderling, 1990b).

Site-directed mutagenesis studies

The regulatory model of Fig. 1 is based largely on reconstitution experiments using proteolyzed kinase and synthetic peptides. We wanted to examine this model using site-directed mutagenesis of selective residues within the multiregulatory domain. The first two residues chosen were Arg^{283} and Thr^{286} since they are the recognition determinant and the regulatory autophosphorylation site, respectively. When Arg^{283} was substituted by either Glu or Gln in synthetic peptide 281–309, the IC_{50} for inhibition of CaM-kinase II increased over 200-fold (Fong and Soderling, 1990), suggesting that Arg^{283} is important for the potency of the autoinhibitory domain. Furthermore, neither of these substituted peptides could be phosphorylated by CaM-kinase II, consistent with the loss of the Arg recognition determinant. When these same mutations were made in the expressed kinase α-subunit, we expected a large increase in the Ca^{2+}-independent activity of the mutants due to the decreased potency of the autoinhibitory domain. However, the Gln^{283} mutant had essentially the same Ca^{2+}-independence ($2.15 \pm 0.62\%$) as the Arg^{283} wild-type ($1.86 \pm 0.72\%$), but the Glu^{283} mutant did show a small but significant ($p < 0.001$) increase in Ca^{2+}-independence to $5.46 \pm 0.90\%$ (Fong and Soderling, 1990). Waxham et al. (1990) mutated Arg^{283} to Ile, and the mutant kinase also was completely Ca^{2+}/CaM-dependent. Although there was autophosphorylation of Thr^{286} in the Gln^{283} and Glu^{283} mutants, the rate was dramatically decreased (Fig. 4). The Ile^{283} mutant did not exhibit any formation of Ca^{2+}-independent activity under autophosphorylation conditions (Waxham et al., 1990). We attribute the much larger effects of substitutions in the synthetic peptides versus mutations in the intact kinase as a difference between intermolecular and intramolecular interactions. There is good evidence (see below) that interaction of the autoinhibitory domain with the catalytic site is intrasubunit. In the mutant kinase there are probably structural constraints which hold the autoinhibitory domain in close proximity to the catalytic site. This may account for the observed autophosphorylation of Thr^{286} in the mutants when the substituted peptides were not phosphorylated at all. Furthermore, Arg^{283} is probably only one of several sites of interaction between the autoinhibitory domain and the catalytic site. When multiple mutations in the vicinity of Arg^{283} are made (His-Arg^{283}-Gln-Glu-Thr^{286} to Asp-Gly^{283}-Glu-Glu-Thr^{286}) to change the positive charges to negative charges, the mutant kinase exhibited 67% Ca^{2+}-independence (Waldmann et al., 1990).

Mutations have also been made at Thr^{286} in the α-subunit. When Ala was substituted for Thr^{286} in peptide 281–309, the IC_{50} was not altered. As expected, the Ala^{286} (Fong et al.,

1989; Waxham et al., 1990) and Leu^{286} (Hanson et al., 1989) mutant kinases had the same basal Ca^{2+}-independence as the Thr^{286} wild-type kinase, and, unlike the wild-type kinase, their Ca^{2+}-independence did not increase when subjected to autophosphorylation even though sites other than residue 286 were autophosphorylated. This result confirms the earlier biochemical studies which indicated that only autophosphorylation of Thr^{286} is responsible for generation of the Ca^{2+}-independence of CaM-kinase II. Most interestingly, when a negative charge was introduced at position 286 by mutation to Asp^{286}, this mutant kinase exhibited about the same level of Ca^{2+}-independence prior to autophosphorylation as the wild-type kinase had after autophosphorylation (Fong et al., 1989; Waldmann et al., 1990). This result suggests that it is largely the negative charge at position 286 in the autophosphorylated kinase which is responsible for the increase in Ca^{2+}-independence.

These results from site-directed mutagenesis strongly support the model of Fig. 1. Additional mutations are in progress to further probe inter-

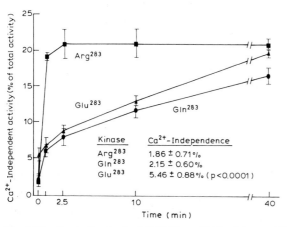 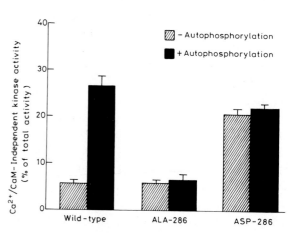

Fig. 4. Site-directed mutants of CaM-kinase II. The left panel illustrates that rates of formation of the Ca^{2+}-independent form (i.e., autophosphorylation of Thr^{286}) of the wild-type (Arg^{283}, ■), Gln^{283} (●) and Glu^{283} (▲) mutants. The Ca^{2+}-independence of these three species prior to autophosphorylation is shown in the inserted table. The right panel depicts the Ca^{2+}-independence of the wild-type (Thr^{286}), Ala^{286} and Asp^{286} mutant kinases prior to (crosshatched bar) and after (solid bar) autophosphorylation. Reproduced by permission from Fong et al. (1989) and Fong and Soderling (1990).

actions between the autoinhibitory domain and the catalytic site. All of our previous work on expressed α-subunit kinases was transcribed and translated in vitro. More recently we have successfully expressed the α-subunit and β-subunit kinases in Sf9 cells using the baculovirus system (Brickey et al., 1990). In agreement with Yamauchi et al. (1989), who expressed the kinases in CHO cells, we find that the expressed α-subunit is an oligomeric kinase of about 650 kDa whereas the expressed β-subunit is primarily a monomeric enzyme. Since the monomeric enzyme is completely Ca^{2+}/CaM-dependent, this indicates that the interaction of the autoinhibitory domain with the catalytic site is intrasubunit. In contrast, the α subunit kinase in E. coli appears to be monomeric (Waxham et al., 1990).

Regulation of CaM-kinase II in the postsynaptic density

Studies on the regulatory properties of CaM-kinase II have utilized the purified cytosolic brain enzyme. It is important to establish whether CaM-kinase II in the PSD also exhibits these regulatory properties, and this has recently been demonstrated (Rich et al., 1989). Formation of the Ca^{2+}-independent form of CaM-kinase II in the PSD through autophosphorylation of Thr^{286} is an attractive model for generating the constitutively-active protein kinase that appears to be important for induction of LTP in hippocampus (Malenka et al., 1989; Malinow et al., 1989). Another potential mechanism to generate a Ca^{2+}-independent form of CaM-kinase II is through limited proteolysis by the Ca^{2+}-dependent protease calpain (Kwiatkowski and King, 1989). This mechanism is attractive since infusion of the protease inhibitor leupeptin prevents induction of LTP (Staubli et al.,1988) and calpain has been localized in the PSD (Perlmutter et al., 1988). Treatment of the isolated PSD with cal-

pain results in a 3–5 fold increase in CaM-kinase II activity due to proteolytic conversion of a small percentage ($< 10\%$) of CaM-kinase II to a soluble, monomeric fragment of about 30 kDa which is fully active in the absence of Ca^{2+} (D.P. Rich et al., 1990). The fact that only a small fraction of the CaM-kinase II in the PSD is solubilized may be important since CaM-kinase II constitutes the major PSD protein and may be required for the structural integrety of the PSD. Since the 30 kDa constitutively-active fragment of CaM-kinase II is now soluble, it would have access to substrates (e.g., receptors and ion channels) other than those in the PSD.

Regulation of CaM-kinase II by gangliosides

Gangliosides, sialic acid containing glycosphingolipids, are found in high concentrations in nerve endings (up to 5–10% of total lipid; Ledeen, 1978) and are thought to regulate several enzymes including protein kinases (Chan, 1988). For this reason we investigated the effect of gangliosides on purified brain CaM-kinase II (Fukunaga et al., 1990a). Gangliosides (GT1b > GD1a > GM1) stimulate CaM-kinase II when assayed in the absence of Ca^{2+}/CaM with little effect (up to 100 μM) on activity in the presence of Ca^{2+}/CaM. Although GT1b stimulates autophosphorylation of CaM-kinase II, there is no formation of the stable Ca^{2+}-independent form since Thr^{286} is not phosphorylated. GT1b appears to bind to the autoinhibitory domain of CaM-kinase II since it can reverse the inhibition by synthetic peptide 281–309 of the proteolyzed catalytic fragment of the kinase. The physiological role of ganglioside stimulation of CaM-kinase II is not clear. If CaM-kinase II is stimulated by gangliosides in vivo, the process would not result in formation of the Ca^{2+}-independent form, as occurs with activation by Ca^{2+}/CaM. These different mechanisms for stimulating CaM-kinase II

176

may be important for different neuronal functions mediated by this multifunctional protein kinase.

Regulation of CaM-kinase II in cultured brain cells

All of the above studies have utilized in vitro approaches. We also wanted to examine regulation of CaM-kinase II in cultured brain cells and chose cerebellar granule cells (Fukunaga et al., 1990b). The hypothesis is that elevations in $[Ca_i^{2+}]$ would trigger autophosphorylation of CaM-kinase II to its Ca^{2+}-independent form which would remain active even when the $[Ca_i^{2+}]$ was resequestered to its basal value (Fig. 5). This could constitute an attractive mechanism for prolonging certain physiological functions in response to transient elevations of $[Ca_i^{2+}]$. The magnitude

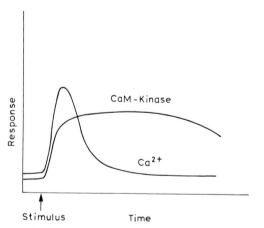

Fig. 5. Model of cellular response of CaM-kinase II to a transient Ca^{2+} signal. A stimulus which triggers a transient elevation in intracellular Ca^{2+} would trigger autophosphorylation of CaM-kinase II and formation of the Ca^{2+}-independent form. This Ca^{2+}-independent form would maintain CaM-kinase II activity, even when intracellular Ca^{2+} returned to its unstimulated value and prolong those physiological functions mediated by this multifunctional protein kinase. The duration of the Ca^{2+}-independent form would depend on the relative activities of the CaM-kinase II autophosphorylation reaction and the opposing protein phosphatases.

and duration of the Ca^{2+}-independent form would depend on the activity of the opposing protein phosphatases, which in turn may depend on the cell type or subcellular localization of CaM-kinase II.

Cultured cells were incubated under various conditions to alter the $[Ca_i^{2+}]$ and then homogenized in the presence of kinase and phosphatase inhibitors to stablize the in situ level of CaM-kinase II autophosphorylation (Fukunaga et al., 1990b,c). When extracellular Ca^{2+} was removed from the cells, only 1–2% of CaM-kinase II was in the Ca^{2+}-independent form, and depolarization with 56 mM K^+ had no effect (Fig. 6). Addition of Krebs-Ringer HEPES buffer containing normal extracellular Ca^{2+} (2.7 mM) resulted in a new steady-state level of 4–5% Ca^{2+}-independent CaM-kinase II, and now K^+ depolarization increased this to about 10% within 30 sec followed by a decline back to 4–5% by 5–10 min (Fig. 6). Ionomycin, a divalent cation ionophore, elicited 10% Ca^{2+}-independence which remained elevated. Inclusion of 5 μM okadaic acid, a cell permeable protein phosphatase inhibitor, in the incubation medium potentiated the magnitude and duration of the Ca^{2+}-independent activity of CaM-kinase II (Fig. 6). Phosphopeptide mapping of CNBr-cleaved ^{32}P-labeled 58–60 kDa subunits of CaM-kinase II revealed that under basal conditions the kinase contains $^{32}PO_4$ in many sites. Agents which promote formation of the Ca^{2+}-independent species of the kinase increase $^{32}PO_4$-incorporation into multiple sites of the kinase (Fig. 7). However, there was a good temporal correlation between ^{32}P-incorporation into CNBr peptide 1, which contains Thr_{287}, and generation of the Ca^{2+}-independent kinase activity.

We have also examined the effects of glutamate treatment of cerebellar granule cells (Fukunaga et al., 1990c). Glutamate is a natural agonist for these cells and can act through either non-NMDA gated ion channels (e.g., kainate or

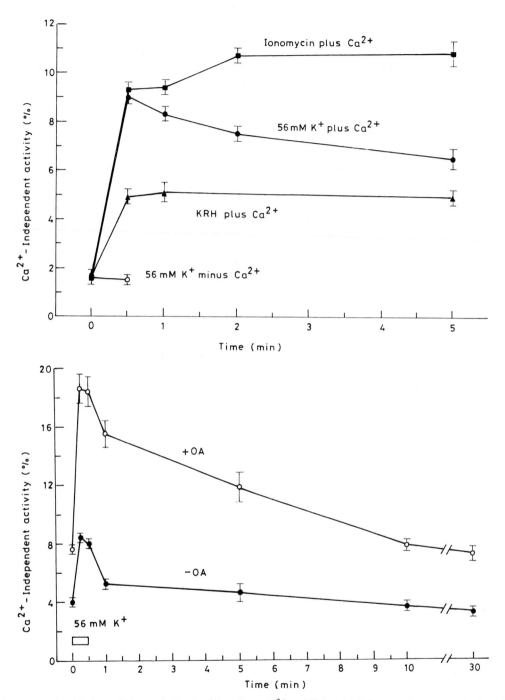

Fig. 6. Response of CaM-kinase II in cerebellar granule cells to Ca^{2+}-mobilizing conditions. In the top panel cultured cerebellar granule cells were preincubated in Krebs-Ringer HEPES buffer (KRH) minus Ca^{2+}. At time 0 the following additions were made: 56 mM K^+ (●), KRH plus 2.7 mM Ca^{2+} alone (▲) plus 56 mM K^+ and 2.7 mM Ca^{2+} (●) or 2.5 μM ionomycin and 2.7 mM Ca^{2+} (■). At the indicated times cells were homogenized and assayed for CaM-kinase II activity in the presence of Ca^{2+}/CaM (total activity) or EGTA (Ca^{2+}-independent activity). In the bottom panel cells were preincubated for 15 min in normal KRH (plus Ca^{2+}) in the absence (●) or presence (0) of 5 μM okadaic acid. At time zero 56 mM K^+ was added for 0.5 min (bar). Cells were homogenized and assayed as in the top panel. Reproduced by permission from Fukunaga et al., 1990b.

178

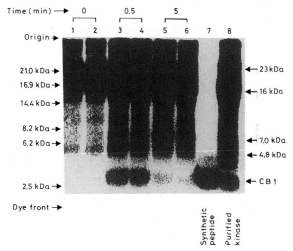

Time (min) → 0   0.5   5
       1 2   3 4   5 6   7 8
Origin →
21.0 kDa →   ← 23 kDa
16.9 kDa →   ← 16 kDa
14.4 kDa →
8.2 kDa →
6.2 kDa →   ← 7.0 kDa
                                  ← 4.8 kDa
2.5 kDa →   ← CB 1
Dye front →

Synthetic peptide
Purified kinase

Fig. 7. Autoradiogram of $^{32}PO_4$-peptide mapping of CaM-kinase II phosphorylated in situ. Granule cells were labeled for 5 h with $^{32}PO_4$ prior to treatment as follows: 45 min in Ca^{2+}-free KRH plus another 15 min with 5 μM okadaic acid. The cells were then incubated for 0, 0.5, or 5 min in KRH plus 2.7 mM Ca^{2+}, 56 mM KCl and 5 μM okadaic acid. Cells were homogenized, CaM-kinase II was immunoprecipitated and subjected to SDS/PAGE. The $^{32}PO_4$ subunits (58–60 kDa) of CaM-kinase II were cut out of the gel, subjected to CNBr hydrolysis and the CNBr peptides separated by urea-SDS/PAGE. Lane 8, CNBr peptides from ^{32}P-labeled purified rat brain CaM-kinase II; lane 7, synthetic peptide corresponding to residues 282–307 (CB1) containing the autophosphorylation site Thr^{286}. Reproduced by permission from Fukunaga et al. (1990b).

quisqualate) or through NMDA gated ion channels (reviewed in Mayer and Westbrook, 1987). The latter category of ion channels have a permeability to Ca^{2+} (MacDermott et al., 1986; Connor et al., 1988) and are therefore of special interest. They are also of great interest because of the requirement for NMDA receptor activation and postsynaptic Ca^{2+} influx in the initiation of long-term potentiation (reviewed in Mayer and Westbrook, 1987; Gustafsson and Wigstrom, 1988). When cerebellar granule cells were treated with 100 μM of L-glutamate, kainate, quisqualate, NMDA, or L-aspartate, only kainate elicited an increase in the Ca^{2+}-independence of CaM-kinase II. Since it is known that the NMDA-gated ion

channel is subject to a voltage-dependent Mg^{2+} blockage (Mayer et al., 1984), we removed extracellular Mg^{2+} and repeated the treatments. Under these conditions all agonists evoked increases (basal = 4.0%; stimulated = 6–8%) in the Ca^{2+}-independence of CaM-kinase II. Since glutamate is the natural agonist, it was utilized for all additional experiments. We found that inclusion of 1 μM glycine in the extracellular incubation medium shifted the dose–response curve to the left such that now 10 μM glutamate was maximally effective rather than 100 μM. This is consistent with the known potentiation by glycine of the NMDA-gated ion channel (Johnson and Ascher, 1987). The increase in Ca^{2+}-independence by 10 μM glutamate plus 1 μM glycine was blocked by specific antagonists (APV and CPP) of the NMDA receptor but not by the non-NMDA antagonist (GAMS) or nitrendipine, a blocker of voltage-dependent Ca^{2+} channels (Fig. 8). In the absence of extracellular Ca^{2+} there was no response of CaM-kinase II to glutamate plus glycine, but with extracellular Ca^{2+} the response was potentiated and prolonged by okadaic acid (Fig. 8). The increase in Ca^{2+}-independence correlated quite well quantitatively and temporally with $^{32}PO_4$ incorporation into Thr^{287} (cerebellum contains predominantly the β subunit kinase) (Fig. 8).

Several aspects of the cerebellar cell study provoke discussion. Firstly, we were surprised that under basal conditions, where intracellular Ca^{2+} is less than 100 nM, the kinase would be partially Ca^{2+}-independent since in vitro studies suggested that CaM-kinase II is less sensitive to Ca^{2+}/CaM (apparent K_a of 20–100 nM CaM) than most other CaM-dependent enzymes (apparent K_a of 1–10 nM CaM). Of course, it is possible there is a gradient (from the cell membrane inward) of $[Ca_i^{2+}]$ and the basal value of 4–5% Ca^{2+}-independence represents partial activation of CaM-kinase II by elevated Ca^{2+} near the membrane. Secondly, the magnitude (10%) and duration (5–10 min) of the increases in Ca^{2+}-independence in response to Ca^{2+} influx and the potentiation by okadaic acid strongly suggests that protein phosphatases are limiting the autophosphorylation reaction. Autophosphorylation in the cell extract with ATP-γ-S produced about 70% Ca^{2+}-independent CaM-kinase II (Fukunaga et al., 1990b). Thiophosphorylated proteins are resistent to protein phosphatases. The fact that okadaic acid, which is a potent inhibitor of protein phosphatases 1 and 2A, only gave partial potentiation suggested the presence of both okadaic acid sensitive and insensitive phosphatases. This was directly demonstrated using the cell extract (Fukunaga et al., 1990b). The identity of the okadaic acid insensitive phosphatase as either type 2C or 2B (calcineurin) has not been established. In summary, these results indicate that formation of the Ca^{2+}-independent species of CaM-kinase II is dynamically regulated in cerebellar granule cells by Ca^{2+}-mobilizing agent and by protein phosphatase activity and is correlated with autophosphorylation of Thr^{287}.

Physiological substrates of CaM-kinase II in brain

Although it is clear from in vitro studies that CaM-kinase II is multifunctional, rigorously established substrates in brain, other than autophosphorylation of the kinase itself as discussed above, are very limited. In large part this is due to the lack of specific probes (inhibitors or activators) of CaM-kinase II that can be used in intact cell studies. This situation may soon change as specific peptide inhibitors of several protein kinases are known, and we are in the process of designing a specific peptide inhibitor of CaM-kinase II (Smith et al., 1990). One might make synthetic genes for these peptides and express them in transfected cells using inducible promoters. Cell permeable organic molecules which selectively inhibit CaM-kinase II also offer great promise for probing the physiological functions of this kinase (Tokumitsu et al., 1990).

Synapsin I is a neuron-specific protein associated with the outer surface of presynaptic vesicles (DeCamilli et al., 1983). It is one of the best

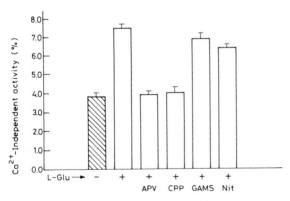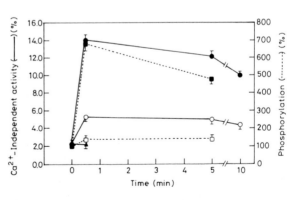

Fig. 8. Response of CaM-kinase II to glutamate and NMDA-antagonists. In the top panel granule cells were incubated in KRH with normal Ca^{2+} in the absence ($-$) or presence ($+$) of 10 μM L-glutamate and 1 μM glycine and either 100 μM NMDA antagonist (APV or CPP), non-NMDA antagonist (GAMS), or the voltage-dependent Ca^{2+} channel blocker nitrendipine (Nit). In the bottom panel cells were preincubated in KRH minus Ca^{2+}. At time zero the following additions were made: 10 μM L-glutamate and 1 μM glycine alone ($\blacktriangle$) or with 2.7 mM Ca^{2+} in the absence ($\bullet$) or presence ($\circ$) of 100 μM APV. The solid line gives Ca^{2+}-independent activity of CaM-kinase II. The dotted lines show $^{32}PO_4$-incorporation in the subunits of CaM-kinase II ($\square$) or CNBr peptide CB1 ($\blacksquare$) which contains Thr^{287}. Reproduced by permission from Fukunaga et al. (1990c).

known in vitro substrates for CaM-kinase II and can be phosphorylated to a molar stoichiometry of about 1.8. Dephospho-synapsin binds to purified synaptic vesicles (Schiebler et al., 1986) and may link synaptic vesicles to the cytoskeleton. Phosphorylation of synapsin I by CaM-kinase II, but not by cAMP-kinase or protein kinase C which phosphorylate different sites, reduces the affinity of synapsin I for the synaptic vesicles. It is proposed that phosphorylation of synapsin I by CaM-kinase II may promote the release of synaptic vesicles from the cytoskeleton for migration to the active zone and exocytosis. Considerable in situ evidence supports this model. Electrical stimulation or K^+ depolarization leads to phosphorylation of synapsin I (Nestler and Greengard, 1982). In isolated synaptosomes K^+ depolarization leads to autophosphorylation of CaM-kinase II, transient formation of its Ca^{2+}-independent form, phosphorylation of synapsin I (Gorelick et al., 1988), and increased neurotransmitter release. This increase in neurotransmitter release can be largely blocked by introduction of a peptide inhibitor of CaM-kinase II into the synaptosome (Nichols et al., 1990).

Another established neuronal substrate for CaM-kinase II is tyrosine hydroxylase, the rate-limiting enzyme in catecholamine biosynthesis. Tyrosine hydroxylase can be phosphorylated by numerous protein kinases in vitro including CaM-kinase II. Phosphorylation by CaM-kinase II increases the V_{max} of tyrosine hydroxylase by promoting its interaction with an activator protein (Atkinson et al., 1987). In PC12 (Yanagihara et al., 1984) and adrenal chromafin cells (Pocotte et al., 1986) depolarizing agents result in a Ca^{2+}-dependent increase in the tyrosine hydroxylase V_{max}. The sites phosphorylated by CaM-kinase II in vitro are transiently phosphorylated in chromaffin cells with temporal correlation to the observed activation in tyrosine hydroxylase (Waymire et al., 1988). Thus, there is good evidence

that CaM-kinase II may be one of several protein kinases that can activate tyrosine hydroxylase in response to different agonists.

There is accumulating evidence for the involvement of CaM-kinase II in the initiation of long-term potentiation and in kindling. CaM-kinase II is an attractive candidate for modulation of synaptic excitability (Lisman and Goldring, 1988) because of its localization in the postsynaptic density (Kennedy et al., 1983) and its potential for prolonged activation through autophosphorylation in response to transient elevations in $[Ca_i^{2+}]$ (see previous sections). It is known that initiation of long-term potentiation requires activation of NMDA-gated ion channels which are permeable to Ca^{2+} (reviewed in Mayer and Westbrook, 1987; Gustafsson and Wigstrom, 1988). When peptide inhibitors of either CaM-kinase II or protein kinase C are microinjected into the postsynaptic cell, they block the acquisition, but not the expression, of LTP (Malenka et al., 1989; Malinow et al., 1989). There is good evidence that initiation of LTP in the CA1 region of hippocampus, where CaM-kinase II is extremely abundant, occurs postsynaptically, but there is controversy concerning whether expression of LTP is presynaptic or postsynaptic (discussed in *Science*, 248: 1603–1605). The putative substrate of CaM-kinase II or protein kinase C in LTP is completely unknown, but the glutamate-gated ion channels (or associated regulatory proteins) which become potentiated in LTP, would be attractive candidates. As these ion channels are cloned (Hollman et al., 1989) the proteins can be directly tested as substrates for multiple protein kinases and effects on ion channel gating properties can be analyzed.

CaM-kinase II may also be involved in the phenomena of kindling, an experimental model of epilepsy (Wada, 1982). When hippocampal membranes were isolated from control and kindled rats, the in vitro autophosphorylation of CaM-kinase II was decreased in membranes from kindled rats (Goldenring et al., 1986). This could be interpreted to suggest that the autophosphorylation state of CaM-kinase II was increased in vivo by the kindling, thus accounting for the decreased subsequent in vitro autophosphorylation. Alternative interpretations are possible which need to be examined.

Perspective

CaM-kinase II is of interest to neuroscientists due to (1) its abundance in neural tissues, (2) its strategic localization both pre- and postsynaptically, (3) its unique regulatory properties allowing for the potential to prolong responses to transient elevations in $[Ca_i^{2+}]$, and (4) the prevalence of signal transduction mechanisms in neural functions. Biochemical and molecular biological investigations from several laboratories over the past ten years have greatly advanced our understanding of the molecular properties of CaM-kinase II. The challenge for the next decade will be to utilize this understanding to further our knowledge of the physiological functions of CaM-kinase II. Several avenues hold great promise. One of these is to develop specific inhibitors or activators of CaM-kinase II which can be used in intact cells. An example of this would be to express synthetic genes encoding a specific peptide inhibitor of CaM-kinase II. We are currently working on the design of a specific peptide inhibitor. A second approach will be the use of transfected cells or transgenic mice containing mutant forms (e.g., constitutively active) of CaM-kinase II. As neuronal specific promoters become available this latter route will become more feasible. It is likely that these methods will establish the importance of CaM-kinase II in multiple neuronal functions (e.g., proto-oncogene regulation) alongside its counterparts cAMP-kinase and protein kinase C.

Acknowledgements

The authors would like to thank Martha Bass for excellent technical assistance and Dr. William Taylor for assistance with the mutagenesis studies. This work was supported by NIH grants GM41292 and NS27037.

References

Atkinson, J., Richtand, N., Schworer, C.M., Kuczenski, R. and Soderling, T.R. (1987) Phosphorylation of purified rat striatal tyrosine hydroxylase by brain calmodulin-dependent protein kinase: effects of an activator protein. *J. Neurochem.*, 49: 1241–1249.

Brickey, D.A., Fong, Y.L. and Soderling, T.R. (1990) Expression and characterization of the α subunit of mouse brain Ca^{2+}/CaM-dependent protein kinase II using the baculovirus expression system. *Fed. Proc.*, 4: A2075.

Bulleit, R.F., Bennett, M.K., Molloy, S.S., Hurley, J.B. and Kennedy, M.B. (1988) Conserved and variable regions in the subunits of brain type II Ca^{2+}/CaM-dependent protein kinase. *Neuron*, 1: 63–72.

Chan, K.F.J. (1988) Ganglioside-modulated protein phosphorylation: Partial purification and characterization of a ganglioside-inhibited protein kinase in brain. *J. Biol. Chem.*, 263: 568–574.

Colbran, R.J., Fong, Y.L., Schworer, C.M. and Soderling, T.R. (1988) Regulatory interactions between the CaM-binding, inhibitory and autophosphorylation domains of Ca^{2+}/CaM-dependent protein kinase II. *J. Biol. Chem.*, 263: 18145–18151.

Colbran, R.J., Schworer, C.M., Hashimoto, Y., Fong, Y.L., Rich, D.P., Smith, M.K. and Soderling, T.R. (1989a) Calcium/calmodulin-dependent protein kinase II. *Biochem. J.*, 258: 313–325.

Colbran, R.J., Smith, M.K., Schworer, C.M., Fong, Y.L. and Soderling, T.R. (1989b) Inhibitory domain of Ca^{2+}/calmodulin-dependent protein kinase II: mechanism of action and regulation by phosphorylation. *J. Biol. Chem.*, 264: 4800–4804.

Colbran, R.J. and Soderling, T.R. (1990a) Ca^{2+}/calmodulin-dependent protein kinase II. *Curr. Topics Cell. Regul.*, 31: in press.

Colbran, R.J. and Soderling, T.R. (1990b) Ca^{2+}/CaM-independent autophos-phorylation sites of Ca^{2+}/CaM-dependent protein kinase II: studies on the effect of phosphorylation of Thr-305/306 and Ser-314 on CaM-binding using synthetic peptides. *J. Biol. Chem.*, 265: 11213–11219.

Connor, J.A., Wadman, W.J., Hockberger, P.E. and Wong, R.K.S. (1988) Sustained dendritic gradients of Ca^{2+} induced by excitatory amino acids in CA1 hippocampal neurons. *Science*, 240: 649–653.

DeCamilli, P., Harris, S.M., Huttner, W.B. and Greengard, P. (1983) Synapsin I (protein I), a nerve terminal-specific phosphoprotein. II. Its specific association with synaptic vesicles demonstrated by immunocytochemistry in agarose-embedded synaptosomes. *J. Cell Biol.*, 96: 1355–1373.

Doskeland, A.P., Schworer, C.M., Doskeland, S.O., Chrisman, T.D., Soderling, T.R., Corbin, J.D. and Flatmark, T. (1984) Phenylalanine 4-monooxygenase. Some aspects of its phosphorylation by a calcium and CaM-dependent protein kinase. *Eur. J. Biochem.*, 145: 31–37.

Erondu, N.E. and Kennedy, M.B. (1985) Regional distribution of type II Ca^{2+}/CaM-dependent protein kinase in rat brain. *J. Neurosci.*, 5: 3270–3277.

Fong, Y.L. and Soderling, T.R. (1990) Studies on the regulatory domain of Ca^{2+}/CaM-dependent protein kinase II: functional analyses of Arg-283 using synthetic inhibitory peptides and site-directed mutagenesis of the α subunit. *J. Biol. Chem.*, 265: 11091–11097.

Fong, Y.L., Taylor, W.L., Means, A.R. and Soderling, T.R. (1989) Studies of the regulatory mechanism of Ca^{2+}/CaM-dependent protein kinase II: mutation of Thr-286 to Ala and Asp. *J. Biol. Chem.*, 264: 16759–16763.

Fukunaga, K., Miyamoto, E. and Soderling, T.R. (1990a) Regulation of Ca^{2+}/CaM-dependent protein kinase II by brain gangliosides. *J. Neurochem.*, 54: 102–109.

Fukunaga, K., Rich, D.P. and Soderling, T.R. (1990b) Generation of the Ca^{2+}-independent form of Ca^{2+}/CaM-dependent protein kinase II in cerebellar granule cells. *J. Biol. Chem.*, 264: 21830–21836.

Fukunaga, K. and Soderling, T.R. (1990c) Activation of Ca^{2+}/CaM-dependent protein kinase II in cerebellar granule cells by N-methyl-D-aspartate receptor activation. *Mol. Cell. Neurosci.*, 1: 133–138.

Goldenring, J.R., Wasterlain, C.G., Oestreicher, A.B., de Graan, P.N.E., Farber, D.B., Glaser, G. and DeLorenzo, R.J. (1986) Kindling induces a long-lasting change in the activity of a hippocampal membrane CaM-dependent protein kinase system. *Brain Res.*, 377: 47–53.

Gorelick, F.S., Wang, J.K., Lai, Y., Nairn, A.C. and Greengard, P. (1988) Autophosphorylation and activation of Ca^{2+}/CaM-dependent protein kinase II in intact nerve terminals. *J. Biol. Chem.*, 263: 17209–17212.

Gustafsson, B. and Wigstrom, H. (1988) Physiological mechanisms underlying long-term potentiation. *Trends Neurosci.*, 11: 156–162.

Hanson, P.I., Kapiloff, M.S., Lou, L.L., Rosenfeld, M.G. and Schulman, H. (1989) Expression of a multifunctional Ca^{2+}/CaM-dependent protein kinase and mutational analysis of its autoregulation. *Neuron*, 3: 59–70.

Hashimoto, Y., Schworer, C.M., Colbran, R.J. and Soderling, T.R. (1987) Autophosphorylation of Ca^{2+}/CaM-depen-

dent protein kinase II: effects on total and Ca^{2+}-independent activities. *J. Biol. Chem.*, 262: 8051–8055.

Hollmann, M., O'Shea-Greenfield, A., Rogers, S.W. and Heinemann, S. (1989) Cloning by functional expression of a member of the glutamate receptor family. *Nature*, 342: 643–648.

Johnson, J.W. and Ascher, P. (1987) Glycine potentiates the NMDA response in cultured mouse brain neurons. *Nature*, 325: 529–531.

Kemp, B.E., Graves, D.J., Benjamini, E. and Krebs, E.G. (1977) Role of multiple basic residues in determining the substrate specificity of cAMP-dependent protein kinase. *J. Biol. Chem.*, 252: 4888–4894.

Kennedy, M.B., Bennett, M.K. and Erondu, N.E. (1983): Biochemical and immuno-chemical evidence that the "major postsynaptic density protein" is a subunit of a CaM-dependent protein kinase. *Proc. Natl. Acad. Sci. USA*, 80: 7357–7361.

Kwiatkowski, A.P., Shell, D.J. and King, M.M. (1988) The role of autophosphorylation in activation of the type II CaM-dependent protein kinase. *J. Biol. Chem.*, 263: 6484–6486.

Kwiatkowski, A.P. and King, M.M. (1989) Autophosphorylation of the type II CaM-dependent protein kinase is essential for a proteolytic fragment with catalytic activity: implications for long-term synaptic potentiation. *Biochemistry*, 28: 5380–5385.

Ledeen, R.W. (1978) Ganglioside structures and distribution: Are they localized at the nerve ending? *J. Supramol. Struct.*, 8: 1–17.

Lin, C.R., Kapiloff, M.S., Durgerian, S., Tatemoto, S., Russo, A.F., Hanson, P., Schulman, H. and Rosenfeld, M.G. (1987) Molecular cloning of a brain-specific Ca^{2+}/CaM-dependent protein kinase. *Proc. Natl. Acad. Sci. USA*, 84: 5962–5966.

Lisman, J.E. and Goldring, M.A. (1988) Feasibility of long-term storage of graded information by the Ca^{2+}/CaM-dependent protein kinase molecules of the postsynaptic density. *Proc. Natl. Acad. Sci. USA*, 85: 5320–5324.

MacDermott, A.B., Mayer, M.L. and Westbrook, G.L. (1986) NMDA-receptor activation increases cytoplasmic calcium in cultured spinal cord neurons. *Nature*, 321: 519–522.

Malenka, R.C., Kauer, J.A., Perkell, D.J., Mauk, M.D., Kelly, P.T., Nicoll, R.A. and Waxham, M.N. (1989) An essential role for postsynaptic CaM and protein kinase activity in long-term potentiation. *Nature*, 340: 554–557.

Malinow, R., Schulman, H. and Tsien, R.W. (1989) Inhibition of postsynaptic protein kinase C or CaM-dependent protein kinase II blocks induction but not expression of long-term potentiation. *Science*, 245: 862–866.

Mayer, M.L., Westbrook, G.L. and Guthrie, P.B. (1984) Voltage-dependent block by Mg^{2+} of NMDA responses in spinal cord neurons. *Nature*, 309: 261–263.

Mayer, M.L. and Westbrook, G.L. (1987) The physiology of

excitatory amino acids in the vertebrate central nervous system. *Prog. Neurobiol.*, 28: 197–276.

Nestler, E.J. and Greengard, P. (1982) Nerve impulses increase the phosphorylation state of protein I in rabbit superior cervical ganglion. *Nature*, 296: 452–454.

Nichols, R.A., Sihra, T.S., Czernik, A.J., Nairn, A.C. and Greengard, P. (1990) Ca^{2+}/CaM-dependent protein kinase II increases glutamate and noradrenaline release from synaptosomes. *Nature*, 343: 647–651.

Patton, B.L., Miller, S.G. and Kennedy, M.B. (1990) Activation of type II Ca^{2+}/CaM-dependent protein kinase by Ca^{2+}/CaM is inhibited by autophosphorylation of threonine within the CaM-binding domain. *J. Biol. Chem.*, 265: 11204–11212.

Payne, M.E., Fong, Y.L., Ono, T., Colbran, R.J., Kemp, B.E., Soderling, T.R. and Means, A.R. (1988) Ca^{2+}/calmodulin-dependent protein kinase II: characterization of distinct CaM-binding and inhibitory domains. *J. Biol. Chem.*, 263: 7190–7195.

Perlmutter, L., Siman, R., Gall, C., Baudry, M. and Lynch, G. (1988) The ultrastructural localization of Ca^{2+}-activated protease calpain in rat brain. *Synapse*, 2: 79–88.

Pocotte, S.L., Holz, R.W. and Ueda, T. (1986) Cholinergic receptor-mediated phosphorylation and activation of tyrosine hydroxylase in cultured bovine adrenal chromaffin cells. *J. Neurochem.*, 46: 610–622.

Rich, D.P., Colbran, R.J., Schworer, C.M. and Soderling, T.R. (1989) Regulatory properties of Ca^{2+}/CaM-dependent protein kinase II in rat brain postsynaptic densities. *J. Neurochem.*, 53: 807–816.

Rich, D.P., Schworer, C.M., Colbran, R.J. and Soderling, T.R. (1990) Proteolytic activation of Ca^{2+}/CaM-dependent protein kinase II: putative function in synaptic plasticity. *Mol. Cell. Neurosci.*, 1: 107–116.

Schiebler, W., Jahn, R., Doucet, J.P., Rothlein, J. and Greengard, P. (1986) Characterization of synapsin I binding to small synaptic vesicles. *J. Biol Chem.*, 261: 8383–8390.

Schulman, H. (1988) The multifunctional Ca^{2+}/CaM-dependent protein kinase. *Adv. Second Messenger Phosphoprotein Res.*, 22: 39–111.

Simmerman, H.K.B., Collins, J.H., Theibert, J.L., Wegener, A.D. and Jones, L.R. (1986) Sequence analysis of phospholamban: identification of phosphorylation sites and two major structural domains. *J. Biol. Chem.*, 261: 13333–13341.

Smith, M.K., Colbran, R.J. and Soderling, T.R. (1990) Specificities of auto-inhibitory domain peptides for four protein kinases: Implications for intact cell studies of protein kinase functions. *J. Biol. Chem.*, 265: 1837–1840.

Soderling, T.R., Schworer, C.M., Payne, M.E., Jett, M.F., Porter, D.K., Atkinson, J.L. and Richtand, N.M. (1986) Ca^{2+}/calmodulin-dependent protein kinase II. In J. Nunez, J.E. Dumont and R.J.B. King (Eds.), *Hormones and Cell Regulation, Colloque INSERM, Vol. 139*, John Libbey Eurotext, London, pp. 141–157.

Soderling, T.R. (1990) Protein kinases: regulation by autoinhibitory domains. *J. Biol. Chem.*, 265: 1823–1826.

Staubli, U., Thibault, O., Baudry, M. and Lynch, G. (1988) Chronic administratioin of a thiol-proteinase inhibitor blocks long-term potentiation of synpatic responses. *Brain Res.*, 444: 153–158.

Tokumitsu, H., Chijiwa, T., Hagiwara, M., Mizutani, A., Terawawa, M. and Hidaka, H. (1990) KN-62, a specific inhibitor of Ca^{2+}/CaM-dependent protein kinase II. *J. Biol. Chem.*, 265: 4315–4320.

Wada, J. (1982) Kindling II. Raven Press, New York.

Waldmann, R., Hanson, P.I. and Schulman, H. (1990) Multifunctional Ca^{2+}/CaM-dependent protein kinase made Ca^{2+}-independent for functional studies. *Biochemistry*, 29: 1679–1684.

Waxham, M.N., Aronowski, J., Westgate, S.A. and Kelly, P.T. (1990) Mutagenesis of Thr-286 in monomeric Ca^{2+}/CaM-dependent protein kinase II eliminates Ca^{2+}-independent activity. *Proc. Natl. Acad. Sci. USA*, 87: 1273–1277.

Waymire, J.C., Johnson, J.P., Hummer-Lickteig, K., Lloyd, A., Vigny, A. and Cravisco, G.L. (1988) Phosphorylation of bovine adrenal chromaffin cell tyrosine hydroxylase: temporal correlation of acetylcholine's effect of site phosphorylation, enzyme activation and catecholamine synthesis. *J. Biol. Chem.*, 263: 12439–12447.

Yamauchi, T., Ohsako, S. and Deguchi, T. (1989) Expression and characterization of CaM-dependent protein kinase II from cloned cDNAs in Chinese hamster ovary cells. *J. Biol. Chem.*, 264: 19108–19116.

Yanagihara, N., Tank, A.W. and Weiner, N. (1984) Relationship between activation and phosphorylation of tyrosine hydroxylase by 56 mM K^+ in PC12 cells in culture. *Mol. Pharmacol.*, 26: 141–147.

SECTION IV

Plasticity and Function of the PKC Substrate B-50 / F1 / GAP-43 / Neuromodulin

W.H. Gispen and A. Routtenberg (Eds.)
Progress in Brain Research, Vol. 89
© 1991 Elsevier Science Publishers B.V.

CHAPTER 13

Protein kinase C substrate B-50 (GAP-43) and neurotransmitter release

P.N.E. De Graan [1], A.B. Oestreicher [1], P. Schotman [2] and L.H. Schrama [2]

[1] *Division of Molecular Neurobiology, Institute of Molecular Biology and Medical Biotechnology, Rudolf Magnus Institute,*
and [2] *Laboratory for Physiological Chemistry, University of Utrecht, Padualaan 8, 3584 CH Utrecht, The Netherlands*

Introduction

It is widely accepted that release of most neurotransmitters from the presynaptic nerve terminal requires fusion of transmitter containing vesicles with the presynaptic membrane, and that Ca^{2+} is an important trigger for this event. However, little is known about the molecular mechanisms underlying Ca^{2+}-induced neurotransmitter release. A major difficulty in resolving these mechanisms has been the inaccessibility of the interior of the presynaptic terminal to experimental manipulations. The small size of most terminals precludes micro-injection and electrophysiological approaches. Thus, most of the information on molecular mechanisms underlying neurotransmitter release is obtained from pharmacological studies.

The pharmacological evidence for a role of protein kinase C (PKC) in neurotransmitter release has been reviewed recently (Nishizuka, 1988; Dekker et al., 1991a) and may be summarized as follows: (i) phorbol esters, which stimulate PKC in vitro enhance depolarization-induced release of various transmitters (Allgaier et al., 1986; Wakade et al., 1986; Versteeg and Florijn, 1987); (ii) phosphorylation of several substrates of PKC is correlated with transmitter release (Dunkley and Robinson, 1986; Wang et al., 1988; Dekker et al., 1989b, 1990a); (iii) polymyxin B, known as an inhibitor of PKC, completely abolishes depolarization-induced transmitter release (Allgaier and Hertting, 1986; Versteeg and Ulenkate, 1987; Dekker et al., 1990b); and (iv) down regulation of PKC in PC12 cells coincides with decreased depolarization-induced release of noradrenaline (NA) (Matthies et al., 1987). Assuming that PKC is involved in the regulation of neurotransmitter release, the question arises which substrate(s) mediate this modulatory role.

One of the well characterized substrates of PKC in neurons is the protein B-50. B-50 is a

---

Correspondence: Dr. P.N.E. De Graan, Division of Molecular Neurobiology, Institute of Molecular Biology and Medical Biotechnology, Rudolf Magnus Institute, Padualaan 8, 3584 CH Utrecht, The Netherlands.

nervous-tissue specific substrate of PKC associated with the cytosolic face of the presynaptic membrane (Zwiers et al., 1980; Sörensen et al., 1981; Kristjansson et al., 1982; Aloyo et al., 1983; Gispen et al., 1985; De Graan et al., 1988, 1989; Van Lookeren Campagne et al., 1989). The B-50 protein is identical to the growth-associated protein GAP-43, to the calmodulin (CaM)-binding protein neuromodulin and to protein F1, which is implicated in long-term potentiation (Basi et al., 1987; Cimler et al., 1987; Karns et al., 1987; Nielander et al., 1987; Rosenthal et al., 1987). In a series of studies we have shown that the degree of PKC-mediated phosphorylation of B-50 in hippocampal slices and synaptosomes is correlated with transmitter release (Dekker et al., 1989b, 1990a,b; Heemskerk et al., 1989a, 1990). Based on these correlative studies and the fact that phorbol esters that stimulate PKC, enhance neurotransmitter release (Allgaier et al., 1986; Wakade et al., 1986; Versteeg and Florijn, 1987) we have suggested that PKC-mediated B-50 phosphorylation may be involved in the regulation of neurotransmitter release.

To investigate a causal relationship between B-50 and neurotransmitter release, a method is required, which renders the presynaptic terminal permeable for ions and macromolecules, without affecting the exocytotic machinery. Several permeation techniques, including erythrocyte ghost fusion (Schweizer et al., 1989), electropermeation (Knight and Scrutton, 1986), detergents (Peppers and Holz, 1986), freeze-thawing (Nichols et al., 1989) or toxins (Ahnert-Hilger et al., 1985a, 1989; Howell and Gomperts, 1987), have been developed to study stimulus-secretion coupling in secretory cells such as mast cells (Howell and Gomperts, 1987; Howell et al., 1987, 1989), chromaffin cells (Kenigsberg and Trifaro, 1985; Lee and Holz, 1986; Bittner and Holz, 1988; Ahnert-Hilger et al., 1989) and PC12 cells (Ahnert-Hilger et al., 1985a; Peppers and Holz, 1986). In these perme-

ated secretory cells, release could be induced by elevation of the Ca^{2+} concentration in the permeation buffer. Ca^{2+}-induced release was found to be dependent on Mg^{2+}, and ATP dependent as well as ATP-independent release have been described (Reynolds et al., 1982; Peppers and Holz, 1986; Ahnert-Hilger et al., 1985a,b, 1987a,b, 1989).

Recently, we have permeated synaptosomes, isolated pinched-off presynaptic nerve terminals, with the toxin streptolysin-O (SL-O; Dekker et al., 1989a). Like in mast cells, chromaffin cells and PC12 cells, NA release from SL-O-permeated synaptosomes could be induced by elevation of external Ca^{2+} (Dekker et al., 1989a). We have used these SL-O-permeated synaptosomes to introduce antibodies to B-50, which are known to inhibit PKC-mediated B-50 phosphorylation. The anti-B-50-IgGs completely inhibited Ca^{2+}-induced release of NA from SL-O-permeated synaptosomes (Dekker et al., 1989a).

In the present paper we summarize recent studies in which we characterized NA release from highly purified SL-O-permeated cortical synaptosomes with respect to permeation conditions, Ca^{2+} sensitivity and ATP dependency and investigated the involvement of PKC and its substrate B-50 in NA release using phorbol esters, protein kinase inhibitors and antibodies to B-50. The results from these experiments indicate that besides phosphorylation by PKC, other properties of B-50, such as its atypical CaM-binding (Andreasen et al., 1983), may be involved in the regulation of NA release. Interestingly, the Ca^{2+} concentrations required to induce release in the permeabilized synaptosomes are much higher than those measured in intact neurons after depolarization. To study the relationship between B-50 phosphorylation and intracellular Ca^{2+} concentrations, we measured changes in intrasynaptosomal Ca^{2+} concentrations induced by K^+ and 4-aminopyridine (4-AP) treatment using the fluorescent probe fura-2. The differential effects of K^+ and 4-AP on B-50 phosphorylation and intrasynaptosomal Ca^{2+} indicate that B-50 phosphorylation is linked to Ca^{2+} influx rather than to Ca^{2+} levels.

NA release from SL-O permeabilized synaptosomes

To further characterize Ca^{2+}-induced NA release we purified synaptosomes according to the method described by Dunkley and Robinson (1986). These synaptosomes were labelled with [^{3}H]NA and treated with different concentrations of SL-O for 5 min in a permeation buffer containing either 10^{-8} or 10^{-5} M Ca^{2+} (Fig. 1; Dekker et al., 1991b). In this purified preparation Ca^{2+}-induced NA release was optimal between 0.2 and 0.4 IU SL-O, and amounted to about 15% of the total NA label in the synaptosomes. The concentration of SL-O required for optimal permeation of these purified synaptosomes is less than 50% of that required for permeation of crude synaptosomes (Fig. 1). Ca^{2+}-induced NA release at optimal SL-O concentrations in purified synaptosomes is at least 3 times higher than in crude synaptosomes.

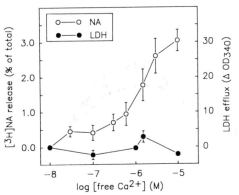

Fig. 2. Ca^{2+} dependency of release of NA and lactate dehydrogenase (LDH) from SL-O-permeated synaptosomes. Crude synaptosomes were permeated with 0.8 IU/ml SL-O at free Ca^{2+} concentration as indicated. The permeation buffer contained 5 mM ATP. Release of NA or LDH was calculated by subtracting the efflux of NA (LDH) from permeated synaptosomes at 10^{-8} M Ca^{2+} from that at the indicated concentration of free Ca^{2+}. Open symbols: NA; filled symbols: LDH. Data are mean $\pm$ SEM of 6–20 observations obtained from 6 independent experiments.

NA release from crude (Fig. 2) and purified synaptosomes was compared with respect to Ca^{2+} sensitivity. Although the magnitude of the response to Ca^{2+} was significantly higher in the purified preparation, the preparations were equally sensitive to Ca^{2+}. Significant stimulation

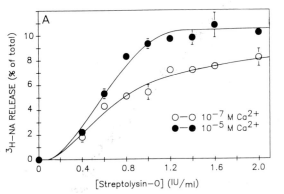

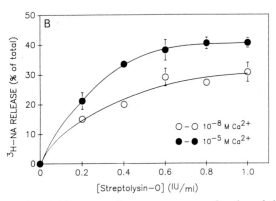

Fig. 1. Ca^{2+}-induced NA release from crude (panel A) and purified (panel B) permeated synaptosomes as a function of the Streptolysin-O (SL-O) concentration. NA efflux is expressed as% of total NA incorporated minus control efflux (in the absence of SL-O). (For experimental details see Dekker et al., 1989a, 1991b.) Bars indicate SEM ($n = 6$). Note the difference in the scale on the Y-axis between panel A and B.

of NA release was found at Ca^{2+} concentrations $> 3 \times 10^{-7}$ M. Control experiments in both preparations showed that the efflux of the cytosolic marker protein lactate dehydrogenase (LDH) is not affected by changing the Ca^{2+} concentration in the permeation buffer.

Based on the work of others (Howell and Gomperts, 1987), we used a buffer system with standard extracellular Na^+ and K^+ concentrations. We tested whether changing the concentrations of these cations towards more intracellular composition would affect the sensitivity of the NA release mechanism for Ca^{2+}. Ca^{2+}-dependent release of NA was not affected by changing the ratio Na^+/K^+ in the buffer, not even by reversing the concentrations. Apparently Na^+ and K^+ can be mutually exchanged and the ratio between these two ions is not relevant to the release process per se.

To investigate the ATP dependency of NA release, the effect of various ATP concentrations on Ca^{2+}-dependent NA release from SL-O-permeated synaptosomes was tested. In the absence of ATP Ca^{2+}-dependent NA release was 1.4% and 2.9% in crude and purified synaptosomes, respectively. Ca^{2+}-dependent release was significantly enhanced at 0.1 mM ATP and was maximal at 2 mM ATP. To control for an effect of ATP on SL-O activity, efflux of LDH was measured. Both at 10^{-7} and at 10^{-5} M Ca^{2+}, ATP had no effect on the efflux of LDH from the SL-O-permeated synaptosomes. We conclude that exocytotic release of NA is largely dependent on ATP. This ATP sensitivity of NA release could indicate the involvement of protein phosphorylation in the release reaction. Ca^{2+}/CaM-dependent kinase as well as PKC-mediated phosphorylation reactions are thought to play a role in transmitter release. The first system is thought to affect release via the neuron-specific vesicle-associated protein synapsin I. This protein could be involved in vesicle-cytoskeletal interactions and its phosphorylation probably liberates exocytotic vesicles from their cytoskeletal environment (Benfenati et al., 1989; Hemmings et al., 1989; Nichols et al., 1990).

NA release from SL-O treated synaptosomes can also be stimulated by phorbol esters. The lowest dose of phorbol 12-myristate 13-acetate (PMA) stimulating NA efflux at 10^{-7} M Ca^{2+} was 10^{-8} M. The effect of phorbol esters was stereospecific as 4α-phorbol 12,13-didecanoate had no effect on NA efflux. Interestingly, the effects of Ca^{2+} and phorbol esters appear to be additive (Dekker et al., 1991b; Hens and De Graan, unpublished), indicating that the two secretagogues induce NA release through at least partially different mechanisms.

To compare phorbol ester and Ca^{2+}-induced NA release we tested the effects of our affinity purified anti-B-50 IgGs, which inhibit B-50 phosphorylation and Ca^{2+}-induced NA release, on phorbol ester-induced release. Much to our surprise the B-50 antibodies only slightly inhibited 10^{-7} M PMA-induced release (14%) at 10^{-7} M Ca^{2+}. At 10^{-5} M Ca^{2+} the antibodies selectively and completely inhibited only that part of the release which can be accounted for by the increase in Ca^{2+} concentration. Total rabbit IgG or heat-inactivated B-50 antibodies, neither affected B-50 phosphorylation nor NA release (Dekker et al., 1991b).

Next we tested the effects of the protein kinase inhibitors H-7 (Hidaka et al., 1984) and polymyxin B (Mazzei et al., 1982) on NA release induced by Ca^{2+} and/or by PMA. These inhibitors both inhibit PKC-mediated B-50 phosphorylation in synaptic plasma membranes (SPM; Dekker et al., 1990b) as well as in SL-O permeated synaptosomes. Polymyxin B (200 IU/ml) inhibited 10^{-7} M PMA-induced NA release by 46%, inhibited the 10^{-5} M Ca^{2+}-induced release by about 64% and inhibited the release induced by 10^{-7} M PMA at 10^{-5} M Ca^{2+} by 56%. At

2000 IU/ml polymyxin B inhibited Ca^{2+} and PMA-induced release almost completely. H-7 (10^{-4} M) inhibited 10^{-7} M PMA-induced release by 57%. However, H-7 reduced 10^{-5} M Ca^{2+}-induced release by only 16% ($P < 0.05$) and the release induced by 10^{-7} M PMA at 10^{-5} M Ca^{2+} only by 47%, thus reducing release to the level obtained at 10^{-5} M Ca^{2+} alone.

In summary, the protein kinase inhibitors polymyxin B (Mazzei et al., 1982) and H-7 (Hidaka et al., 1984) and B-50 antibodies inhibit PKC-mediated B-50 phosphorylation in permeated synaptosomes, showing that they indeed interfere with the PKC system. Ca^{2+}-induced release can be inhibited by antibodies to B-50 and by polymyxin B, but not by H-7. PMA-induced release can be inhibited by polymyxin B and H-7, but only marginally by B-50 antibodies. Our effects of polymyxin B and H-7 are in agreement with those obtained by others in intact synaptosomes and hippocampal slices. In both preparations, polymyxin B completely inhibits evoked release of various transmitters (Allgaier and Hertting, 1986; Versteeg and Ulenkate, 1987) and H-7 has only marginal effects (Daschmann et al., 1989), whereas phorbol ester-induced transmitter release could be inhibited by all PKC inhibitors tested by these authors. Thus, evoked and Ca^{2+}-induced release are not affected by the most specific PKC inhibitor, H-7, whereas phorbol ester-induced release is completely blocked. It is possible that the inhibition of neurotransmitter release by polymyxin B, a much less specific kinase inhibitor (Dekker et al., 1990b) is elicited through non-PKC mediated pathways. These data are in line with the hypothesis that Ca^{2+} and phorbol ester induce release through different mechanisms. This hypothesis is further supported by the finding that B-50 antibodies inhibit Ca^{2+} but hardly affect PMA-induced release.

The precise mechanisms of Ca^{2+} and PMA-induced release and their relationship remain to be elucidated. PMA-induced release probably involves translocation of PKC to the membrane, but it is not clear to what extent PKC translocation contributes to evoked NA release under physiological conditions. Although our studies clearly show an involvement of PKC substrate B-50 in Ca^{2+}-induced NA release from permeabilized synaptosomes, they do not provide conclusive evidence for a role of PKC-mediated B-50 phosphorylation in release. Our results leave open the possibility that other properties of B-50, which are affected by the B-50 antibodies or polymyxin B, also contribute to the release process. An interesting possibility is the CaM-binding property of B-50. It has been shown by Storm and co-workers that purified B-50 has a higher affinity for CaM in the absence of Ca^{2+} than in the presence (Andreasen et al., 1983; Cimler et al., 1987) and that prephosphorylation of purified B-50 by PKC inhibits CaM binding (Alexander et al., 1987, 1988). Based on these in vitro studies it has been suggested (Alexander et al., 1987) that in vivo B-50 may act as a local CaM store. CaM may be dissociated from B-50 after a depolarization-induced rise in intracellular Ca^{2+} and subsequently activate cellular processes involved in transmitter release, for instance CaM-dependent kinases and phosphorylation of synapsin I (Nichols et al., 1990). Once CaM and B-50 have been dissociated, B-50 may be phosphorylated by PKC, thus preventing reassociation of CaM and prolonging CaM action. In subsequent studies we investigated the binding of CaM to endogenous B-50 in different membrane systems.

Binding of CaM to endogenous B-50 in SPM and growth cones

In contrast to other CaM-binding proteins, purified B-50 as well as B-50 solubilized from cortical membranes, has a higher affinity for CaM in the absence of free Ca^{2+} than in the presence

(Andreasen et al., 1983). Studies on purified B-50 confirmed that the dissociation constant for the CaM/B-50 complex is lowest in the presence of excess Ca^{2+} chelator (Alexander et al., 1987). Phosphorylation of B-50 by PKC reduces the affinity for CaM (Alexander et al., 1987), probably because the CaM-binding domain in B-50 (Alexander et al., 1988) is localized adjacent to the only PKC phosphorylation site (Nielander et al., 1990). Based on these studies with purified B-50, Storm and coworkers (Liu and Storm, 1990) proposed that in the growth cone B-50 regulates local CaM concentrations, which are important for growth cone mobility. However, direct evidence for the binding of CaM to endogenous B-50 in native neuronal membranes was lacking. Therefore we studied the interaction between CaM and B-50 in native synaptosomal and growth cone membranes in the presence and absence of free Ca^{2+} in order to assess the physiological relevance of this interaction.

The CaM/B-50 interaction was detected using the homo-bifunctional crosslinker disuccinimidyl suberate (DSS) to form a covalent B-50/CaM complex suitable for detection on SDS-PAGE and Western blots (De Graan et al., 1990). First, we investigated whether conditions can be found, which allow the crosslinking of B-50 and CaM. Indeed under appropriate conditions, incubation of purified B-50 and CaM with DSS and without Ca^{2+}, results in the formation of a protein complex with an apparent molecular weight (M_r) of about 70 kDa on 11% SDS-PAGE, a value which is close to the sum of the M_r values of B-50 and CaM. Complex formation requires the presence of B-50, CaM and DSS in the incubation mixture, showing that the complex is not a multimer of B-50 or CaM. No difference in complex formation could be detected between rat and bovine CaM. The fact that the protein complex can only be detected in the absence of free Ca^{2+} indicates that only under Ca^{2+}-free conditions B-50 and

CaM are closely associated. This is in line with findings of Storm and co-workers (Andreasen et al., 1983; Alexander et al., 1987), who showed with different techniques that purified B-50 exhibits a higher affinity for CaM in the absence of Ca^{2+}. Based on the intensity of the silver staining of the complex and a 1:1 stoichiometry (Alexander et al., 1987) crosslinking efficiency appears to be about 15% as estimated by densitometry. Higher concentrations of CaM do not increase crosslinking yield and higher concentrations of DSS cannot be used due to its limited solubility. From these experiments we conclude that under suitable conditions DSS can be used to detect the binding of CaM to B-50.

Because of the abundance of proteins in native SPM in the 70 kDa range it is essential to show that a complex formed by DSS treatment contains B-50 as well as CaM. Therefore, we investigated complex formation between purified B-50 and CaM on Western blots with anti-B-50 and anti-CaM IgGs (Fig. 3). Both antibodies crossreact with the 70 kDa complex, which is only detectable under EGTA conditions in the presence of all reactants (B-50, CaM and DSS). The B-50 antibody reveals several faint crossreacting bands in the 100 kDa range, which are present under both Ca^{2+} and EGTA conditions. These bands probably represent B-50 multimers, which are not detectable in the absence of DSS (Fig. 3, lanes 2 and 7). Preimmune serum or second antibody controls did not reveal any staining. Thus, under the conditions used both IgGs recognize the complex.

Using the same protocol, complex formation between B-50 in native SPM and purified CaM can be detected with anti-B-50 IgG only in the absence of Ca^{2+} (Fig. 4, right panel). Complex formation in SPM can also be detected without exogenous CaM, but the staining is much less pronounced. The SPM complex has a M_r identical to the complex formed from purified B-50 and

CaM. Similar results were obtained with washed growth cone membranes. The 70 kDa complex not only contains B-50, but also CaM (Fig. 4, left panel) as it crossreacts with the CaM antibodies. Under Ca^{2+} conditions a number of bands, including a 72 kDa band which also faintly stained under EGTA conditions, crossreacts with the CaM antiserum. These bands represent CaM-binding proteins, since preabsorption of the antibody with CaM abolished all immunostaining. Recent results in SPM and SL-O permeated synaptosomes (De Graan, unpublished) show that the Ca^{2+} sensitivity of the complex formation closely resembles that of NA release from permeated synaptosomes. Moreover, these results (De Graan, unpublished) indicate that increasing the ionic strength of the incubation buffer to 150 mM K^+ or Na^+, does not affect the Ca^{2+} sensitivity of complex formation. These results are in contradiction with results by Alexander et al. (1987) who showed that the binding of AEDANS-CaM to purified B-50 lost its sensitivity to Ca^{2+} at high ionic strength. The major difference between the two experiments is that our data were obtained using endogenous B-50 in its membrane attached environment.

In conclusion, our data show that with the crosslinker DSS a complex can be detected between CaM and endogenous B-50 in native SPM and growth cone membranes, and that this com-

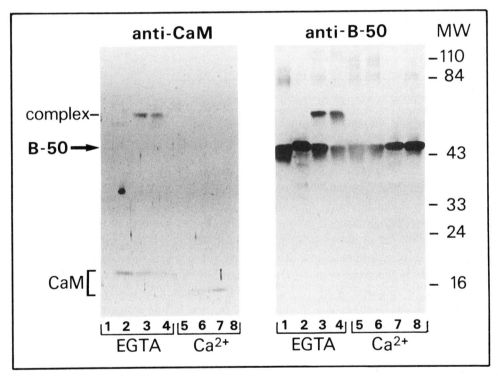

Fig. 3. Crosslinking of purified B-50 and CaM detected on Western blots with B-50 and CaM antibodies. Dephospho-B-50 (0.15 μg) and CaM (0.2 μg) were crosslinked with 1 mM DSS in the absence (EGTA: lanes 1–4) or presence (Ca^{2+}: lanes 5–8) of Ca^{2+}. Proteins were separated on 11% SDS-PAGE. Western blots were stained with anti-B-50 IgG (dilution 1:2000) or anti-CaM IgG (dilution 1:100). Lanes 1 and 8: B-50 + DSS; lanes 2 and 7: B-50 + CaM; lanes 3–6: B-50 + CaM + DSS. MW: molecular weight standards. Note the characteristic Ca^{2+}-induced migration shift of CaM. The faint B-50 staining in lanes 4–6 is an artefact due to incomplete protein transfer to the blot in that region and was not seen in similar experiments.

194

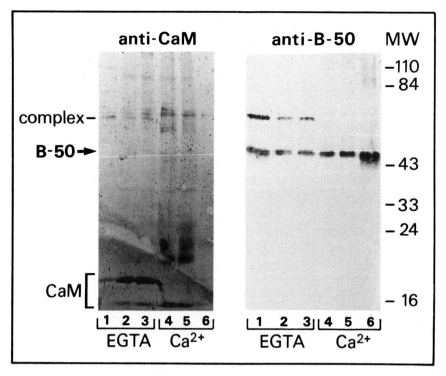

Fig. 4. Complex formation between CaM and endogenous B-50 in native SPM. SPM (10 μg) or dephospho-B-50 (0.15 μg) were preincubated with 0.2 μg CaM in the absence (EGTA: lanes 1–3) or presence (Ca^{2+}: lanes 4–6) of Ca^{2+} and subsequently treated with 1 mM DSS. Samples were analyzed on Western blots as in Fig. 2 with anti-B-50 or anti-CaM IgG. Lanes 1 and 6: B-50 + CaM + DSS; lanes 2–5: SPM + CaM + DSS.

plex formation is regulated by free Ca^{2+} concentrations which occur under physiological conditions in a neuron. These results strongly suggest that the binding of CaM to B-50 in the presynaptic and growth cone membrane is of physiological relevance. One possibility is that B-50 serves as a local CaM store, which could be released upon an increase in intracellular free Ca^{2+} (either directly or indirectly through PKC activation and B-50 phosphorylation) and could subsequently activate CaM-dependent processes involved in regulating growth cone motility (see Liu and Storm, 1990) or transmitter release (Nichols et al., 1990). Alternatively, the binding of CaM and phosphorylation of B-50 could regulate B-50 function in signal transduction, neurotransmitter release or

neurite outgrowth. These two possibilities are not mutually exclusive.

Interestingly, the concentrations of Ca^{2+} required to induce NA release and dissociation of the B-50/CaM complex are much higher than those measured in synaptosomes using the fluorescent Ca^{2+} indicator fura-2. The Ca^{2+} sensitivity of NA release from permeated synaptosomes was similar to that observed for catecholamine secretion from permeated PC12 cells and chromaffin cells and to histamine secretion from permeated mast cells (Ahnert-Hilger et al., 1985a; Lee and Holz, 1986; Howell and Gomperts, 1987). Free Ca^{2+} levels below 10^{-7} M did not induce release in any of the systems, but levels of 10^{-6} to 10^{-5} M Ca^{2+} were found to be effective. The high Ca^{2+} concentration necessary to induce release contrasts with the overall Ca^{2+} levels measured after K^+ depolarization of intact synaptosomes, which do not exceed the 0.5×10^{-6} M level (Verhage et al., 1988). There are two explanations for this apparent contradiction: (i) local high levels of Ca^{2+} are required to induce transmitter release (Smith and Augustine, 1988), and (ii) high concentrations of Ca^{2+} are necessary in the permeated system to overcome SL-O-induced damage to the release machinery, or to overcome dilution of cytosolic components of the release system. Extremely high local Ca^{2+} concentrations would occur just below the plasma membrane immediately following depolarization when Ca^{2+} enters through voltage sensitive Ca^{2+} channels (VSCC). If B-50 is involved in the release mechanism the prediction would be that for instance changes in the degree of phosphorylation would parallel a Ca^{2+} influx rather than an overall increase in the Ca^{2+} concentration. In a series of experiments using different depolarization techniques in intact synaptosomes we studied the relationship between the intrasynaptosomal Ca^{2+} concentration, NA release and B-50 phosphorylation.

Involvement of extracellular Ca^{2+} in neurotransmitter release and B-50 phosphorylation

Two secretagogues have been employed to investigate the dependence of B-50 phosphorylation on influx of extracellular Ca^{2+}, high extracellular K^+ and the convulsant drug 4-aminopyridine (4-AP). K^+ depolarizes the membrane thereby opening Ca^{2+} channels, allowing influx of extracellular Ca^{2+}. In excitable tissues 4-AP specifically blocks voltage-dependent K^+ channels known to carry the A-current (Segal et al., 1984, Rogawski, 1985). These channels are thought to play an important role in neuronal activity by regulating spike frequency (Segal et al., 1984;

Rogawski, 1985). Although the drug is widely applied as a blocker of K^+ channels in electrophysiological experiments, little information is available on its impact on biochemical processes inside the cell. One of the most striking features of 4-AP is that the drug is capable of stimulating release of many neurotransmitters in the PNS and the CNS. In brain, stimulatory effects of 4-AP on Ca^{2+}-dependent transmitter release have been reported both in slices (Löffelholz and Weide, 1982; Doležal and Tuček, 1983; Foldes et al., 1988; Heemskerk et al., 1989a) and in synaptosomes (Tapia and Sitges, 1982; Tibbs et al., 1989a).

Recently, we have found that 4-AP potently stimulated $[^3H]$-NA release from hippocampal slices (Heemskerk et al., 1989a). Concomitantly, 4-AP stimulated the phosphorylation of B-50 in these slices in a time and concentration-dependent manner. Moreover, this effect of 4-AP was found to be PKC-mediated (Heemskerk et al., 1989b). We were interested in the presynaptic mechanism of transmitter release stimulated by 4-AP. Depolarization of synaptosomes with 30 mM K^+ has been shown to increase PKC activity (Wu et al., 1982; Rodnight and Perrett, 1986; Diaz-Guerra et al., 1988; Wang et al., 1988) and B-50 phosphorylation (Dekker et al., 1990a). Therefore, we investigated the effects of 4-AP on the phosphorylation of B-50 in synaptosomes and compared these to the effects of K^+ depolarization on B-50 phosphorylation, in order to gain more insight into the (presynaptic) mechanism of action of 4-AP.

4-AP-induced changes in presynaptic protein phosphorylation were investigated in $[^{32}P]$-orthophosphate labeled synaptosomes. Since B-50 phosphorylation stimulated by depolarization with 30 mM K^+ is only observed in the presence of extracellular Ca^{2+} (Dekker et al., 1990b), we also investigated the effects of K^+ depolarization and 4-AP on B-50 phosphorylation at low extracellu-

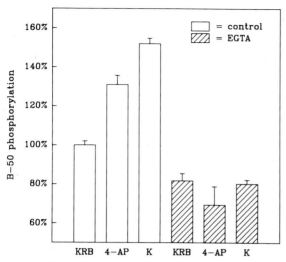

Fig. 5. Ca^{2+} dependency of stimulation of B-50 phosphorylation by 4-AP or depolarization with K^+. $[^{32}P]$-Orthophosphate labeled synaptosomes were incubated in KRB (control) or in the presence of 100 μM 4-AP (for 2 min) or 30 mM K^+ (for 15 s). The $[^{32}P]$-incorporation in B-50 was analyzed by 11% SDS-PAGE and autoradiography. Data are expressed as% ($\pm$SEM) of control incubations without 4-AP (5 mM K^+). Total incubation was 10 min, under control (2 mM Ca^{2+}) or low extracellular Ca^{2+} (with EGTA added).

lar Ca^{2+} concentrations ($< 10^{-7}$ M). $[^{32}P]$-orthophosphate labeled synaptosomes were incubated for 10 min in KRB with Ca^{2+} (2 mM) or with low extracellular Ca^{2+} (with EGTA added) and 4-AP was added 2 min before the reaction was stopped. Parallel samples were depolarized with 30 mM K^+ for 15 s under these two Ca^{2+} conditions, before terminating the reaction. As can be seen in Fig. 4, B-50 phosphorylation in KRB with Ca^{2+} was stimulated by 100 μM 4-AP as well as by depolarization. Chelation of extracellular Ca^{2+} with EGTA decreased phosphorylation of B-50 to $82 \pm 5\%$ of control incubations. Under these EGTA conditions 4-AP was not able to change B-50 phosphorylation significantly, nor was B-50 phosphorylation significantly increased by depolarization (Fig. 5).

Although many authors have suggested that 4-AP could have an effect on Ca^{2+} entry either

directly (Rogawski and Barker, 1983) or indirectly (Lundh, 1978; Thesleff, 1980), there are only a few studies investigating the effects of aminopyridines on Ca^{2+} homeostasis or Ca^{2+} influx in isolated presynaptic terminals. Using the fluorescent dye fura-2, two recent studies reported that 4-AP is able to elevate the Ca^{2+} concentration in synaptosomes (Gibson and Manger, 1988; Tibbs et al., 1989b). Studies in synaptosomes have shown that $^{45}Ca^{2+}$ uptake, under non-depolarizing conditions, was enhanced by 4-AP (Pasantes-Morales and Arzate, 1981; Tibbs et al., 1989b) or by the structurally related 3,4-diaminopyridine (Peterson and Gibson, 1983, 1985). However, it cannot be concluded from these studies whether Ca^{2+} influx through Ca^{2+} channels was affected by 4-AP, since intracellular redistribution of $^{45}Ca^{2+}$ might have occurred during the relatively long incubations used (1–20 min) in these experiments. Other studies, determining $^{45}Ca^{2+}$ uptake after seconds did not provide conclusive evidence either (Agoston et al., 1983; Tapia et al., 1985). Therefore, we determined the effects of 4-AP and K^+ depolarization on $[Ca^{2+}]_i$ in synaptosomes loaded with fura-2 under the same experimental conditions used for studying the effects of the two secretagogues on B-50 phosphorylation. This way, we hoped to find a relationship between Ca^{2+} entry into the synaptosome and the observed increases in B-50 phosphorylation induced by 4-AP and K^+ depolarization.

$[Ca^{2+}]_i$ was determined as described previously (Verhage et al., 1988). Drugs (or KRB as control) were added at $t = 300$ s (or at $t = 240$ and 480 s in case of two additions). $[Ca^{2+}]_i$ (in nM) was averaged over fixed periods of time (of 175 s starting 60 s after every drug addition). In addition the slope of the increase of $[Ca^{2+}]_i$ in time (in nM/s) was determined. The mean $[Ca^{2+}]_i$ was calculated at a time central to each period (i.e. at $t = 148$ s (period I), $t = 388$ s (period II) and $t = 628$ s (period III) in case of two addi-

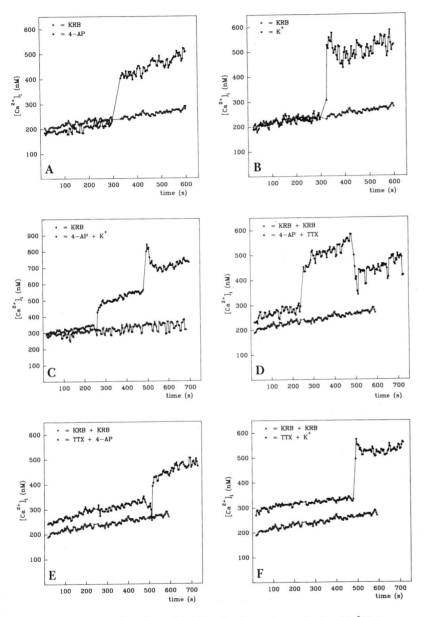

Fig. 6. Time course of the effects of 4-AP or depolarization on the free $[Ca^{2+}]_i$ in synaptosomes, determined with fura-2. In every experiment R_{max} and R_{min} were determined and corrections for possible effects on autofluorescence were made. All open circles refer to $[Ca^{2+}]_i$ in the presence of buffer. Panel A: 100 μM 4-AP was added at $t = 300$ s. Panel B: 30 mM K^+ was added at $t = 300$ s. Panel C: 100 μM 4-AP was added at $t = 240$ s, 30 mM K^+ was added at $t = 480$ s. Panel D: 100 μM 4-AP was added at $t = 240$ s, 1 μM TTX was added at $t = 480$ s. Panel E: 1 μM TTX was added at $t = 240$ s, 100 μM 4-AP was added at $t = 480$ s. F: 1 μM TTX was added at $t = 240$ s, 30 mM K^+ was added at $t = 480$ s.

198

tions). The values thus obtained for periods II and III (expressing $[Ca^{2+}]_i$ levels after the first or second addition, respectively) were divided by their own basal (control) values at the respective times, extrapolated from the first period, to yield the percentage changes. The mean values of three independent synaptosome preparations were averaged and expressed as the mean $\pm$ SEM.

Synaptosomes isolated on Percoll/sucrose gradients and loaded with fura-2, showed a basal $[Ca^{2+}]_i$, at the start of the experiment, of 239 ± 30 nM. During the incubation at 37°C this level slowly increased linearly to 363 ± 31 nM at 750 s (untreated control incubations, Fig. 6C). After addition of 4-AP (100 μM) $[Ca^{2+}]_i$ rose within a few seconds to 481 ± 29 s (Fig. 6A), corresponding to an increase of $158\% \pm 9$. This effect of 4-AP increased with the concentrations tested (10, 50 μM and 1 mM, results not shown). Depolarization of the synaptosomes with 30 mM K^+ elevated $[Ca^{2+}]_i$ to 474 ± 29 nM at $t = 448$ s (Fig. 6B), corresponding to an increase of $170\% \pm 9$. In most experiments a transient elevation of $[Ca^{2+}]_i$ was observed in the first min after addition of K^+. We did not observe such initial transient elevations in any of the experiments using 4-AP. Addition of 4-AP as well as K^+ depolarization resulted in a larger increase of $[Ca^{2+}]_i$ in time (reflected by a larger slope) of $184\% \pm 6$ and $256\% \pm 54$, respectively.

K^+-induced depolarization of the synaptosomes treated with 100 μM 4-AP resulted in an additional rise of $[Ca^{2+}]_i$ above the level obtained with 4-AP alone. After an initial transient elevation the mean $[Ca^{2+}]_i$ at $t = 628$ s was 686 ± 59 nM (Fig. 6C), corresponding to an increase to $180\% \pm 8$ of the extrapolated basal level at $t = 628$ s.

In order to investigate the possible involvement of Na^+ channels in 4-AP and K^+-induced effects we used the specific Na^+ channel blocker tetrodotoxin (TTX). Addition of TTX to the

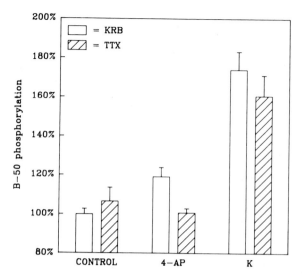

Fig. 7. Effect of TTX (1 μM) on B-50 phosphorylation stimulated by 4-AP (2 min, 100 μM) or depolarization (15 s, 30 mM K^+). B-50 phosphorylation was determined as described in the legend of Fig. 5. Data are expressed as percentual change over untreated (100$\pm$3%) or TTX-treated (107$\pm$7%) controls.

synaptosomes before addition of 4-AP or in the presence of 4-AP resulted in a decrease of $[Ca^{2+}]_i$ (Fig. 6D,E) to a level at $t = 628$ s (69% $\pm$ 6 of $[Ca^{2+}]_i$ level in the presence of 4-AP). TTX did not affect the slope of $[Ca^{2+}]_i$ in time in the presence of 4-AP. In contrast, the increase of $[Ca^{2+}]_i$ by K^+ depolarization was not prevented by TTX (Fig. 6F).

To investigate B-50 phosphorylation under these conditions, $[^{32}P]$orthophosphate labelled synaptosomes were incubated for 2 min with 4-AP (100 μM) or depolarized for 15 s with 30 mM K^+. 4-AP as well as depolarization resulted in an increase in B-50 phosphorylation of 20 and 50–90%, respectively (Figs. 5, 7, 8, 9 open bars).

The presence of TTX (1 μM) for 10 min did not change basal B-50 phosphorylation significantly. However, the stimulation of B-50 phosphorylation by 4-AP was abolished in the presence of TTX, in contrast to the effect of depolarization with 30 mM K^+ (Fig. 4). Involvement of

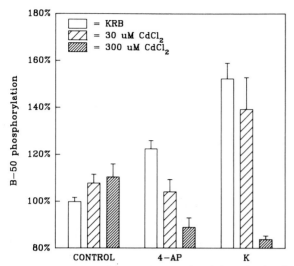

Fig. 8. Effects of CdCl$_2$ on B-50 phosphorylation stimulated by 4-AP (2 min, 100 μM) or depolarization (15 s, 30 mM K$^+$). B-50 phosphorylation was determined as described in the legend of Fig. 5. Data are expressed as percentual change over untreated (100$\pm$3%) or CdCl$_2$-treated (30 μM: 108$\pm$4% or 300 μM: 110 $\pm$6%) controls.

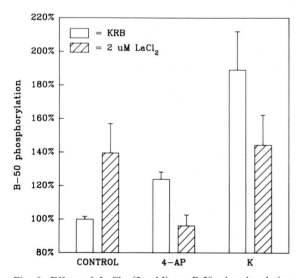

Fig. 9. Effect of LaCl$_3$ (2 μM) on B-50 phosphorylation stimulated by 4-AP (2 min, 100 μM) or depolarization (15 s, 30 mM K$^+$). B-50 phosphorylation was determined as described in the legend of Fig. 5. Data are expressed as percentual change over untreated (100$\pm$2%) or LaCl$_3$-treated (139$\pm$18%) controls.

Na$^+$ channels in the effects of 4-AP was further suggested by results from experiments where the extracellular concentration of Na$^+$ was lowered from 124 mM to 44 mM by equimolar replacement with choline. Under these conditions B-50 phosphorylation was 126% $\pm$ 8 of untreated control incubation after 10 min of incubation, while addition of 4-AP did not increase B-50 phosphorylation further. In contrast, depolarization with 30 mM K$^+$ stimulated B-50 phosphorylation to a level similar to that observed upon depolarization in control KRB (results not shown).

We tested whether the stimulation of B-50 phosphorylation by 4-AP or depolarization was sensitive to the inorganic Ca^{2+} channel antagonists CdCl$_2$ (Miller, 1987; Guan et al., 1988; Tsien et al., 1988) and LaCl$_3$ (Nachshen and Blaustein, 1980; Okada et al., 1989). In the presence of Cd^{2+} (30–300 μM CdCl$_2$) B-50 phosphorylation was no longer stimulated by 4-AP, as can be seen in Fig. 8. The response to K$^+$ depolarization was completely inhibited by 300 μM CdCl$_2$, but 30 μM was not sufficient to inhibit the depolarization evoked B-50 phosphorylation. Basal B-50 phosphorylation was not affected by Cd^{2+} alone. The effect of 4-AP on B-50 phosphorylation was also completely abolished by another inorganic Ca^{2+} channel antagonist (LaCl$_3$, 2 μM), however, in the presence of LaCl$_3$ alone the basal phosphorylation of B-50 was enhanced (Fig. 9).

As both K$^+$ depolarization and 4-AP stimulated B-50 phosphorylation in a Ca^{2+}-dependent manner, we investigated whether these stimulatory effects were additive. In Fig. 10 the time course of the effects of prolonged incubation with K$^+$ or 100 μM 4-AP on B-50 phosphorylation are shown. During incubation of synaptosomes with 30 mM K$^+$, B-50 phosphorylation is transiently enhanced, with a maximum at 15 s returning to control levels within 5 min (see also Dekker et al., 1990a). In contrast to this, incubation in the presence of 100 μM 4-AP and 30 mM K$^+$, transiently

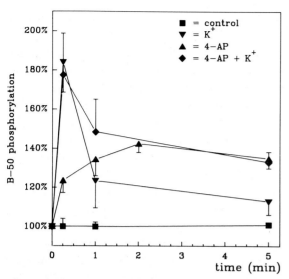

Fig. 10. Effects of 4-AP and K$^+$ depolarization on the phosphorylation of B-50. B-50 phosphorylation was determined as described in the legend of Fig. 5. The time course of B-50 phosphorylation is shown after incubation in the presence (▲) or absence (■) of 100 μM 4-AP under normal (5 mM K$^+$) conditions or in the presence (♦) or absence (▼) of 100 μM 4-AP under depolarized conditions (30 mM K$^+$). Data are expressed as% (±SEM) of control incubations (5 mM K$^+$) without 4-AP.

enhanced B-50 phosphorylation, but remained significantly elevated above control levels ($P < 0.05$) after 5 min.

Comparing the effects of 4-AP and K$^+$ depolarization on B-50 phosphorylation and changes in [Ca^{2+}]$_i$ in isolated synaptosomes we reach the following conclusions. 4-AP induced a concentration-dependent elevation of [Ca^{2+}]$_i$ within seconds and 100 μM 4-AP elevated [Ca^{2+}]$_i$ to almost the same level as evoked by 30 mM K$^+$ (481 and 474 nM, respectively). Although both treatments elevated [Ca^{2+}]$_i$ to the same extent, the underlying mechanisms may differ. In contrast to 4-AP, K$^+$ depolarization induced an initial large and transient elevation of [Ca^{2+}]$_i$. K$^+$ depolarization in the presence of 4-AP resulted in an additional rise of [Ca^{2+}]$_i$. Moreover, the attenuation

of the effect of 4-AP on [Ca^{2+}]$_i$ by TTX is in sharp contrast with the fact that TTX does not affect the increase of [Ca^{2+}]$_i$ evoked by 30 mM K$^+$. These data are in line with recent observations by Nicholls and co-workers in guinea pig cortical synaptosomes (Tibbs et al., 1989a,b).

The observation that the increase in [Ca^{2+}]$_i$ due to 4-AP could be lowered by TTX, indicates that synaptosomes are capable of actively controlling their [Ca^{2+}]$_i$, even in the presence of 4-AP. Therefore the increase of [Ca^{2+}]$_i$ due to 4-AP is most likely the net result of influx and extrusion, and not a passive accumulation of Ca^{2+}.

Theoretically, the elevation of [Ca^{2+}]$_i$ by 4-AP could be the result of Ca^{2+} entry or by (inositol phosphate-mediated) release of Ca^{2+} from intracellular stores. However, since TTX is a specific blocker of Na$^+$ channels (Moore et al., 1967; Ulbricht, 1974; Catterall, 1980), Ca^{2+} entry through voltage-sensitive channels in the plasma membrane triggered by repetitive spiking seems to be the more likely explanation. This is supported by the observation that 4-AP does not affect PPI turnover in hippocampal slices or cortical synaptosomes (data not shown).

It has been calculated that 30 mM K$^+$ depolarizes the synaptosomes by approximately 45 mV (Suszkiw et al., 1989), which will activate VSCC without the involvement of Na$^+$ channel activation (Adam-Vizi and Ligeti, 1986). In contrast to this large depolarization, even 1 mM 4-AP has been shown to cause 8–15 mV depolarization at most (Agoston et al., 1983; Tibbs et al., 1989a; McMahon et al., 1989). Our observation that the elevation of [Ca^{2+}]$_i$ by 4-AP is sensitive to TTX, indicates the need for Na$^+$ channel activity for its action and is in close agreement with recent data reported by others (Nicholls et al., 1989; Tibbs et al., 1989b). In our experiments, TTX was able to attenuate the effect of 4-AP even when 4-AP was already present, suggesting that Na$^+$ channels are continuously active in the presence of 4-AP.

These results suggest a different activation pattern of Ca^{2+} channels in synaptosomes by 4-AP compared to that observed during prolonged depolarization with 30 mM K^+.

Differences in underlying mechanisms between K^+ depolarization and 4-AP were also evident from the effects of 4-AP or K^+ depolarization on B-50 phosphorylation. 4-AP as well as K^+ depolarization stimulate B-50 phosphorylation in a Ca^{2+}-dependent manner. Chemical depolarization of synaptosomes with K^+ stimulates B-50 phosphorylation within 15 s to a maximal level, whereas stimulation of B-50 phosphorylation by 4-AP is much slower, reaching its maximal level not before 2 min (Fig. 10). Moreover, stimulation of B-50 phosphorylation in the presence of 30 mM K^+ seems to be a transient phenomenon, while 4-AP induces a sustained stimulation of B-50 phosphorylation. Since the effects of 4-AP and 30 mM K^+ were not additive, they are apparently not independent (see Fig. 10). This implies that one of the steps leading from Ca^{2+} entry to B-50 phosphorylation was already maximally stimulated by 30 mM K^+. The increase in $[Ca^{2+}]_i$ due to 4-AP and due to K^+ depolarization are faster than the stimulation of B-50 phosphorylation, suggesting that the increase in $[Ca^{2+}]_i$ precedes the enhancement of phosphorylation. The continued presence of K^+ does induce a sustained elevation of $[Ca^{2+}]_i$, but the degree of phosphorylation of B-50 returns to basal levels within 2 min (Fig. 10). The effects of 4-AP and K^+ depolarization on $[Ca^{2+}]_i$ do seem to be independent, since K^+ depolarization induces an additional rise in $[Ca^{2+}]_i$ when 4-AP has already elevated $[Ca^{2+}]_i$. These data suggested to us, that the phosphorylation of B-50 is not merely a reflection of the $[Ca^{2+}]_i$, but that the influx of Ca^{2+} through the membrane determines the degree of phosphorylation of B-50. Moreover, in experiments with the Ca^{2+} ionophore A23187, inducing a large elevation of $[Ca^{2+}]_i$, a relatively small increase in B-50 phosphorylation was detected only after 5 min (Dekker et al., 1990a).

Thus it seems that B-50 phosphorylation might be stimulated by local Ca^{2+} influx through ion channels. The observations that the effects of depolarization or 4-AP on B-50 phosphorylation are abolished by lowering the extracellular Ca^{2+} concentration with EGTA and attenuated by several ion channel antagonists are consistent with this hypothesis.

The fast and large increase in B-50 phosphorylation upon depolarization could reflect a large Ca^{2+} influx through VSCC opened immediately after the application of 30 mM K^+. In contrast, 4-AP might cause a continuous enhancement of Ca^{2+} cycling across the plasma membrane, possibly by repetitive depolarization and repolarization, as pointed out already by Tibbs et al. (1989b). 4-AP is known to block transient K^+ channels which are important in controlling excitability at voltages near the resting membrane potential, thereby preventing small depolarizing stimuli to reach threshold and trigger action potentials (Rogawski, 1985; Storm, 1988). A block of such a presynaptic K^+ current by 4-AP would result in the removal of this dampening effect on the membrane potential. This might lead to larger fluctuations, repetitive opening of voltage-sensitive Na^+ channels and subsequently to enhanced Ca^{2+} influx. If this is the case, stimulation by 4-AP probably resembles electrical stimulation more than K^+ depolarization does, as proposed by Tibbs et al. (1989b).

Taken together our results indicate that B-50 phosphorylation in mature presynaptic nerve terminals is associated with Ca^{2+} influx through voltage-sensitive channels. This influx of Ca^{2+} could then directly stimulate B-50 phosphorylation, without the need for receptor-mediated polyphosphoinositide turnover in order to generate diacylglycerol and inositol trisphosphate. We believe that PKC phosphorylating B-50 is already membrane associated, since in permeated synaptosomes we are able to phosphorylate B-50 after changing the Ca^{2+} concentration from 10^{-8} to 10^{-5} M (Dekker et al., 1989a). In the non-permeated synaptosome the influx of Ca^{2+} through VSCC might then be the physiological stimulus for B-50 phosphorylation and subsequent neurotransmitter release. Since there is some evidence for more than one type of Ca^{2+} channel in the presynaptic nerve terminal with different activation/inactivation characteristics (Turner and Goldin, 1985; Lemos and Nowycky, 1989; Martínez-Serrano et al., 1989), it remains to be shown which of these are involved in the effects of 4-AP and K^+ depolarization. The use of $CdCl_2$ and $LaCl_3$ as relatively unspecific Ca^{2+} channel antagonists does not allow any speculation yet about which Ca^{2+} channel subtype might be involved.

Summarizing, the following model can be proposed to describe a possible mechanism by which 4-AP stimulated B-50 phosphorylation and transmitter release (Fig. 11). The application of 4-AP, blocking an I_A-like K^+ current (I), results in the removal of the dampening effect of this current

on the membrane potential, leading to repetitive opening of voltage-dependent Na^+ channels (sensitive to TTX, II) and Ca^{2+} influx (III). Incubation with 30 mM K^+ would depolarize enough to bypass the Na^+ channels and enhance Ca^{2+} influx more directly (IV). Ca^{2+} would stimulate PKC activity, B-50 phosphorylation (V) and evoke transmitter release (VI).

Concluding remarks

The data presented in this chapter point to an important function of B-50 in neurotransmitter release. The active principles in this event appear to be B-50, PKC, a B-50 phosphatase, Ca^{2+} and CaM. A model, proposing the role of B-50 and all its properties in neurotransmitter release is presented in Fig. 12. The key event leading to release is the influx of Ca^{2+} probably through VSCCs. Depolarization- or 4-AP-induced Ca^{2+} influx would dissociate CaM and B-50. B-50, once liberated from CaM can be phosphorylated by PKC. It is at present not clear whether Ca^{2+} directly activates membrane-bound PKC, or whether a translocation of PKC from the cytosol

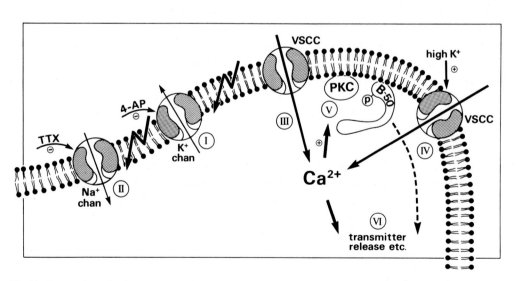

Fig. 11. Proposed mechanism of action of 4-AP in the presynaptic terminal.

is required. Once CaM and B-50 have been dissociated, B-50 may be phosphorylated by PKC preventing reassociation of B-50 and CaM, leaving CaM available for ongoing release. The phosphorylation state of B-50 may also be an important factor in regulating release. CaM may activate cellular processes involved in transmitter release, for instance activation of CaM-dependent kinases and phosphorylation of synapsin I (Nichols et al., 1990). This hypothesis implies the existence of various conformations of B-50 e.g. being (de)phosphorylated or binding CaM. The polyclonal antibodies used to interfere with NA release in permeated synaptosomes recognize all three B-50 configurations. Thus, besides interfering with B-50 phosphorylation, they may also interfere with CaM availability for neurotransmitter release. The nature and role of B-50 phos

phatase(s) remain to be established. Several phosphatases are capable of dephosphorylating B-50 in vitro (phosphatase 2B; Liu and Storm, 1989; Schrama et al., 1989), and phosphatase 1 and 2A (Dokas et al., chapter 3, this volume) and a B-50 phosphatase isolated from rat SPM (Dokas et al., 1990). To what extent regulation of phosphatase activity may contribute to the mechanism of transmitter release remains to be determined.

In conclusion, B-50 is an important molecule in release of NA from the presynaptic terminal. Whether B-50 is involved in the regulation of release of other transmitters has to be investigated. To obtain more insight into the involvement of B-50 in release, all molecular properties of B-50 (including phosphorylation, CaM binding and probably others) and their interactions should be considered.

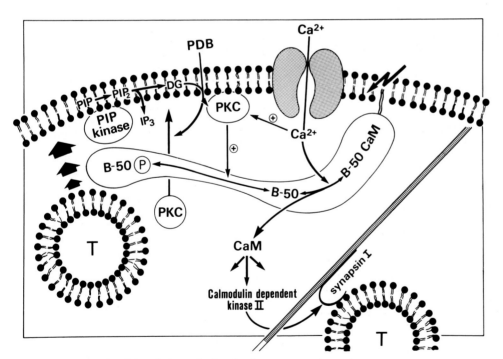

Fig. 12. Properties of B-50 in relation to its function in neurotransmitter release.

Acknowledgements

We thank Marina de Wit and Ank Frankhuyzen for their excellent experimental assistance, Dr. W.H. Ghijsen for his expertise in the fura-2 measurements and Drs. Lodewijk Dekker, Frank Heemskerk and Jac Hens for stimulating discussions. Paul van der Most is gratefully acknowledged for his artwork. This work was supported by grants 900–548–082 and 900–548–113 of the Netherlands Organization for Scientific Research (NWO), and by CLEO-TNO grant A66 of the Dutch Epilepsy Foundation.

References

Adam-Vizi, V. and Ligeti, E. (1986) Calcium uptake of rat brain synaptosomes as a function of membrane potential under different depolarizing conditions. *J. Physiol. (London)*, 372: 363–377.

Agoston, D., Hargittai, P. and Nagy, A. (1983) Effects of 4-aminopyridine in calcium movements and changes of membrane potential in pinched-off nerve terminals from rat cerebral cortex. *J. Neurochem.*, 41: 745–751.

Ahnert-Hilger, G., Bhakdi, S. and Gratzl, M. (1985a) Minimal requirements for exocytosis. *J. Biol. Chem.*, 260: 12730–12734.

Ahnert-Hilger, G., Bhakdi, S. and Gratzl, M. (1985b) α-Toxin permeabilized rat pheochromo cytoma cells: a new approach to investigate stimulus-secretion coupling. *Neurosci. Lett.*, 58: 107–110.

Ahnert-Hilger, G. and Gratzl, M. (1987a) Further characterization of dopamine release by permeabilized PC12 cells. *J. Neurochem.*, 49: 764–770.

Ahnert-Hilger, G., Bräutigam, M. and Gratzl, M. (1987b) Ca^{2+}-stimulated catecholamine release from α-toxin-permeabilized PC12 cells: biochemical evidence for exocytosis and its modulation by PKC and G proteins. *Biochemistry*, 26: 7842–7848.

Ahnert-Hilger G., Bader M.F., Bhakdi S. and Gratzl, M. (1989) Introduction of macro-molecules into rat bovine adrenal medullary chromaffin cells and rat pheochromocytoma cells (PC12) by permeabilization with streptolysin O: inhibitory effect of tetanus toxin on catecholamine secretion. *J. Neurochem.*, 52: 1751–1758.

Alexander K.A., Cimler B.M., Meier K.E. and Storm D.R. (1987) Regulation of calmodulin binding to P-57. *J. Biol. Chem.*, 262: 6108–6113.

Alexander, K.A., Wakim, B.T., Doyle, G.S., Walsh, K.A. and

Storm, D.R. (1988) Identification and characterization of the calmodulin-binding domain of neuromodulin, a neuron-specific calmodulin-binding protein. *J. Biol. Chem.*, 263: 7544–7549.

Allgaier, C. and Hertting, G. (1986) Polymyxin B, a selective inhibitor of protein kinase C, diminishes the release of noradrenaline and the enhancement of release caused by phorbol 12,13-dibutyrate. *Naunyn-Schmiedeberg's Arch. Pharmacol.*, 334: 218–221.

Allgaier, C., Von Kügelgen, O. and Hertting, G. (1986) Enhancement of noradrenaline release by 12-O-tetradecanoyl phorbol-13-acetate, an activator of protein kinase C. *Eur. J. Pharmacol.*, 129: 389–392.

Aloyo, V.J., Zwiers, H. and Gispen, W.H. (1983) Phosphorylation of B-50 protein by calcium-activated, phospholipid-dependent protein kinase and B-50 protein kinase. *J. Neurochem.*, 41: 649–653.

Andreasen, T.J., Luetje, C.W., Heideman, W. and Storm, D.R. (1983) Purification of a novel calmodulin binding protein from bovine cerebral cortex membranes. *Biochemistry*, 22: 4615–4618.

Basi, G.S., Jacobson, R.D., Virag, I., Schilling, J. and Skene, J.H.P. (1987) Primary structure and transcriptional regulation of GAP-43, a protein associated with nerve growth. *Cell*, 49: 785–791

Benfenati, F., Bähler, M., Jahn, R. and Greengard, P. (1989) Interactions of synapsin I with small synaptic vesicles: distinct sites in synapsin I bind to vesicle phospholipids and vesicle proteins. *J. Cell Biol.*, 108: 1863–1872.

Bittner, M.A. and Holz, R.W. (1988) Effects of tetanus toxin on catecholamine release from intact and digitonin-permeabilized chromaffin cells. *J. Neurochem.*, 51: 451–456.

Catterall, W.A. (1980) The molecular basis of neuronal excitability. *Annu. Rev. Toxicol.*, 20: 15–43.

Cimler, B.M., Giebelhaus, D.H., Wakim, B.T., Storm, D.R. and Moon, R.T. (1987) Characterization of murine cDNAs encoding P-57, a neural-specific calmodulin-binding protein. *J. Biol. Chem.*, 262: 12158–12163.

Daschmann, B., Allgaier, C., Nakov, R. and Hertting, G. (1988) Staurosporine counteracts the phorbol ester-induced enhancement of neurotransmitter release in hippocampus. *Arch. Int. Pharmacodyn. Ther.*, 296: 232–245.

De Graan, P.N.E., Heemskerk, F.M.J., Dekker, L.V., Melchers, B.P.C., Gianotti, C. and Schrama, L.H. (1988) Phorbol esters induce long- and short-term enhancement of B-50/GAP-43 phosphorylation in rat hippocampal slices. *Neurosci. Res. Commun.*, 3: 175–182.

De Graan, P.N.E., Oestreicher, A.B., Dekker, L.V., Van der Voorn, L. and Gispen, W.H. (1989) Determination of changes in the phosphorylation state of the neuron specific protein kinase C substrate B-50 (GAP-43). *J. Neurochem.*, 52: 17–23.

De Graan, P.N.E., Oestreicher, A.B., De Wit, M., Kroef, M., Schrama, L.H. and Gispen, W.H. (1990) Evidence for the binding of calmodulin to endogenous B-50 (GAP-43) in native synaptosomal plasma membranes. *J. Neurochem.*, 55: 2139–2141.

Dekker, L.V., De Graan, P.N.E., Oestreicher, A.B., Versteeg, D.H.G. and Gispen, W.H. (1989a) Inhibition of noradrenaline release by antibodies to B-50 (GAP-43). *Nature*, 342: 74–76.

Dekker, L.V., De Graan, P.N.E., Versteeg, D.H.G., Oestreicher, A.B. and Gispen, W.H. (1989b) Phosphorylation of B-50 (GAP-43) is correlated with neurotransmitter release in rat hippocampal slices. *J. Neurochem.*, 52: 24–30.

Dekker, L.V., De Graan, P.N.E., De Wit, M., Hens, J.J.H. and Gispen, W.H. (1990a) Depolarization-induced phosphorylation of the protein kinase C substrate B-50 (GAP-43) in rat cortical synaptosomes. *J. Neurochem.*, 54: 1645–1652.

Dekker, L.V., De Graan, P.N.E., Spierenburg, H., De Wit, M., Versteeg, D.H.G. and Gispen, W.H. (1990b) Evidence for a relationship between B-50 (GAP-43) and [³H]-noradrenaline release in rat brain synaptosomes. *Eur. J. Pharmacol.*, 188: 113–122.

Dekker, L.V., De Graan, P.N.E. and Gispen, W.H. (1991a) Transmitter release: target of regulation by protein kinase C? In W.H. Gispen and A. Routtenberg (Eds.), *Protein Kinase C and its Brain Substrates: Role in Growth and Plasticity, Progr. Brain Res., vol. 89*, Elsevier, Amsterdam, pp. 209–233.

Dekker, L.V., De Graan, P.N.E., Pijnappel, P., Oestreicher, A.B. and Gispen, W.H. (1991b) Noradrenaline release from streptolysin-O-permeated rat cortical synaptosomes. *J. Neurochem.*, 56: 1146–1153.

Diaz-Guerra, M.J.M., Sánchez-Prieto, J., Bosca, L., Pocock, J., Barrie, A. and Nicholls, D. (1988) Phorbol ester translocation of protein kinase C in guinea-pig synaptosomes and the potentiation of calcium-dependent glutamate release. *Biochim. Biophys. Acta*, 970: 157–165.

Dokas, L.A., Pisano, M.R., Schrama, L.H., Zwiers, H. and Gispen, W.H. (1990) Dephosphorylation of B-50 in synaptic plasma membranes. *Brian Res. Bull.*, 24: 321–329.

Doležal, V. and Tuçek, S. (1983) The effects of 4-aminopyridine and tetrodotoxin on the release of acetylcholine from rat striatal slices. *Naunyn-Schiedeberg's Arch. Pharmacol.*, 323: 90–95.

Dunkley, P.R. and Robinson, P.J. (1986) Depolarization-dependent protein phosphorylation in synaptosomes: mechanisms and significance. *Progr. Brain Res.*, 69: 273–294.

Foldes, F.F., Ludvig, N., Nagashima, H. and Vizi, E.S. (1988) The influence of aminopyridines on Ca²⁺-dependent evoked release of acetylcholine from rat cortex slices. *Neurochem. Res.*, 13: 761–764.

Gibson, G.E. and Manger, T. (1988) Changes in cytosolic free calcium with 1,2,3,4-tetra-hydro-5-aminoacridine, 4-aminopyridine and 3,4-diaminopyridine. *Biochem. Pharmacol.*, 37: 4191–4196.

Gispen, W.H., Leunissen, J.L.M., Oestreicher, A.B., Verkleij, A.J. and Zwiers, H. (1985) Presynaptic localization of B-50 phosphoprotein: the (ACTH)-sensitive protein kinase substrate involved in rat brain polyphosphoinositide metabolism. *Brain Res.*, 328: 381–385.

Guan, Y.-Y., Quastel, D.M.J. and Saint, D.A. (1988) Single Ca²⁺ entry and transmitter release systems at the neuromuscular synapse. *Synapse*, 2: 558–564.

Heemskerk, F.M.J., Schrama, L.H., Gianotti, C., Spierenburg, H., Versteeg, D.H.G., De Graan, P.N.E. and Gispen, W.H. (1989a) 4-Aminopyridine stimulates B-50/GAP-43 phosphorylation and [³H]-noradrenaline release in rat hippocampal slices. *J. Neurochem.*, 54: 863–869.

Heemskerk, F.M.J., Schrama, L.H. and Gispen, W.H. (1989b) Activation of protein kinase C by 4-aminopyridine dependent on Na⁺ channel activity in rat hippocampal slices. *Neurosci. Lett.* 106: 315–321.

Heemskerk, F.M.J., Schrama, L.H., De Graan, P.N.E. and Gispen, W.H. (1990) 4-Aminopyridine stimulates B-50 (GAP-43) phosphorylation in rat synaptosomes. *J. Mol. Neurosci.*, 2: 11–17.

Hemmings, H.C., Nairn, A.C., McGuinness, T.L., Huganir, R.L. and Greengard, P. (1989) Role of protein phosphorylation in neuronal signal transduction. *FASEB J.*, 3: 1583–1592.

Hidaka, H., Inagaki, M., Kawamoto, S. and Sasaki, Y. (1984) Isoquinolinesulfonamides, novel and potent inhibitors of cyclic nucleotide dependent protein kinase and protein kinase C. *Biochemistry*, 23: 5036–5041.

Howell, T.W. and Gomperts, B.D. (1987) Rat mast cells permeabilized with streptolysin O secrete histamine in response to Ca²⁺ at concentrations buffered in the micromolar range. *Biochim. Biophys. Acta*, 927: 177–183.

Howell, T.W., Cockcroft, S. and Gomperts, B.D. (1987) Essential synergy between Ca²⁺ and guanine nucleotides in exocytotic secretion from permeabilized rat mast cells. *J. Cell Biol.*, 105: 191–197.

Howell, T.W., Kramer, Y.M. and Gomperts, B.D. (1989) Protein phosphorylation and the dependence on Ca²⁺ and GTP-γ-S for exocytosis from permeabilized mast cells. *Cell Signalling*, 1: 157–163.

Karns, L.R., Ng, S.-C., Freeman, J.A. and Fishman, M.C. (1987) Cloning of complementary DNA for GAP-43, a neuronal growth-related protein. *Science*, 236: 597–600.

Kenigsberg, R.L. and Trifaro, J.M. (1985) Micro-injection of calmodulin antibodies into cultured chromaffin cells blocks catecholamine release in response to stimulation. *Neuroscience*, 14: 335–347.

206

Knight, D.E. and Scrutton, M.C. (1986) Gaining access to the cytosol: the technique and some applications of electropermeabilization. *Biochem. J.*, 234: 497–506.

Kristjansson, G.I., Zwiers, H., Oestreicher, A.B. and Gispen, W.H. (1982) Evidence that the synaptic phosphoprotein B-50 is localized exclusively in nerve tissue. *J. Neurochem.*, 39: 371–378.

Lee, S.A. and Holz, R.W. (1986) Protein phosphorylation and secretion in digitonin permeabilized adrenal chromaffin cells. *J. Biol. Chem.*, 261: 17089–17098.

Lemos, J.R. and Nowycky, M.C. (1989) Two types of calcium channels coexist in peptide-releasing vertebrate nerve terminals. *Neuron*, 2: 1419–1426.

Liu, Y. and Storm, D.R. (1989) Dephosphorylation of neuromodulin by calcineurin. *J. Biol. Chem.*, 264: 12800–12804.

Liu, Y. and Storm, D.R. (1990) Regulation of free calmodulin levels by neuromodulin: neuron growth and regeneration. *Trends Pharmacol. Sci.*, 11: 107–111.

Löffelholz, K. and Weide, W. (1982) Aminopyridines and the release of acetylcholine. *Trends Pharmacol. Sci.*, 4: 147–149.

Lundh, H. (1978) Effects of 4-aminopyridine on neuromuscular transmission. *Brain Res.*, 153: 307–318.

Martínez-Serrano, A., Bogónez, E., Vitórica, J. and Satrústegui, J. (1989) Reduction of K^+-stimulated $^{45}Ca^{2+}$ influx in synaptosomes with age involves inactivating and non-inactivating calcium channels and is correlated with temporal modifications in protein dephosphorylation. *J. Neurochem.*, 52: 576–584.

Matthies, H.J.G., Palfrey, H.C., Hirning, L.D. and Miller, R.J. (1987) Down regulation of protein kinase C in neuronal cells: effects on neurotransmitter release. *J. Neurosci.*, 7: 1198–1206.

Mazzei, G.J., Katoh, N. and Kuo, J.F. (1982) Polymyxin B is a more selective inhibitor for phospholipid-sensitive Ca^{2+}-dependent protein kinase than for calmodulin-sensitive Ca^{2+}-dependent protein kinase. *Biochem. Biophys. Res. Commun.*, 109: 1129–1133.

McMahon, H.T., Barrie, A.P., Lowe, M. and Nicholls, D.G. (1989) Glutamate release from guinea pig synaptosomes: stimulation by reuptake-induced depolarization. *J. Neurochem.*, 53: 71–79.

Miller, R.J. (1987) Multiple calcium channels and neuronal function. *Science*, 235: 46–52.

Moore, J.W., Blaustein, M.P, Andersson, N.C. and Narahshi, T. (1967) Basis of tetrodotoxin's selectivity in blockage of squid axons. *J. Gen. Physiol.*, 50: 1401–1411.

Nachshen, D.A. and Blaustein, M.P. (1980) Some properties of potassium-stimulated calcium influx in presynaptic nerve endings. *J. Gen. Physiol.*, 76: 709–728.

Nicholls, D.G., Tibbs, G. and Barrie, A.P. (1989) Cytosolic free calcium in synaptosomes and its coupling to glutamate exocytosis. *J. Neurochem.*, 52 Suppl.: 47D.

Nichols, R.A., Wu, W.C.S., Haycock, J.W. and Greengard, P. (1989) Introduction of impermeant molecules into synaptosomes using freeze/thaw permeabilization. *J. Neurochem.*, 52: 521–529.

Nichols, R.A., Sihra, T.S., Czernik, A.J., Nairn, A. and Greengard, P. (1990) Calcium/calmodulin-dependent protein kinase II increases glutamate and noradrenaline release from synaptosomes. *Nature*, 343: 647–651.

Nielander, H.B., Schrama, L.H., Van Rozen, A.J., Kasperaitis, M., Oestreicher, A.B., De Graan, P.N.E., Gispen, W.H. and Schotman, P. (1987) Primary structure of the neuron-specific phosphoprotein B-50 is identical to growth-associated protein GAP-43. *Neurosci. Res. Commun.*, 1: 163–172.

Nielander, H.B., Schrama, L.H., Van Rozen, A.J., Kasperaitis, M., Oestreicher, A.B., Gispen, W.H. and Schotman, P. (1990) Mutation of serine 41 in the neuron-specific protein B-50 (GAP-43) prohibits phosphorylation by protein kinase C. *J. Neurochem.*, 55: 1442–1445.

Nishizuka, Y. (1988) The molecular heterogeneity of protein kinase C and its implications for cellular regulation. *Nature*, 334: 661–665.

Okada, M., Mine, K., Iwasaki, K. and Fujiwara, M. (1989) Is the augmentation of K^+-evoked intrasynaptosomal Ca^{2+} concentration due to the influx of Ca^{2+} in rat brain synaptosomes? *J. Neurochem.*, 52: 1837–1842.

Pasantes-Morales, H. and Arzate, M.E. (1981) Effect of taurine on seizures induced by 4-aminopyridine. *J. Neurosci. Res.*, 6: 465–474.

Peppers, S.C. and Holz, R.W. (1986) Catecholamine secretion from digitonin-treated PC12 cells. *J. Biol. Chem.*, 261: 14665–14669.

Peterson, C. and Gibson, G.E. (1983) Aging and 3,4-diaminopyridine alter synaptosomal calcium uptake. *J. Biol. Chem.*, 258: 11482–11486.

Peterson, C. and Gibson, G.E. (1985) Synaptosomal calcium metabolism during hypoxia and 3,4-diaminopyridine treatment. *J. Neurochem.*, 42: 248–253.

Reynolds E.E., Melega W.P. and Howard B.D. (1982) Adenosine 5′-triphosphate independent secretion from PC12 pheochromocytoma cells. *Biochemistry*, 21: 4795–4799.

Rodnight, R. and Perrett, C. (1986) Protein phosphorylation and synaptic transmission: receptor-mediated modulation of protein kinase C in a rat brain fraction enriched in synaptosomes. *J. Physiol.*, 81: 340–348.

Rogawski, M.A. (1985) The A-current: how ubiquitous a feature of excitable cells is it? *Trends Neurosci.*, 5: 214–219.

Rogawski, M.A. and Barker, J.L. (1983) Effects of 4-aminopyridine on calcium action potentials and calcium current under voltage clamp in spinal neurons. *Brain Res.*, 280: 180–185.

Rosenthal, A., Chan, S.Y., Henzel, W., Haskell, C., Kuang, W.-J., Chen, E., Wilcox, J.N., Ullrich, A., Goeddel, D.V. and Routtenberg, A. (1987) Primary structure and mRNA localization of protein F1, a growth-related protein kinase C substrate associated with synaptic plasticity. *EMBO J.*, 6: 3641–3646.

Schrama, L.H., Heemskerk, F.M.J. and De Graan, P.N.E. (1989) Dephosphorylation of protein kinase C phosphorylated B-50/GAP-43 by the calmodulin-dependent phosphatase calcineurin. *Neurosci. Res. Commun.*, 5: 141–147.

Schweizer, F.E., Schäfer, T., Tapparelli, C., Grob, M., Karli, U.O., Heumann, R., Thoenen, H., Bookman, R.J. and Burger, M.M. (1989) Inhibition of exocytosis by intracellularly applied antibodies against a chromaffin granule-binding protein. *Nature*, 339: 709–712.

Segal, M., Rogawski, M.A. and Barker, J.L. (1984) A transient potassium conductance regulates the excitability of cultured hippocampal and spinal neurons. *J. Neurosci.*, 4: 604–609.

Smith, S.J. and Augustine, G.J. (1988) Calcium ions, active zones and synaptic transmitter release. *Trends Neurosci.*, 11: 458–464.

Sörensen, R.G., Kleine, L.P. and Mahler, H.R. (1981) Presynaptic localization of phosphoprotein B-50. *Brain Res. Bull.*, 7: 57–61.

Storm, J.F. (1988) Temporal integration by a slowly inactivating K^+ current in hippocampal neurons. *Nature*, 336: 379–381.

Suszkiw, J.B., Murawsky, M.M. and Shi, M. (1989) Further characterization of phasic calcium influx in rat cerebrocortical synaptosomes: inferences regarding calcium channel type(s) in nerve endings. *J. Neurochem.*, 52: 1260–1269.

Tapia, R. and Sitges, M. (1982) Effect of 4-aminopyridine on transmitter release in synaptosomes. *Brain Res.*, 250: 291–299.

Tapia, R., Sitges, M. and Morales, E. (1985) Mechanism of the calcium-dependent stimulation of transmitter release by 4-aminopyridine in synaptosomes. *Brain Res.*, 361: 373–382.

Thesleff, S. (1980) Aminopyridines and synaptic transmission. *Neuroscience*, 5: 1413–1419.

Tibbs, G.R, Dolly, J.O. and Nicholls, D.G. (1989a) Dendrotoxin, 4-aminopyridine and β-bungarotoxin act at common loci but by two distinct mechanisms to induce Ca^{2+}-dependent release of glutamate from guinea-pig cerebrocortical synaptosomes. *J. Neurochem.*, 52: 201–206.

Tibbs, G.R., Barrie, A.P., Van Mieghem, F.J.E., McMahon, H.T. and Nicholls, D.G. (1989b) Repetitive action potentials in isolated nerve terminals in the presence of 4-aminopyridine: effects on cytosolic free Ca^{2+} and glutamate release. *J. Neurochem.*, 53: 1693–1699.

Tsien, R.W., Lipscombe, D., Madison, D.V., Bley, K.R. and Fox, A.P. (1988) Multiple types of neuronal calcium channels and their selective modulation. *Trends Neurosci.*, 11: 431–438.

Turner, T.J. and Goldin, S.M. (1985) Calcium channels in rat brain synaptosomes: identification and pharmacological characterization. High affinity blockade by organic Ca^{2+} channel blockers. *J. Neurosci.*, 5: 841–849.

Ulbricht, W. (1974) Drugs to explore the ionic channels in the axon membrane. In L. Jaenicke (Ed.), *Biochemistry of Sensory Functions*, Springer Verlag, Berlin, pp. 351–365.

Van Lookeren Campagne, M., Oestreicher, A.B., Van Bergen en Henegouwen, P.M.P. and Gispen, W.H. (1989) Ultrastructural immunocytochemical localization of B-50/GAP-43, a protein kinase C substrate, in isolated presynaptic nerve terminals and neuronal growth cones. *J. Neurocytol.*, 18: 479–489.

Verhage, M., Besselsen, E., Lopes da Silva, F.H. and Ghijsen, W.E.J.M. (1988) Evaluation of the Ca^{2+} concentration in purified nerve terminals: relationship between Ca^{2+} homeostasis and synaptosomal preparation. *J. Neurochem.*, 51: 1667–1674.

Versteeg, D.H.G. and Florijn, W.J. (1987) Phorbol 12,13-dibutyrate enhances electrically stimulated neuromessenger release from rat dorsal hippocampal slices in vitro. *Life Sci.*, 40: 1237–1243.

Versteeg, D.H.G. and Ulenkate, H.J.L.M. (1987) Basal and electrically stimulated release of [3H]-noradrenaline and [3H]-dopamine from rat amygdala slices in vitro: effects of 4ß-phorbol 12,13-dibutyrate, 4α-phorbol 12,13-didecanoate and polymyxin B. *Brain Res.*, 416: 343–348.

Wakade, A.R., Malhotra, R.K. and Wakade, T.D. (1986) Phorbol ester, an activator of protein kinase C, enhances calcium-dependent release of sympathetic neurotransmitter. *Naunyn-Schmiedeberg's Arch. Pharmacol.*, 331: 122–124.

Wang, J.K.T., Walaas, S.I. and Greengard, P. (1988) Protein phosphorylation in nerve terminals: comparison of calcium/calmodulin-dependent and calmodulin/diacylglycerol-dependent systems. *J. Neurosci.*, 8: 281–288.

Wu, W.C-S., Walaas, S.I., Nairn, A.C. and Greengard, P. (1982) Calcium/phospholipid regulates phosphorylation of a Mr '87 k' substrate protein in rat brain synaptosomes. *Proc. Natl. Acad. Sci. USA*, 79: 5249–5253.

Zwiers, H., Schotman, P. and Gispen, W.H. (1980) Purification and some characteristics of an ACTH-sensitive protein kinase and its substrate protein in rat brain membranes. *J. Neurochem.*, 34, 1689–1699.

W.H. Gispen and A. Routtenberg (Eds.)
Progress in Brain Research, Vol. 89
© 1991 Elsevier Science Publishers B.V.

CHAPTER 14

Transmitter release: target of regulation by protein kinase C?

L.V. Dekker *, P.N.E. De Graan and W.H. Gispen

Division of Molecular Neurobiology, Rudolf Magnus Institute and Institute of Molecular Biology and Medical Biotechnology, University of Utrecht, Padualaan 8, 3584 CH Utrecht, The Netherlands

Introduction

Chemical neurotransmission involves the presynaptic release of transmitter substances that act on postsynaptic receptors. The most widely accepted hypothesis concerning the mechanism of transmitter release states that a depolarization-induced rise in intracellular Ca^{2+} triggers extrusion of prepackaged transmitter from vesicles localized in the nerve terminal via an exocytotic pathway (Augustine et al., 1987; Smith and Augustine, 1988; Thorn et al., 1988).

Our knowledge on the regulation of exocytosis during synaptic transmitter release is still very limited. The importance of $[Ca^{2+}]_i$ as initiator of release seems unquestioned. Elevation of the Ca^{2+} levels in the terminal occurs after depolarization of the membrane and subsequent influx of extracellular Ca^{2+} via voltage-sensitive Ca^{2+} channels. Probably in this way very high Ca^{2+} levels are generated at the very site of the influx. Once present in the synaptosol, the Ca^{2+} signal is further transmitted via a putative Ca^{2+} receptor molecule. After the stimulus, Ca^{2+} returns to its original level by processes such as binding to cytosolic Ca^{2+}-binding proteins (calmodulin, parvalbumin, calbindin and calcineurin), uptake in intracellular organelles and transport over the plasma membrane. In spite of the existence of these general views on the regulation of the intracellular free Ca^{2+} concentration, many details remain to be resolved, particularly in the nerve terminal. This aspect of stimulus-secretion coupling is largely out of scope of this paper (for reviews, see Augustine et al., 1987; Kaczmarek and Levitan, 1987; Blaustein, 1988; Thorn et al., 1988).

Only little information is available on the nature of the putative Ca^{2+} receptor or, more in general, on the molecular events occurring between the depolarization-induced elevation of the intracellular concentration of Ca^{2+} and the actual fusion of transmitter vesicles with the presynaptic plasma membrane. Ca^{2+}-binding proteins

---

* *Present address*: Sandoz, Institute for Medical Research, 5 Gower Place, WC1E 6BN London, U.K.
Correspondence: Dr. P.N.E. De Graan, Division of Molecular Neurobiology, Rudolf Magnus Institute, University of Utrecht, Padualaan 8, 3584 CH Utrecht, The Netherlands.

localized in the vesicle membrane, the cytoskeleton and the presynaptic plasma membrane, as well as the cytoskeletal elements themselves are thought to be involved in transducing the Ca^{2+} signal (Smith and Augustine, 1988). The functional state of these proteins may be regulated by processes such as proteolytic cleavage, phosphorylation and dephosphorylation (Augustine et al., 1987; Thorn et al., 1988).

Many investigators have emphasized the role of protein kinases in the molecular mechanism of neurotransmission (Nestler and Greengard, 1984; Gispen and Routtenberg, 1986; Kaczmarek and Levitan, 1987). It has been shown that phosphorylation of specific proteins in the brain is highly correlated with the process of transmitter release (see for instance, Dunkley and Robinson, 1986). Evidence for a role of protein kinases in neuronal signal transduction has also been obtained from studies involving the intracellular injection of purified protein kinases or specific protein kinase inhibitors into identified cells (Kandel and Schwartz, 1982; Llinas et al., 1985; Paupardin-Tritsch et al., 1986; Rodnight and Perrett, 1986; Hammond et al., 1987; Kaczmarek and Levitan, 1987; Hemmings et al., 1989). For transmitter release, Ca^{2+}-dependent protein kinases are of particular interest and several of these are present in the brain. They are roughly divided into Ca^{2+}/calmodulin-dependent and Ca^{2+}/phospholipid-dependent protein kinases (Nestler and Greengard, 1984).

In this paper we will evaluate the involvement of the Ca^{2+}/phospholipid-dependent protein kinase C (PKC) in transmitter release in the nervous system. We will mainly discuss pharmacological and biochemical evidence, indicating the involvement of PKC and its neuron-specific substrate B-50 (GAP-43) in the molecular mechanism of stimulus secretion coupling during synaptic transmitter release. Data obtained from experiments on secretion and exocytosis in non-neuro-nal preparations, for instance platelets or glandular tissue, will not be considered.

PKC

The first report on PKC by Nishizuka and coworkers (Inoue et al., 1977), describes a widely distributed, proteolytically activated protein kinase. Following the finding that PKC is activated by Ca^{2+} and phospholipids (Takai et al., 1979) and that diacylglycerol (DG) profoundly enhanced the Ca^{2+} sensitivity of the molecule (Kishimoto et al., 1980), a role for PKC in cellular signal transduction was postulated. Furthermore, it became clear that PKC is the major receptor for tumor-promoting phorbol esters which activate PKC in a way similar to DG (Castagna et al., 1982; Niedel et al., 1983; Leach et al., 1983).

Recent evidence suggests that PKC is not a single molecular entity but that it consists of a family of proteins which differ considerably both in sequence and in chromatographic behaviour (Coussens et al., 1986; Parker et al., 1986; Huang et al., 1986, 1987; Knopf et al., 1986; Ohno et al., 1987; Ono et al., 1987a; Kikkawa et al., 1987; Nishizuka, 1988; Parker et al., 1989). The various subtypes show a very heterogeneous tissue distribution. The α-subtype (also known as isozyme type III, identified by its chromatographic behaviour) has a universal distribution whereas the β-subtype (isozyme type II), further divided into β_I and β_{II}, is enriched in brain and spleen and the γ-subtype (isozyme type I) is exclusively present in the brain (Knopf et al., 1986; Ohno et al., 1987; Brandt et al., 1987; Huang et al., 1987; Yoshida et al., 1988). In addition four other subtypes have been identified (defined as ϵ-, ϵ'-, δ- and ζ-subtypes), which do not appear to be related to any of the chromatographically separated isozymes (Ono et al., 1987b, 1988, 1989; Nishizuka et al., 1988; Schaap et al., 1989, 1990; Parker et al., 1989).

A large number of studies have been devoted to the localization of PKC subtypes in the brain, both at the level of protein (using mono- and polyclonal antibodies) and at the level of mRNA. These in general revealed a heterogenous distribution for all different PKC subtypes (for review see Huang, 1989). A presynaptic localization of PKC, a prerequisite for a putative function of PKC in the (presynaptic) process of neurotransmitter release, has been shown by immunocytochemical and by biochemical procedures. Immuno electron microscopical evidence suggests the presence of the γ-subtype in the axonal terminal of cerebellar Purkinje cells and the presence of the β_{II}-subtype in nerve endings, terminating at dendrites and cell bodies of Purkinje cells in the cerebellum. Neither the γ-subtype nor the $\beta_{I,II}$-subtypes have been detected in presynaptic terminals of cerebral neocortex neurons (Ase et al., 1988; Hidaka et al., 1988; Kose et al., 1988; Tsujino et al., 1990).

Phosphorylation studies in various subcellular fractions have revealed an active PKC present in the synaptic terminal together with PKC substrates serving as phosphate acceptors (Dunkley and Robinson, 1986; Wang et al., 1988; see also section "Substrates of PKC in the brain" in this chapter). A large portion of brain PKC is associated with synaptic membranes, in contrast to most other tissues where the enzyme is present mainly in the soluble fraction in its inactive form (Kikkawa et al., 1982; Wolf et al., 1985; Yoshida et al., 1988; Lester, 1989). Interestingly, the brain specific isozyme I (or γ-subtype) is recovered predominantly in the particulate fraction of the brain and cannot be extracted with Ca^{2+} chelators (Yoshida et al., 1988; Lester, 1989), suggesting that it is tightly bound to the membrane. Furthermore, in vitro PKC exists in at least two different membrane-associated states, which differ in their activation by Ca^{2+} and phorbol esters (Burgoyne, 1989).

PKC activity has originally been defined as depending on Ca^{2+} and phospholipids (Takai et al., 1979). DG and phorbol esters enhance the enzyme's affinity for Ca^{2+} (Kishimoto et al., 1980; Castagna et al., 1982; Nishizuka, 1984, 1986, 1988; Huang et al., 1988b) and can thus render it fully active without a net increase in Ca^{2+} concentration. Thus, in vitro the activity of the enzyme is dependent on Ca^{2+}, but under some conditions physiologically independent of Ca^{2+}. The enzyme could also be activated by a synergistic action of an increase in Ca^{2+} and the generation of DG. In view of the apparent heterogeneity of PKC a reinvestigation of the kinetic properties of the enzyme subtypes has been performed. In the absence of phosphatidylserine (PS) and DG, or in the presence of only DG, the α-, β-, and γ-subtype do not show any activity, irrespective of the Ca^{2+} concentration (Nishizuka, 1988; Huang et al., 1988b; Shearman et al., 1989). In the presence of only PS, all three subtypes respond to Ca^{2+} at concentrations higher than 10 μM (Nishizuka, 1988; Shearman et al., 1989). However, in the presence of PS and DG the various subtypes respond differently to Ca^{2+}. β_{I}- and β_{II}-subtypes show substantial activity in the absence of Ca^{2+}, whereas the α- and γ-subtype require μM levels of Ca^{2+} for their activation in vitro (Nishizuka, 1988; Shearman et al., 1989). In contrast to these subtypes, the ϵ-subtype (artificially expressed in a baculovirus expression system) is independent of Ca^{2+} (Schaap and Parker, 1990). The ζ-subtype (expressed in COS cells) also has considerable activity in the absence of Ca^{2+} (Ono et al., 1989). In the presence of DG and/or PS this enzyme subtype is fully activated and it shows no Ca^{2+} dependency under any of these conditions (Ono et al., 1989). The differences in Ca^{2+} sensitivity of the α, β and γ PKC subtypes suggested above have not been found by Huang et al. (1988b) nor by Marais and Parker (1989). These investigators used the mixed micellar assay to study PKC activity in vitro. Finally the various PKC subtypes differ in their response to arachidonic acid (Sekigushi et al., 1987; Shearman et al., 1989).

The discovery of considerable molecular heterogeneity within the PKC family, coinciding with a difference in localization and activation of the various subtypes of PKC, creates the possibility of a very complex regulation of the enzyme. Furthermore, compartmentalization of PKC-activating compounds and PKC substrates could contribute to an increasing refinement of messages to be transmitted by PKC. This makes PKC an ideal candidate molecule for the multifactorial regulation of a complicated process such as the release of neurotransmitter from the presynaptic nerve terminal.

PKC and neurotransmitter release

The indications for a role of PKC in transmitter release are manifold. However, none of the individual studies seems to justify a conclusive statement on the way in which PKC is involved in this process. In the following section, the experimental data from the various approaches used to investigate the involvement of PKC in neurotransmitter release, have been summarized.

Studies using phorbol esters

Tumor-promoting phorbol esters have a molecular structure that in part is very similar to that of DG and activate PKC directly both in vitro and in vivo (Castagna et al., 1982; Niedel et al., 1983; Leach et al., 1983). Like DG, phorbol esters dramatically increase the affinity of the enzyme for Ca^{2+}, which results in the full activation of PKC at physiological Ca^{2+} concentrations (Nishizuka, 1984). Furthermore they induce translocation of PKC from the cytosolic compartment to the plasma membrane (Kraft and Anderson, 1983; Wolf et al., 1985). Phorbol esters are widely used to study the involvement of PKC in cellular processes. Modulation by phorbol esters of a physiological process is usually interpreted as an involvement of PKC in this process. Based on the stimulatory effect of phorbol esters, a role for PKC in stimulus-secretion coupling has been postulated in a variety of cell types (Kikkawa et al., 1986). Table I summarizes evidence for a role of PKC in transmitter release from nervous tissues obtained by the use of phorbol esters. From this evidence it is clear that the effect of phorbol esters on release is not restricted to a particular kind of transmitter nor to a specific brain area or neuronal preparation. Furthermore, in all the studies presented in Table I the effect of phorbol esters on release was stereo-specific and phorbol ester isomers that did not activate PKC in vitro had no effect on transmitter release. From this observation it is generally concluded that PKC and not, for instance, a change in membrane configuration is responsible for the action of phorbol esters in release. However, some caution must be used in interpreting the data. After physiological activation of PKC by DG, DG is rapidly degraded and thus only transiently present in membranes (Nishizuka, 1984). In contrast, phorbol esters are hardly degraded and remain in the membrane. A sustained presence of phorbol esters has been shown to induce down-regulation of PKC rather than activation (see for instance, Matthies et al., 1987 and references herein). Finally, the concentration of phorbol esters should be given special consideration when attempting to evaluate the exact contribution of PKC to the physiological process. Particularly at higher (μM) concentrations, PKC is not necessarily the sole target of the phorbol ester (Kikkawa et al., 1986).

The actual point of action of PKC in stimulating neurotransmitter release, is at present unknown. However, some possibilities can be excluded in advance. Phorbol esters do not seem to act on transmitter re-uptake system. It has been

TABLE I

Transmitter release from neuronal tissues stimulated by porbol esters

Brain area	Transmitter	Stimulus	Reference
Cerebral cortex	ACh	PDB, PMA	Nichols et al. (1987)
	CCK	PDB, PMA	Allard and Beinfeld (1988)
	DA	TPA	Zurgil and Zisapel (1985)
			Zurgil et al. (1986)
	Glutamate	PMA, PDD	Diaz-Guerra et al. (1988)
	5HT	PMA, PDB	Wang and Friedman (1987)
	NE	TPA	Shuntoh et al. (1988)
		PDB, PMA	Nichols et al. (1987)
Hippocampus	ACh	PDB	Versteeg and Florijn (1987)
			Allgaier et al. (1988)
			Daschmann et al. (1988)
	CCK	PMA, PDB	Allard and Beinfeld (1988)
	Glutamate	PDA	Malenka et al. (1987)
		OAG	Lynch et al. (1986)
	5HT	PDB	Versteeg and Florijn (1987)
			Daschmann et al. (1988)
		TPA, PDB	Feuerstein et al. (1987)
	NE	PDB	Versteeg and Florijn (1987)
			Allgaier and Hertting (1986)
			Allgaier et al. (1987)
			Huang et al. (1988a)
			Daschmann et al. (1988)
		TPA	Allgaier et al. (1986)
Caudate putamen	CCK	PMA, PDB	Allard and Beinfeld (1988)
Caudate nucleus	ACh	TPA	Tanaka et al. (1986)
	GABA	PDB	Bartmann et al. (1989)
Striatum	DA	TPA, PDB, PDA	Shu and Selmanoff (1988)
		TPA	Chandler and Leslie (1989)
	GABA	PDB	Weiss et al. (1989)
Median eminence	DA	TPA, PDB, PDA	Shu and Selmanoff (1988)
Amygdala	DA	PDB	Versteeg and Ulenkate (1987)
	NE	PDB	Versteeg and Ulenkate (1987)
Sympathetic neurons	NE	PDB, TPA	Malhotra et al. (1988)
Ileum nerve endings	ACh	TPA	Tanaka et al. (1984)
	GABA	TPA	Shuntoh et al. (1989)
PC12 cells	DA	TPA, OAG	Pozzan et al. (1984)
	NE	TPA, PDB	Matthies et al. (1987)
Neuromuscular Junction	ACh	TPA	Haimann et al. (1987)
			Eusebi et al. (1986)
			Publicover (1985)
			Shapira et al. (1987)
		PDB, OAG	Murphy and Smith (1987)
Arterie	NE	PMA	Balfagon et al. (1989)
Sinus node	NE	TPA	Shuntoh and Tanaka (1986)

OAG: 1,2-oleoylacetylglycerol; PDA: phorbol 12,13-diacetate; PDB: phorbol 12,13-dibutyrate; PDD: phorbol 12,13-didecanoate; PMA, TPA: 12-O-tetradecanoyl-phorbol-13-acetate.

shown that in the presence of re-uptake inhibitors for acetylcholine (ACh), noradrenaline (NA) (Allgaier et al., 1987, 1988), 5-hydroxytryptamine (5-HT) (Feuerstein et al., 1987) and γ-amino butyric acid (GABA) (Bartmann et al., 1989), phorbol esters still enhance the release of these transmitters. Moreover, as most of these experiments have been performed in a superfusion assay system, it is assumed that transmitter re-uptake makes no important contribution to the phenomena observed.

Phorbol esters could affect the release inhibitory system by modulation of auto-receptor activity. At least in the case of release inhibition by muscarinic (Versteeg and Florijn, 1987; Allgaier et al., 1988), serotonergic (Feuerstein et al., 1987; Wang and Friedman, 1987), noradrenergic (Allgaier and Hertting, 1986; Allgaier et al., 1986,1987; Versteeg and Florijn, 1987), and GABAergic (Limberger et al., 1986) autoreceptors this does not seem to happen. Occupation of these receptors by application of their agonists or antagonists results in modulation of release of the respective transmitters completely independent of the stimulatory effect of phorbol esters on release.

Phorbol esters could modulate release of transmitter by increasing Ca^{2+} levels in the cell and a number of studies have been performed to obtain more insight into this possibility. Several authors have reported that the stimulation of transmitter release by phorbol esters occurs only in the presence of extracellular Ca^{2+} (e.g. Wakade et al., 1986; Allgaier et al., 1987) and is blocked by Cd^{2+} (Huang et al., 1988a), whereas according to others even in the absence of extracellular Ca^{2+} (Murphy and Smith, 1987; Pozzan et al., 1984; Chandler and Leslie, 1989) or in the presence of Ca^{2+} channel blockers (Zurgil and Zisapel, 1985) phorbol esters have a stimulatory effect on release. It has been suggested that phorbol esters enhance influx of extracellular Ca^{2+}

and that this is in fact the trigger for release (Peterfreund and Vale, 1983; Wakade et al., 1986; Zurgil et al., 1986; Malhotra et al., 1988). Studies using $^{45}Ca^{2+}$ have shown that phorbol esters enhance the uptake of $^{45}Ca^{2+}$ in cultured chicken brain neurons (Zurgil et al., 1986; Malhotra et al., 1988). However, in rat striatal synaptosomes $^{45}Ca^{2+}$ uptake was not affected (Chandler and Leslie, 1989) nor was $[Ca^{2+}]_i$ (measured using fluorescent dyes; ibidem). In guinea pig cortical synaptosomes (Diaz-Guerra et al., 1988), in rat hippocampal synaptosomes (Ghijsen and Verhage, personal communication) and in PC12 cells (Pozzan et al., 1984) phorbol esters enhance transmitter release without inducing a rise in intracellular Ca^{2+} (both measured using fluorescent dyes). To add to the confusion, phorbol esters appear to have an inhibitory effect on $^{45}Ca^{2+}$ uptake in PC12 cells (Harris et al., 1986; DiVirgilio et al., 1986; Messing et al., 1986).

At the electrophysiological level modulation of Ca^{2+} channel activity by phorbol esters has been reported for chicken dorsal root ganglion cells and for hippocampal neurons (Rane and Dunlap, 1986; Malenka et al., 1986; Doerner et al., 1988). In both cases, however, the effect of phorbol ester treatment was inhibitory rather than stimulatory, which is in contrast with the hypothesis that phorbol esters enhance transmitter release by elevating the intracellular concentration of Ca^{2+}. Furthermore in various invertebrate systems, phorbol esters enhance activity of Ca^{2+} channels either by stimulation of the channel itself (DeRiemer et al., 1985; Farley and Auerbach, 1986) or by recruitment of previously inactive channels (Strong et al., 1987). The Ca^{2+} influx could be affected by PKC indirectly via modulation of the action potential, which in itself is determined by activation of various ion channels. A blockade of K^+ channels by phorbol esters, for instance, has been reported for numerous preparations and this could be responsible for a broadened action potential and hence enhanced release of transmitter (Baraban et al., 1985; Farley and Auerbach, 1986; Grega et al., 1987; Doerner et al., 1988). However, using ion channel blockers it has been demonstrated that K^+ channels are not mediating the effect of phorbol esters on transmitter release in rabbit hippocampal slices (Huang et al., 1988a).

The effects of phorbol esters on Ca^{2+} levels in the cell are hard to interpret and very much depend on the approaches that have been used. Electrophysiological studies reveal that Ca^{2+} channels can be modulated by phorbol esters but, at least in vertebrates, this is an inhibitory rather than a stimulatory effect and thus conceptually hard to reconcile with a phorbol ester-induced stimulation of transmitter release. Also from biochemical studies no single option emerges regarding the effects of phorbol esters on Ca^{2+} levels. However, in isolated nerve terminals, synaptosomes, Ca^{2+} levels do not seem to be affected by phorbol esters (Diaz-Guerra et al., 1988; Chandler and Leslie, 1989). This is an important observation as the synaptic nerve terminal is the relevant site of action of the stimulatory effect of phorbol esters on release.

In most studies, described in Table I, the effects of phorbol esters on transmitter release from polarized cells occur only at high concentrations (10^{-6} M) and are considerably smaller than those on transmitter release from depolarized cells (see for instance Versteeg and Ulenkate, 1987). This questions the assumption that PKC acts as a transducer of the initial Ca^{2+} signal towards the final release event. For, in that case phorbol esters, activating PKC, would bypass the Ca^{2+} signal and affect transmitter release in polarized tissues to the same degree as Ca^{2+} elevates transmitter release. It rather implies that activation of PKC is a modulatory event in transmitter release, potentiating the initial effect of Ca^{2+}.

In conclusion, studies using phorbol esters strongly indicate that PKC is involved in release of transmitter in the CNS. It is not clear at which level PKC is involved though it seems to affect the presynaptic release system neither by elevating $[Ca^{2+}]_i$, nor by affecting transmitter reuptake nor by modulation of presynaptic auto-inhibition.

Studies using PKC inhibitors

If transmitter release involves the activation of protein kinases, then inhibition of these by protein kinase inhibitors should attenuate release. Most widely used in transmitter release studies are the inhibitors polymyxin B, 1-(5-isoquinolinyl-sulfonyl)-2-methylpiperazine (H-7) and staurosporine. In vitro studies have shown that these compounds inhibit PKC via a different mechanism. Polymyxin B probably acts by competing with the binding of the kinase to PS (Mazzei et al., 1982). The inhibition of PKC by H-7 occurs at the catalytic site of the enzyme, by a competition with ATP (Hidaka et al., 1984; Kawamoto and Hidaka, 1984; Hidaka and Hagiwara, 1987). Staurosporine does not compete with activators of PKC such as PS, Ca^{2+}, ATP, DG or substrate (Tamaoki et al., 1986). It has been reported that staurosporine apart from inhibiting PKC has a stimulatory effect on the translocation of PKC (Wolf and Baggiolini, 1988).

The effects of H-7, staurosporine and polymyxin B on transmitter release from neuronal preparations are summarized in Table II. Some general conclusions can be drawn. In all studies the effect of phorbol esters on depolarization-evoked release is completely antagonized by the inhibitors indicating that the effects of the phorbol esters are truly PKC-mediated. On the other hand, the inhibition of depolarization-evoked release itself is less pronounced and seems to be different for the various inhibitors. Polymyxin B is the most reliable compound with respect to the inhibition of release. In most reports it was found to reduce the release almost to basal levels (Versteeg and Ulenkate, 1987; Allgaier and Hertting, 1986; Bartmann et al., 1989). The effects of H-7 and staurosporine on depolarization-evoked release are usually very small and in some cases even absent (Daschmann et al., 1988; Bartmann et al., 1989).

It should be emphasized that, in general, there is a lot of debate on the use of these inhibitors (Garland et al., 1987). A major issue in this respect is their usually poor specificity. For instance staurosporine though very effective in inhibiting PKC ($IC_{50} = 2.7$ nM; Tamaoki et al., 1986) inhibits cAMP-dependent protein kinase with the same efficiency (Tamaoki et al., 1986). H-7 inhibits PKC with an IC_{50} value of 6 μM (Hidaka et al., 1984). It affects myosin light chain kinase, cAMP- and cGMP-dependent protein kinases with an IC_{50} of approximately 97, 3.0 and 5.8 μM, respectively. Polymyxin B inhibits PKC with an IC_{50} value of 6–8 μM and myosin light chain kinase with an IC_{50} of 80–100 μM and has no effect on cyclic nucleotide-dependent kinases (Mazzei et al., 1982). The specificity of the PKC inhibitors as mentioned above is, in most cases determined in an in vitro assay system in which neither the substrate nor the PKC subtype, relevant to the in vivo situation, has been used. The existence of a neuron-specific subtype of PKC (see section "PKC" in this chapter) already indicates that such data should be carefully considered. Furthermore, it has been shown that in brain synaptosomal plasma membranes polymyxin B is not a specific inhibitor of PKC. In these membranes, calmodulin-dependent autophosphorylation of the 50 kDa subunit of calmodulin-dependent kinase II is much more sensitive to treatment with polymyxin B than the phosphorylation of the nervous tissue-specific PKC substrate B-50 (Dekker et al., 1990a).

The fact that various PKC inhibitors affect transmitter release differentially requires some special consideration. The different PKC subtypes show slight differences in biochemical properties (see section "PKC" in this chapter). It is possible that only a certain subtype(s) of PKC is (are) involved in the mechanism of release and that there is a difference in the sensitivity of the subtypes for the inhibitors used. The intracellular distribution of PKC (subtypes) could also determine the sensitivity of the enzyme for the various inhibitors. It has already been mentioned that PKC exists in a cytosolic as well as a membrane-bound configuration. The brain-specific γ-subtype seems to be tightly bound to the membrane and cannot be extracted with Ca^{2+} chelators (Yoshida et al., 1988; Lester, 1989). Moreover, in vitro PKC exists in at least two different membrane-associated states, which differ in their activation by Ca^{2+} and phorbol esters (Burgoyne, 1989). In keeping with these considerations, it is still a striking finding that all inhibitors very effectively antagonize phorbol ester-stimulated release and have differential effects on depolarization-induced release. This could imply that PKC is not as much involved in the stimulus secretion coupling chain itself but merely acts as a modulator of this chain after it has been activated by elevation of the levels of Ca^{2+} in the synaptic terminal.

An alternative approach to inhibit PKC has been applied by Matthies et al. (1987), who used PKC-depleted PC12 cells and sympathetic neurons to study the involvement of PKC in release. Loss of PKC activity was induced by long-term phorbol ester treatment which causes down regulation of PKC to undetectable levels (see references in Matthies et al., 1987). This treatment has no effect on the activity of other protein kinases (cAMP- and Ca^{2+}/calmodulin-dependent kinases). Concomitantly with the loss of PKC activity a reduction in the extent of depolarization-induced NA release was observed, but not a total loss of release, indicating that a PKC-dependent and a PKC-independent/Ca^{2+}-dependent release component is present in these cells.

Depolarization and translocation of PKC

There is general consensus in the literature that phorbol esters induce translocation of PKC from the cytosolic compartment to the membrane (Kraft and Anderson, 1983; Wolf et al., 1985). Experiments, trying to establish a translocation of PKC during transmitter release have provided very conflicting evidence. Diaz-Guerra et al. (1988) reported that a 10 min exposure of guinea pig synaptosomes to 30 mM KCl had no effect on subcellular distribution of PKC, while treatment with phorbol ester did. They concluded that translocation of PKC is not a prerequisite for Ca^{2+}-dependent transmitter release. Similarly, Zatz reported in two separate papers that KCl depolarization of hippocampal slices (Zatz, 1986) and of anterior pituitary tumor cells (Zatz et al., 1987) failed to induce translocation of PKC. Again phorbol ester treatment could induce translocation of PKC in these cells. Wakade et al. (1988) studied activation of PKC by high KCl treatment in chicken sympathetic neurons in culture. They reported that PKC activity in the membrane fraction, but not in the cytosolic fraction, is enhanced by KCl treatment and thus no physical translocation of the enzyme is occurring. They interpreted their results as a depolarization-induced activation of the pro-enzyme present in the membrane. This observation indicates that the intracellular topography of the enzyme could cause a difference in sensitivity to the depolarizing stimulus.

A small depolarization-induced translocation of cerebral cortex PKC from cytosol to membrane has been reported by Friedman and Wang (1989). The effect of depolarization on PKC translocation was dependent on the age of the rat; no depolarization-induced translocation was observed in 24-month-old rats (Friedman and Wang, 1989). K^+ depolarization or Ca^{2+} ionophore A23187 enhanced the amount of tightly bound chelator stable PKC in rat pinealocyte membranes by a mechanism which is dependent on extracellular Ca^{2+}, but not on generation of DG (Ho et al., 1988). It is suggested that a change in $[Ca^{2+}]_i$ is the major perturbation involved in translocation of PKC. However, the authors do not rule out the possibility that DG is involved in the translocation as basal levels could be high enough to support and/or stabilize the translocation of PKC caused by $[Ca^{2+}]_i$ elevating agents.

It is hard to obtain a conclusive picture from the data on the subcellular distribution of PKC after a depolarizing stimulus. This in contrast to the data on PKC translocation after phorbol ester treatment of the various tissues. The difference between the two stimuli gives rise to the notion that depolarization and phorbol esters affect PKC in a different way. However, some methodological reservations have to be made too. Measurement of PKC translocation usually involves a subcellular fractionation and an extraction step followed by the determination of PKC activity in the "subcellular extracts". Phorbol esters induce a relatively stable membrane translocation of PKC, which is not affected by any of these processing steps. As far as transmitter release is concerned, it is not clear whether the action of PKC in this process involves a change in the intracellular distribution of the enzyme. Furthermore, translocation of PKC during transmitter release, if occurring, is not necessarily as stable as phorbol ester-induced translocation and could get lost during the sample processing.

In fact, a subdivision in PKC translocating agents has been suggested, one acting through an increase in Ca^{2+} influx (for instance K^+ depolarization), the second depending on both Ca^{2+} and DG (receptor-operating agents such as NA and

ACh) and the third consisting of phorbol esters and DG, which cause a direct translocation of PKC (Ho et al., 1988). In addition, one could speculate that different subcellular species of PKC do not respond similarly to these stimuli and that, for instance, membrane-bound PKC is directly activated by a depolarizing stimulus whereas soluble PKC is not. If membrane-bound PKC is indeed the species which is relevant to the release process, then a translocation of PKC from the cytosol to the membrane can, of course, not be measured. Probably an immunocytochemical approach, using subtype-specific antibodies, could shed more light on this problem.

Depolarization-induced production of DG

A physiological activator of PKC is DG. DG can activate PKC at resting Ca^{2+} levels. PKC can also be activated by a synergistic action of both DG and Ca^{2+}. DG is produced after phosphodiesteratic cleavage of phosphoinositides in the cellular membrane (Berridge, 1987) leading to the concomitant formation of inositol phosphates. Generation of inositol phosphates thus indicates that breakdown of phosphoinositides occurs and that circumstances are created in which PKC activation may occur. There is general consensus that the binding of several ligands to their receptors finally results in the breakdown of phosphoinositides. In the nervous system, but also in other tissues, ACh, NA, 5-HT, histamine, glutamate and various peptides induce breakdown of phosphoinositides (Michell, 1975; Fisher and Agranoff, 1987). More specific for nervous tissue is the observation that depolarizing conditions, e.g. K^+ or veratridine, elicit generation of inositol phosphates (Pickard and Hawthorne, 1977; Griffin and Hawthorne, 1978; Hawthorne and Pickard, 1979; Fisher and Agranoff, 1987; Kendall and Nahorski, 1984; Gusovsky and Daly, 1988; Eberhard and Holz, 1988; Challiss et al., 1988; Audi-

ger et al., 1988). There is discussion concerning the mechanism that underlies this phenomenon. A depolarization-induced rise in Ca^{2+} could directly act on presynaptic phospholipase C or alternatively transmitters, released from the presynaptic terminal after depolarization, could, via a receptor-mediated event, activate postsynaptic phospholipase C. Furthermore, it is possible that glial phospholipase C is involved. In general, depolarization-induced inositol phosphate production cannot be blocked by antagonists of muscarinic, adrenergic, serotonergic or histaminergic receptors (Bone and Michell, 1985; Kendall and Nahorski, 1985, 1987). This suggests that a non-receptor-mediated presynaptic mechanism is involved. However, it is also possible that an unknown transmitter, for instance a peptide, induces the effect (Bone and Michell, 1985). It has been shown that K^+-induced inositol phosphate accumulation is enhanced by a voltage-sensitive Ca^{2+} channel activator and suppressed stereospecifically by dihydropyridine antagonists (Kendall and Nahorski, 1985; Rooney and Nahorski, 1986). Thus, there could be a direct action of Ca^{2+} on phosphoinositide-specific phospholipase C activity after entry through voltage-sensitive Ca^{2+} channels. A strong indication for presynaptic phospholipase C activity has been provided by Audiger et al. (1988) who showed that in rat brain synaptosomes, depolarization as well as muscarinic agonists could generate formation of inositol phosphates. These two stimuli differed with respect to the type of inositol phosphate that was generated but it is likely that in both cases DG is produced and PKC activation may occur.

In conclusion it seems that in the presynaptic nerve terminal, not only ligand-stimulated but also direct depolarization-stimulated inositol phosphate production occurs. This implies that depolarization will also initiate the concomitant production of DG, which could result in an activation of PKC under these conditions.

Substrates of PKC in the brain

A more close examination of the molecular events occurring after activation of PKC, the most important being the phosphorylation of its substrates, would increase our knowledge on the involvement of PKC in transmitter release. Various substrates of PKC exist in neuronal preparations determined on the basis of phorbol ester-stimulated phosphorylation. In cultured dissociated rat embryo neurons, phorbol esters enhance the phosphorylation of a 43 kDa protein (Burgess et al., 1986; Zurgil et al., 1986) as well as phosphorylation of a 47, a 75 and a 83 kDa protein (Burgess et al., 1986). Rodnight and Perrett (1986) showed that in synaptosomes, phosphorylation of a 45 and a 82 kDa protein are enhanced by phorbol ester treatment. Wang et al. (1988) reported phorbol ester-induced phosphorylation of 45 and 87 kDa proteins.

It is very likely that the PKC substrates, identified in these earlier studies are similar to the two well-defined proteins that are known as substrates of PKC in the brain: the 87 kDa protein (also known as MARCKS) and the B-50 protein (also known as GAP43; F1; pp46; neuromodulin). Both are potential candidates for mediating the effects of PKC in neurotransmitter release. In the next section the properties of these two proteins will be described in more detail.

The 87 kDa protein

The 87 kDa protein (also known as myristoylated alanine-rich C kinase substrate: MARCKS) is present in monkey, human, mouse and bovine brain tissue and in torpedo electric organs (Albert et al., 1986); mRNA has been found in rat, chicken and bovine brain, while the brains of other organisms have not been tested (Stumpo et al., 1989). The protein is not specific for the brain, as protein and mRNA have been identified in non-neuronal rat and bovine tissue with high

amounts in spleen and lung and lower amounts in testis, pancreas, kidney, adrenal and liver (Albert et al., 1986; Stumpo et al., 1989). In the brain it is enriched in certain brain regions; it is most easily detected in axons, axon terminals, dendrites of small diameter and dendritic spines. Furthermore it is present in distinct populations of glial cells (Quimet et al., 1990).

Based on stokes radius, sedimentation coefficient and partial specific volume, an extremely elongated molecular shape and a molecular weight of 68,000 was predicted (Albert et al., 1987). However, sequence information revealed a molecular weight of 31,000 or 32,000 (Stumpo et al., 1989). On sodiumdodecylsulfate (SDS) gels MARCKS has an apparent molecular weight of 87,000. The protein is extremely rich in alanine residues and relatively rich in glycine, proline and glutamate. Other than the amino-terminal methionine, there are no methionine residues (Albert et al., 1987; Stumpo et al., 1989). In broken cell preparations MARCKS has been shown to be a major substrate of PKC (Wu et al., 1982). In intact nerve terminals as well as in neuronal cell cultures phorbol esters enhance the phosphorylation of this protein (Rodnight and Perrett, 1986; Burgess et al., 1986; Wang et al., 1988). Depolarization of intact nerve terminals with either high potassium or veratridine has also been shown to enhance phosphorylation of MARCKS (Wu et al., 1982; Wang et al., 1988). A role for the phosphorylation of this protein in cellular physiology has not yet been established. Interestingly, phosphorylation of MARCKS is correlated with its translocation from the membrane to the cytosol (Wang et al., 1989). This occurs in isolated membrane fractions as well as in intact synaptosomes (Wang et al., 1989). Again, the meaning of this phenomenon is not known.

Despite the abundant characterization of this protein, its function remains to be elucidated. It has been suggested that it plays a role in transmitter release, but only results of correlative studies are available at the moment (Nichols et al., 1987).

B-50 (GAP-43)

The B-50 protein has been intensively investigated during the last ten years. Primary structure (Nielander et al., 1987) and immunological characteristics (Gispen et al., 1986a) of B-50 are identical to those of several proteins that have been independently investigated by others. Among these are protein F1 (Rosenthal et al., 1987), GAP-43 (Basi et al., 1987; Karns et al., 1987) and P-57/neuromodulin (Cimler et al., 1987). These proteins have been implicated in synaptic potentiation, neuronal growth and calmodulin-binding, respectively (see reviews by Benowitz and Routtenberg, 1987, and Skene, 1989). B-50 has an elongated molecular shape with a molecular weight of 25 700 (Masure et al., 1986). Sequence information revealed a molecular weight of 23 600 (Basi et al., 1987; Nielander et al., 1987). On SDS gels, the migration of the protein is anomalous, revealing a molecular weight of 43 000–57 000, depending on the gel concentration (Schrama et al., 1987). B-50 has an isoelectric point of 4.5 (Zwiers et al., 1980). The protein is rich in glycine, glutamate and aspartate residues and relatively rich in alanine and histidine (Zwiers et al., 1980; Basi et al., 1987; Karns et al., 1987; Nielander et al., 1987; Rosenthal et al., 1987). No hydrophobic domains are present, in spite of its predominant localization at synaptic membranes (Sörensen et al., 1981; Van Lookeren Campagne et al., 1989).

B-50 is a nervous tissue-specific protein that is widely present in the nerves of different species in the animal kingdom (Kristjansson et al., 1982; Oestreicher et al., 1984). In mature rat brain B-50 immunoreactivity is found in cerebral cortex, hippocampus and septum in high concentrations (Oestreicher et al., 1981, 1986; Oestreicher and

Gispen, 1986; Benowitz et al., 1988). B-50 has, in general, a distribution that coincides with regions of high synaptic density (Oestreicher et al., 1981; Oestreicher and Gispen, 1986). In the hippocampus B-50 is present in the neuropil region, whereas cell bodies of pyramidal and granule cells retain little immunostaining (Oestreicher et al., 1981). At the subsynaptic level the main source of B-50 immunoreactivity and of B-50 phosphorylation is the plasma membrane of the presynaptic nerve terminal (Sörensen et al., 1981; Gispen et al., 1985; Van Lookeren Campagne et al., 1989). Pre-embedding labelling studies showed that B-50 is associated with the inner leaflet of the plasma membrane (Gispen et al., 1985; Van Hooff et al., 1989b). Recently it has been reported that B-50 is anchored to the membrane by palmitate residues that are esterified to the two N-terminal cysteine residues (Skene and Virag, 1989; Zuber et al., 1989).

Extensive evidence exists indicating that B-50 is a substrate of PKC. Purified B-50 kinase and purified PKC show the same biochemical characteristics (Aloyo et al., 1983). Purified B-50 can be phosphorylated in vitro by PKC, but not by cAMP-dependent protein kinases (Aloyo et al., 1983). The only PKC phosphorylation site is serine residue 41 in the molecule (Coggins and Zwiers, 1989; Nielander et al., 1990). In synaptosomal plasma membranes and growth-cone membranes, B-50 is a major substrate of exogenously added PKC (De Graan et al., 1988a; Van Hooff et al., 1988). The phorbol esters 4β-phorbol 12,13-dibutyrate (PDB) and phorbol 12-myristate 13-acetate (PMA) increase B-50 phosphorylation in these membranes (Eichberg et al., 1986; De Graan et al., 1988a). Furthermore, phorbol esters enhance the degree of phosphorylation of B-50 in intact preparations like hippocampal slices, synaptosomes and growth cones (De Graan et al., 1988b, 1989; Dekker et al., 1989a, 1990a,b; Van Hooff et al., 1989a). The DG derivative DOG, which is considered as a more "natural" activator of PKC than phorbol esters, enhances B-50 phosphorylation in synaptosomes (Dekker et al., 1990b) and intact growth cones (Van Hooff et al., 1989a). Moreover, like activation of PKC in vitro, phorbol ester-induced B-50 phosphorylation in synaptosomes is stereospecific and can be inhibited by the PKC inhibitors polymyxin B, H-7 and staurosporine (Dekker et al., 1990a,b). In hippocampal slices, phorbol ester-induced B-50 phosphorylation is catalyzed at the peptide fragment which contains the serine 41 residue (Heemskerk et al., 1989). These results convincingly show that in synaptosomal plasma membranes as well as in intact nerve tissue preparations, B-50 phosphorylation is catalyzed by PKC.

B-50 phosphorylation and release

In view of the possible involvement of PKC in transmitter release and of the presynaptic localization of the PKC substrate B-50, we started a series of studies to investigate the relationship between B-50 phosphorylation and transmitter release.

Depolarization-induced B-50 phosphorylation

Stimulation of transmitter release from rat brain synaptosomes by high K^+, 4-aminopyridine or veratridine, induces a concomitant phosphorylation of B-50 (Fig. 1; Dekker et al., 1990a,b; Heemskerk et al., 1990). We more closely investigated K^+-induced phosphorylation of B-50. Like K^+-evoked release, K^+-evoked B-50 phosphorylation is dependent on extracellular Ca^{2+} (Dekker et al., 1989a, 1990b). Most likely both processes are activated by an influx of extracellular Ca^{2+} into the synaptic terminal. K^+-evoked B-50 phosphorylation depends on PKC activity as it is inhibited by polymyxin B, H-7 and staurosporine, 3 inhibitors of PKC (Dekker et al., 1989a, 1990a,b). In synaptosomes, K^+ and veratridine-induced

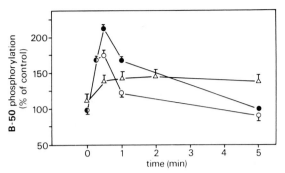

Fig. 1. Enhancement of B-50 phosphorylation by high potassium, veratridine and 4-aminopyridine. Cortical synaptosomes were treated with 20 mM K^+ (●), with 10 μM veratridine (○) or with 0.1 mM 4-aminopyridine (△) for the indicated period. B-50 phosphorylation was measured by immunoprecipitation. High potassium and veratridine induce a transient enhancement of B-50 phosphorylation. 4-Aminopyridine gradually enhances B-50 phosphorylation. For experimental details see Dekker et al. (1990a,b) and Heemskerk et al. (1990).

phosphorylation of B-50 is transient, that is, it is rapidly induced and returns to basal levels within 5 min (Dekker et al., 1990a,b). Such a time dependency profile has also been reported for the amount of transmitter released per unit of time from synaptosomes after depolarization (Cotman et al., 1976; Nicholls et al., 1987) and for the amount of Ca^{2+} entering the synaptosomes per unit of time (DiVirgilio et al., 1987), indicating that Ca^{2+} influx, PKC-mediated B-50 phosphorylation and transmitter release may be closely related.

Interference with anti-B-50 IgGs

The clear correlation that has been found between B-50 phosphorylation and transmitter release, did not answer the question to which extent B-50 phosphorylation is essential to this process. To address this issue, specific B-50 antibodies, which inhibit the phosphorylation of B-50, were tested on transmitter release. As B-50 is localized at the inner leaflet of the synaptic plasma membrane, exposed towards the cytosol (Gispen et al., 1985; Van Hooff et al., 1989b), the antibodies needed to be introduced into the synaptosomes. To introduce the antibodies, a streptolysin-O permeation technique was used, originally developed to study Ca^{2+}-induced vesicular histamine release from mast cells (Howell and Gomperts, 1987). The effect of K^+ depolarization on NA release can be mimicked in SL-O-permeated synaptosomes by increasing the Ca^{2+} concentration in the incubation buffer from 10^{-8} to 10^{-5} (Fig. 2; Dekker et al., 1989b). The effect of Ca^{2+} cannot be due to an efflux of cytosolic NA, since Ca^{2+} did not stimulate the SL-O-induced efflux of the cytosolic marker protein lactate dehydrogenase (LDH). This confirms that the Ca^{2+}-dependent part of SL-O-induced NA efflux is vesicular NA release.

Next, permeated synaptosomes were incubated with anti-B-50 IgGs concomitantly with the Ca^{2+} trigger and NA release and B-50 phosphorylation were measured. Ca^{2+}-induced release of NA was inhibited by the anti-B-50 IgGs in a concentration-dependent way (Fig. 2; Dekker et al., 1989b), as was B-50 phosphorylation (Fig. 3; Dekker et al., 1989b). Control IgGs and heat-inactivated anti-B-50 IgGs were ineffective. Neither control nor anti-B-50 IgGs affected NA efflux at 10^{-8} and 10^{-7} M Ca^{2+}, showing that the antibodies did not influence the Ca^{2+}-independent efflux. Surprisingly, we found that, in contrast to Ca^{2+}-induced NA release, phorbol ester-induced release was not inhibited by the anti-B-50 antibodies. Though at present we don't have a rigorous explanation for this, it may indicate that phorbol esters and Ca^{2+} have a different effect on transmitter release.

From these studies we conclude that B-50 plays a crucial role in the molecular mechanism of NA release from rat cerebral cortex synaptosomes. Since immunocytochemical studies indicate that B-50 does not co-localize with one particular type

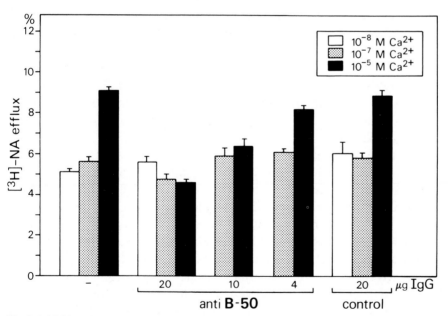

Fig. 2. Inhibition of NA release by antibodies to B-50. Cortical synaptosomes were permeated with streptolysin-O at 10^{-8}, 10^{-7} or 10^{-5} M Ca^{2+} in the presence of anti-B-50 or control IgGs. Anti-B-50 IgGs inhibit Ca^{2+}-dependent NA release in a concentration-dependent way whereas control IgGs are ineffective. For experimental details see Dekker et al. (1989b).

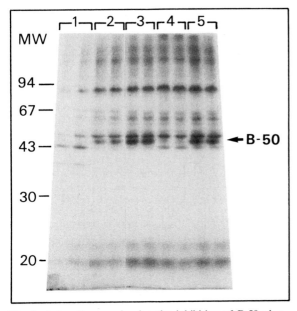

Fig. 3. Autoradiogram showing the inhibition of B-50 phosphorylation by antibodies to B-50. Cortical synaptosomes were treated with 0 (lanes 1), 0.4 (lanes 2) or 0.8 (lanes 3) IU/ml streptolysin-O before phosphorylation with radiolabelled ATP. The effect of anti B-50 IgGs (lanes 4) or control IgGs (lanes 5) was tested at 0.8 IU/ml streptolysin-O concentration. Positions of molecular weight markers are shown on the left. For experimental details see Dekker et al. (1989b)

of neurotransmitter (Gispen et al., 1985; Benowitz et al., 1988), B-50 may be more generally involved in transmitter release.

PKC, B-50 phosphorylation and release: conclusions

Several prerequisites have been fulfilled to propose an involvement of PKC in transmitter release in the central nervous system. First, PKC and PKC substrates are present in the presynaptic nerve terminal, the site where the actual release process takes place. At present the nature of the subtype(s) involved and its (their) subsynaptic localization are unknown. Second, activators of PKC, such as Ca^{2+} and probably also DG, are present in the synaptic terminal at the time

transmitter release occurs. It is not known whether Ca^{2+} and DG are both required during release or that for instance a rise in Ca^{2+} is sufficient to activate PKC. Similarly, it is unclear whether a translocation of PKC is involved in its activation during transmitter release. Thirdly, pharmacological stimulation of PKC by phorbol esters enhances depolarization-evoked transmitter release. This effect of phorbol esters could not be explained by an action on transmitter re-uptake, auto-inhibition or synaptosolic Ca^{2+} levels and indicates that PKC is involved in presynaptic stimulus-secretion coupling. Furthermore, down-regulation of PKC by long-term phorbol esters treatment has an inhibitory effect on release. Some PKC inhibitors also inhibit release, though from this line of research no conclusive evidence has been obtained. Finally, phosphorylation of various endogenous PKC substrates is correlated with the process of transmitter release. Both the B-50 protein and the 87 kDa protein are phosphorylated in response to stimuli which also affect transmitter release.

As for the B-50 protein, it has been established that blocking of its phosphorylation by specific antibodies, is accompanied by complete abolishment of NA release from permeated isolated nerve terminals, thus providing a strong indication for an involvement of B-50 phosphorylation in transmitter release. However, it is also possible that the B-50 antibodies block properties of B-50, other than its phosphorylation or in conjunction with its phosphorylation, which are of importance to transmitter release. In favor of a direct involvement of PKC-mediated phosphorylation of B-50 in transmitter release, is the following evidence: (i) phorbol esters that directly activate PKC and induce B-50 phosphorylation, enhance the release of a variety of neurotransmitters (see Table I), and this enhancement can be antagonized by the PKC inhibitors polymyxin B, staurosporine and H-7 (see Table II); (ii) both

TABLE II

Effects of PKC inhibitors on transmitter release evoked in the absence and in the presence of phorbol esters

Brain area	Trans-mitter	Stimulus	Inhibitor		Reference
Cerebral cortex	5HT	K	H-7	+	Wang and Friedman (1987)
		K + PDB	H-7	+ +	Wang and Friedman (1987)
	NA	K	PMB	−	Shuntoh et al. (1988)
			Stau	−	Shuntoh et al. (1988)
		K + TPA	PMB	+ +	Shuntoh et al. (1988)
			Stau	+ +	Shuntoh et al. (1988)
Hippocampus	Ach	K	Stau	+	Daschmann et al. (1988)
			H-7	+	Daschmann et al. (1988)
			PMB	+ +	Daschmann et al. (1988)
		El	Stau	−	Daschmann et al. (1988)
			PMB	+ +	Allgaier et al. (1988)
		K + PDB	Stau	+ +	Daschmann et al. (1988)
			H-7	+ +	Daschmann et al. (1988)
			PMB	+ +	Daschmann et al. (1988)
		El-PDB	Stau	+ +	Daschmann et al. (1988)
			PMB	+ +	Allgaier et al. (1988)
	5HT	K	Stau	−	Daschmann et al. (1988)
			PMB	+ +	Daschmann et al. (1988)
		El	Stau	−	Daschmann et al. (1988)
			PMB	+ +	Feuerstein et al. (1987)
		K + PDB	Stau	+ +	Daschmann et al. (1988)
			PMB	+ +	Daschmann et al. (1988)
		El + PDB	Stau	+ +	Daschmann et al. (1988)
			PMB	+ +	Feuerstein et al. (1987)
	Na	K	Stau	−	Daschmann et al. (1988)
			H-7	−	Daschmann et al. (1988)
			PMB	+ +	Daschmann et al. (1988)
		El	Stau	−	Daschmann et al. (1988)
			PMB	+ +	Allgaier et al. (1987)
		K + PDB	Stau	+ +	Daschmann et al. (1988)
			H-7	+ +	Daschmann et al. (1988)
			PMB	+ +	Daschmann et al. (1988)
		El + PDB	Stau	+ +	Daschmann et al. (1988)
			PMB	+ +	Allgaier et al. (1987)
Amygdala	DA	El	PMB	+ +	Versteeg and Ulenkate (1987)
		El + PDB	PMB	+ +	Versteeg and Ulenkate (1987)
	NA	El	PMB	+ +	Versteeg and Ulenkate (1987)
		El + PDB	PMB	+ +	Versteeg and Ulenkate (1987)
Striatum	ACh	K	PMB	+ +	Tanaka et al. (1986)
		K + TPA	PMB	+ +	Tanaka et al. (1986)
			H-7	+ +	Tanaka et al. (1986)
	DA	K	Sphi.	+ +	Shu and Selmanoff (1988)
		K + TPA	Sphi.	+ +	Shu and Selmanoff (1988)
Caudate nucleus	GABA	El	Stau	+	Bartmann et al. (1989)
		El + PDB	Stau	+ +	Bartmann et al. (1989)

K: Stimulation by high potassium; El: Electrical stimulation; tra: transmitter; PDB: phorbol 12,13-dibutyrate; TPA; 12-*O*-tetra-decanoyl-phorbol-13-acetate; H-7: 1-(5-isoquinolinesulfonyl)-2-methylpiperazine; PMB: polymyxin B; Sphi: sphingosine; Stau: staurosporine; − / + / + +: 0/0–20%/20–100% inhibition by the PKC inhibitor.

B-50 phosphorylation and neurotransmitter release can be enhanced by high K^+, veratridine and 4-aminopyridine (Fig. 1); (iii) the protein kinase inhibitor polymyxin B inhibits K^+-induced B-50 phosphorylation and K^+-induced neurotransmitter release and finally (iv) anti-B-50 IgGs inhibit B-50 phosphorylation as well as Ca^{2+}-dependent transmitter release (Figs. 2 and 3). However, in disagreement with a direct role for PKC-mediated B-50 phosphorylation in neurotransmitter release are the observations that (i) phorbol esters enhance B-50 phosphorylation but not transmitter release in polarized tissue; (ii) antibodies to B-50 only partially interfere with transmitter release induced by phorbol esters, which stimulate PKC-mediated phosphorylation of B-50; and (iii) the protein kinase inhibitors H-7 and staurosporine, both inhibiting B-50 phosphorylation, do not inhibit Ca^{2+}-dependent transmitter release.

In evaluating these data, it should be emphasized that the precise mechanisms of Ca^{2+} and phorbol ester-induced transmitter release remain to be elucidated. It may very well be that phorbol esters and Ca^{2+} affect neurotransmitter release in a different way, thus giving rise to the differential effects of inhibitors and antibodies on Ca^{2+} and phorbol ester-induced release. Alternatively, as suggested, the present results leave open the possibility that properties of B-50 other than its PKC-mediated phosphorylation are contributing to the transmitter release process and that these are affected by antibodies and inhibitors.

Such properties could involve the calmodulin-binding capacity of B-50. It has been shown that B-50 binds calmodulin in vitro under low Ca^{2+} conditions and releases calmodulin when Ca^{2+} is raised (Andreasen et al., 1983; Cimler et al., 1985; See chapter 13 by De Graan et al., this volume). The phosphorylation site and the calmodulin-binding site of B-50 are very close to each other (Alexander et al., 1988) and prephos-

phorylation of B-50 by PKC inhibits its property to bind calmodulin (Alexander et al., 1987). The data suggest that B-50 acts as a membrane-bound calmodulin store, releasing calmodulin after elevation of the Ca^{2+} levels or upon phosphorylation. A depolarization-induced rise of the intracellular concentration of Ca^{2+} in vivo would dissociate calmodulin and B-50. B-50, once liberated from calmodulin can be phosphorylated by PKC. PKC may be activated by Ca^{2+} entering the synaptic terminal after the depolarization. It is not clear whether such activation should involve the translocation of PKC from the cytosol to the membrane or whether Ca^{2+} directly activates membrane-bound PKC. Alternatively, Ca^{2+} may activate phospholipase C generating DG which could induce activation of PKC (Eberhard and Holz, 1988). PKC-mediated phosphorylation could prevent calmodulin from re-associating with B-50.

Phosphorylation of B-50 has also been reported to be a modulatory event in the activity of PIP kinase, the enzyme which converts the membrane lipid phosphatidylinositol 4-phosphate (PIP) to phosphatidylinositol 4,5-bisphosphate (PIP_2), (Gispen et al., 1986b). Specific blocking of B-50 phosphorylation in synaptosomal plasma membranes by adreno-corticotropic hormone$_{1-24}$ or by B-50 antibodies, induced a stimulation of PIP_2 production (Oestreicher et al., 1983). Enhancement of B-50 phosphorylation by addition of purified PKC to the synaptosomal plasma membranes, resulted in an inhibition of PIP kinase activity (De Graan et al., 1988a). Thus B-50 phosphorylation, modulating PIP kinase activity, could regulate the amount of PIP_2 which is available in the nerve terminal. The amount of PIP_2 seems to affect NA release in chromaffin cells probably by interfering with the cytoskeleton (Forscher, 1989).

Recently it has been suggested that B-50 could act as a modulator of G_o activity (Strittmatter et al., 1990). It is at present not clear whether phosphorylation of B-50 by PKC is of any relevance to its G_o modulating effects. Ample evidence exists suggesting the involvement of G_o in secretory processes in mast cells (Howell et al., 1987) and it would be very interesting to investigate whether the apparent necessity of B-50 to secretion in the nerve terminal is related to modulation of G_o activity.

In conclusion, there is ample evidence suggesting an involvement of PKC in stimulus-secretion coupling during synaptic transmitter release. In general this role seems to be modulatory rather than essential. An intriguing possibility is that PKC affects the ability of calmodulin-binding proteins to bind calmodulin, a property which in itself could be modulated by the concentration of intracellular Ca^{2+} in the synaptic terminal. This would imply that the molecular mechanism of synaptic transmitter release involves a coordinated activation of PKC-dependent, Ca^{2+}-dependent and calmodulin-dependent processes.

Acknowledgements

We thank Marina de Wit and Ank Frankhuyzen for their excellent technical assistance, Dr. A.B. Oestreicher for the anti-B-50 antibody and Dr. D.H.G. Versteeg for valuable discussions. This work was supported by grant 900-548-082 of the Netherlands Organization for Scientific Research (NWO).

References

Albert, K.A., Walaas, S.I., Wang, J.K.T. and Greengard, P. (1986) Wide-spread occurrence of "87 kDa", a major specific substrate of protein kinase C. *Proc. Natl. Acad. Sci. USA*, 83: 2822–2826.

Albert, K.A., Nairn, A.C. and Greengard, P. (1987) The 87-kDa protein, a major specific substrate for protein kinase C: purification from bovine brain and characterization. *Proc. Natl. Acad. Sci. USA*, 84: 7046–7050.

Alexander, K.A., Cimler, B.M., Meier, K.E. and Storm, D.R. (1987) Regulation of calmodulin binding to P-57. *J. Biol. Chem.*, 262: 6108–6113.

Alexander, K.A., Wakim, B.T., Doyle, G.S., Walsh, K.A. and Storm, D.R. (1988) Identification and characterization of the calmodulin-binding domain of neuromodulin, a neurospecific calmodulin-binding protein. *J. Biol. Chem.*, 263: 7544–7549.

Allard, L.R. and Beinfeld, M.C. (1988) Phorbol esters stimulate the potassium-induced release of cholecystokinin from slices of cerebral cortex, caudato-putamen and hippocampus incubated in vitro. *Biochem. Biophys. Res. Commun.*, 153: 372–376.

Allgaier, C., Von Kügelgen, O. and Hertting, G. (1986) Enhancement of noradrenaline release by 12-O-tetradecanoyl phorbol-13-acetate, an activator of protein kinase C. *Eur. J. Pharmacol.*, 129: 389–392.

Allgaier, C. and Hertting, G. (1986) Polymyxin B, a selective inhibitor of protein kinase C, diminishes the release of noradrenaline and the enhancement of release caused by phorbol 12,13-dibutyrate. *Naunyn-Schmiedeberg's Arch. Pharmacol.*, 334: 218–221.

Allgaier, C., Hertting, G., Huang, H.Y. and Jackisch, R. (1987) Protein kinase C activation and α_2-autoreceptor-modulated release of noradrenaline. *Br. J. Pharmacol.*, 92: 161–172.

Allgaier, C., Daschmann, B., Huang, H.Y. and Hertting, G. (1988) Protein kinase C and presynaptic modulation of acetylcholine release in rabbit hippocampus. *Br. J. Pharmacol.*, 93: 525–534.

Aloyo, V.J., Zwiers, H. and Gispen, W.H. (1983) Phosphorylation of B-50 protein by calcium-activated, phospholipid-dependent protein kinase and B-50 protein kinase. *J. Neurochem.*, 41: 649–653.

Andreasen, T.J., Luetje, C.W., Heideman, W. and Storm, D.R. (1983) Purification of a novel calmodulin binding protein from bovine cerebral cortex membranes. *Biochemistry*, 22: 4615–4618.

Ase, K., Saito, N., Shearman, M.S., Kikkawa, U., Ono, Y., Igarashi, K., Tanaka, C. and Nishizuka, Y. (1988) Distinct cellular expression of β_I- and β_{II}-subspecies of protein kinase C in rat cerebellum. *J. Neurosci.*, 8: 3850–3856.

Audiger, S.M.P., Wang, J.K.T. and Greengard, P. (1988) Membrane depolarization and carbamoylcholine stimulate phosphatidylinositol turnover in intact nerve terminals. *Proc. Natl. Acad. Sci. USA* 85: 2859–2863.

Augustine, G.J., Charlton, M.P. and Smith, S.J. (1987) Calcium action in synaptic transmitter release. *Annu. Rev. Neurosci.*, 10: 633–693.

Balfagon, G., De Sagarra, M.R., Barrus, M.T., Arrivas, S., Capilla M.I. and Marin, J. (1989) Effect of phorbol esters on noradrenaline release from cerebral arteries. *Brain Res.*, 477: 196–201.

Baraban, J.M., Snyder, S.H. and Alger, B.E. (1985) Protein

kinase C regulates ionic conductance in hippocampal pyramidal neurons: electrophysiological effects of phorbol esters. *Proc. Natl. Acad. Sci. USA*, 82: 2538–2542.

Bartmann, P., Jackisch, R., Hertting, G. and Allgaier, C. (1989) A role for protein kinase C in the electrically evoked release of [^{3}H]γ-aminobutyric acid in rabbit caudate nucleus. *Naunyn-Schmiedeberg's Arch. Pharmacol.*, 339: 302–305.

Basi, G.S., Jacobson, R.D., Virag, I., Schilling, J. and Skene, J.H.P. (1987) Primary structure and transcriptional regulation of GAP-43, a protein associated with nerve growth. *Cell*, 49: 785–791

Benowitz, L.I. and Routtenberg, A. (1987) A membrane phosphoprotein associated with neural development, axonal regeneration, phospholipid metabolism, and synaptic plasticity. *Trends Neurosci.*, 10: 527–532.

Benowitz, L.I., Apostolides, P.J., Perrone-Bizzozero, N.I., Finklestein, S.P. and Zwiers, H. (1988) Anatomical distribution of the growth-associated protein GAP-43/B-50 in the adult rat brain. *J. Neurosci.*, 8: 339–352.

Berridge, M.J. (1987) Inositol trisphosphate and diacylglycerol: two interacting second messengers. *Annu. Rev. Biochem.*, 56: 159–193.

Blaustein, M.P. (1988) Ca^{2+} buffering and transport in neurons. *Trends Neurosci.*, 11: 438–443.

Bone, E.A. and Michell, R.H. (1985) Accumulation of inositol phosphates in sympathetic ganglia. *Biochem. J.*, 227: 263–269.

Brandt, S., Niedel, J.E., Bell, R.M. and Young III W.S. (1987) Distinct patterns of expression of different protein kinase C mRNAs in rat tissues. *Cell*, 49: 57–63.

Burgess, S.K., Sanhyoun, N., Blanchard, S.G., LeVine III H., Chang K.J. and Cuatrecasas, P. (1986) Phorbol ester receptors and protein kinase C in primary neuronal cultures: development and stimulation of endogenous phosphorylation. *J. Cell Biol.*, 102: 312–319.

Burgoyne, R.D. (1989) A role for membrane-inserted protein kinase C in cellular memory. *Trends Biochem. Sci.*, 14: 87–88.

Castagna, M., Takai, Y., Kaibuchi, K., Sano, K., Kikkawa, V. and Nishizuka, Y. (1982) Direct activation of calcium-activated, phospholipid-dependent protein kinase by tumor-promoting phorbol esters. *J. Biol. Chem.*, 257: 7847–7851

Challiss, R.A.J., Batty, I.H. and Nahorski, S.R. (1988) Mass measurements of inositol(1,4,5)trisphosphate in rat cerebral cortex slices using a radioreceptor assay: effects of neurotransmitters and depolarization. *Biochem. Biophys. Res. Commun.*, 157: 684–691.

Chandler, L.J. and Leslie, S.W. (1989) Protein kinase C activation enhances K$^+$-stimulated endogenous dopamine release from rat striatal synaptosomes in the absence of an increase in cytosolic Ca^{2+}. *J. Neurochem.*, 52: 1905–1912.

Cimler, B.M., Andreasen, T.J., Andreasen, K.I. and Storm, D.R. (1985) P-57 is a neural specific calmodulin binding protein. *J. Biol. Chem.*, 260: 10784–10788.

Cimler, B.M., Giebelhaus, D.H., Wakim, B.T., Storm, D.R. and Moon, R.T. (1987) Characterization of murine cDNAs encoding P-57, a neural-specific calmodulin-binding protein. *J. Biol. Chem.*, 262, 12158–12163.

Coggins, P.J. and Zwiers, H. (1989) Evidence for a single protein kinase C-mediated phosphorylation site in rat brain protein B-50. *J. Neurochem.*, 53: 1895–1901.

Cotman, C.W., Haycock, J.W. and White, W.F. (1976) Stimulus-secretion coupling processes in brain: analysis of noradrenaline and gamma-aminobutyric acid. *J. Physiol. (London)*, 254: 475–505.

Coussens, L., Parker, P.J., Rhee, L., Yang-Feng, T.L., Chen, E., Waterfield, M.D., Francke, U. and Ullrich, A. (1986) Multiple, distinct forms of bovine and human protein kinase C suggests diversity in cellular signalling pathways. *Science*, 233: 859–866.

Daschmann, B., Allgaier, C., Nakov, R. and Hertting, G. (1988) Staurosporine counteracts the phorbol ester-induced enhancement of neurotransmitter release in hippocampus. *Arch. Int. Pharmacodyn. Ther.*, 296: 232–245.

De Graan, P.N.E., Dekker, L.V., De Wit, M., Schrama, L.H. and Gispen W.H. (1988a) Modulation of B-50 phosphorylation and polyphosphoinositide metabolism in synaptic plasma membranes by protein kinase C, phorbol diesters and ACTH. *J. Receptor Res.*, 8: 345–361.

De Graan, P.N.E., Heemskerk, F.M.J., Dekker, L.V., Melchers, B.P.C., Gianotti, C. and Schrama, L.H. (1988b) Phorbol esters induce long- and short-term enhancement of B-50/GAP-43 phosphorylation in hippocampal slices. *Neurosci. Res. Commun.*, 3: 175–182.

De Graan, P.N.E., Dekker, L.V., Oestreicher, A.B., Van Der Voorn, L. and Gispen, W.H. (1989) Determination of changes in the phosphorylation state of the neuron-specific protein kinase C substrate B-50 (GAP-43). *J. Neurochem.*, 52: 17–23.

Dekker, L.V., De Graan, P.N.E., Versteeg, D.H.G., Oestreicher, A.B. and Gispen, W.H. (1989a) Phosphorylation of B-50 (GAP-43) is correlated with neurotransmitter release in rat hippocampal slices. *J. Neurochem.*, 52: 24–30.

Dekker, L.V., De Graan, P.N.E., Oestreicher, A.B., Versteeg, D.H.G. and Gispen, W.H. (1989b) Inhibition of noradrenaline release by antibodies to B-50 (GAP-43). *Nature*, 342: 74–76.

Dekker, L.V., De Graan, P.N.E., Spierenburg, H., De Wit, M., Versteeg D.H.G. and Gispen, W.H. (1990a) Evidence for a relationship between B-50 (GAP-43) and [^{3}H]noradrenaline release in rat brain synaptosomes. *Eur. J. Pharmacol.*, 188: 113–122.

Dekker, L.V., De Graan, P.N.E., De Wit, M., Hens, J.J.H. and Gispen W.H. (1990b) Depolarization-induced phosphorylation of the protein kinase C substrate B-50 (GAP-43) in rat cortical synaptosomes. *J. Neurochem.*, 54: 1645–1652.

DeRiemer, S.A., Strong, J.A., Albert, K.A., Greengard, P. and Kaczmarek, L.K. (1985) Enhancement of calcium current in Aplysia neurons by phorbol ester and protein kinase C. *Nature*, 313: 313–316.

Diaz-Guerra, M.J.M., Sanchez-Prieto, J., Bosca, L., Pocock, J., Barrie, A. and Nicholls, D. (1988) Phorbol ester translocation of protein kinase C in guinea-pig synaptosomes and the potentiation of calcium-dependent glutamate release. *Biochim. Biophys. Acta*, 970: 157–165.

DiVirgilio, F., Pozzan, T., Wollheim, C.B., Vicentini, L.M. and Meldolesi, J. (1986) Tumour promoter phorbol myristate acetate inhibits Ca^{2+} influx through voltage-gated Ca^{2+} channels in two secretory cell lines, PC12 and RINm5F. *J. Biol. Chem.*, 261: 32–35.

DiVirgilio, F., Milani, D., Leon, A., Meldolesi, J. and Pozzan, T. (1987) Voltage-dependent activation and inactivation of calcium channels in PC12 cells. *J. Biol. Chem.*, 262: 9189–9195.

Doerner, D., Pitler, T.A. and Alger, B.E. (1988) Protein kinase C activators block specific calcium and potassium current components in isolated hippocampal neurons. *J. Neurosci.*, 8: 4069–4078.

Dunkley, P.R. and Robinson, P.J. (1986) Depolarization-dependent protein phosphorylation in synaptosomes: mechanisms and significance. *Progr. Brain Res.*, 69: 273–294.

Eberhard, D.A. and Holz, R.W. (1988) Intracellular Ca^{2+} activates phospholipase C. *Trends Neurosci.*, 11: 517–520.

Eichberg, J., De Graan, P.N.E., Schrama, L.H. and Gispen, W.H. (1986) Dioctanoylglycerol and phorbol diesters enhance phosphorylation of phosphoprotein B-50 in native synaptic plasma membranes. *Biochem. Biophys. Res. Commun.*, 136: 1007–1012.

Eusebi, F., Molinaro, M. and Caratsch, C.G. (1986) Effects of phorbol ester on spontaneous transmitter release at frog neuromuscular junction. *Pflügers Arch.*, 406: 181–183.

Farley, J. and Auerbach, S. (1986) Protein kinase C activation induces conductance changes in *Hermissenda* photoreceptors like those seen in associative learning. *Nature*, 319: 220–223.

Feuerstein, T.J., Allgaier, C. and Hertting, G. (1987) Possible involvement of protein kinase C (PKC) in the regulation of electrically evoked serotonin (5-HT) release from rabbit hippocampal slices. *Eur. J. Pharmacol.*, 139: 267–272.

Fisher, S.K. and Agranoff, B.W. (1987) Receptor activation and inositol lipid hydrolysis in neural tissues. *J. Neurochem.*, 48: 999–1017.

Forscher (1989) Calcium and polyphosphoinositide control of cytoskeletal dynamics. *Trends Neurosci.*, 12: 468–473.

Friedman, E. and Wang, H.-Y. (1989) Effect of age on brain cortical protein kinase C and its mediation of 5-hydroxytryptamine release. *J. Neurochem.*, 52: 187–192.

Garland, L.G., Bonser, R.W. and Thompson, N.T. (1987) Protein kinase C inhibitors are not selective. *Trends Pharmacol. Sci.*, 8: 334.

Gispen, W.H., Leunissen, J.L.M., Oestreicher, A.B., Verkleij, A.J. and Zwiers, H. (1985) Presynaptic localization of B-50 phosphoprotein: the (ACTH)-sensitive protein kinase substrate involved in rat brain polyphosphoinositide metabolism. *Brain Res.*, 328: 381–385.

Gispen, W.H. and Routtenberg, A. (Eds.) (1986) *Phosphoproteins in Neuronal Function,* Elsevier, Amsterdam.

Gispen, W.H., De Graan, P.N.E., Chan, S.Y. and Routtenberg, A. (1986a) Comparison between the neural acidic proteins B-50 and F1. *Progr. Brain Res.*, 69: 383–386.

Gispen, W.H., De Graan, P.N.E., Schrama, L.H. and Eichberg, J. (1986b) Phosphoprotein B-50 and polyphosphoinositide-dependent signal transduction in brain. In: Horrocks, L.A., Freysz, L. and Tofano, G. (Eds.), *Phospholipids in the Nervous System: Biochemical and Molecular Pharmacology*, pp. 31–41, Liviana Press, Padova.

Grega, D.S., Werz, M.A. and MacDonald, R.L. (1987) Forskolin and phorbol esters reduce the same potassium conductance of mouse neurons in culture. *Science,* 235: 345–348.

Griffin, H.D. and Hawthorne, J.N. (1978) Calcium-activated hydrolysis of phosphatidyl-*myo*-inositol 4-phosphate and phosphatidyl-*myo*-inositol 4,5-bisphosphate in guinea-pig synaptosomes. *Biochem. J.*, 176: 541–552.

Gusovsky, F. and Daly, J.W. (1988) Formation of second messengers in response to activation of ion channels in excitable cells. *Cell. Mol. Neurobiol.*, 8: 157–169.

Haimann, C., Meldolesi, J. and Ceccarelli, B. (1987) The phorbol ester, 12-O-tetradecanoyl-phorbol-13-acetate, enhances the evoked quanta release of acetylcholine at the frog neuromuscular junction. *Pflügers Arch.*, 408: 27–31.

Hammond, C., Paupardin-Tritsch, D., Nairn, A.C., Greengard, P. and Gerschenfeld, H.M. (1987) Cholecystokinin induces a decrease in Ca^{2+} current in snail neurons that appears to be mediated by protein kinase C. *Nature*, 325: 809–811.

Harris, K.M., Kongsamut, S. and Miller, R.J. (1986) Protein kinase C mediated regulation of calcium channels in PC-12 pheochromocytoma cells. *Biochem. Biophys. Res. Commun.*, 134: 1298–1305.

Hawthorne, J.N. and Pickard, M.R. (1979) Phospholipids in synaptic function. *J. Neurochem.*, 32: 5–14.

Heemskerk, F.M.J., Schrama, L.H. and Gispen, W.H. (1989) Activation of protein kinase C by 4-aminopyridine dependent on Na^+ channel activity in rat hippocampal slices. *Neurosci. Lett.*, 106: 315–321.

Heemskerk, F.M.J., Schrama, L.H., Gianotti, C., Spierenburg, H., Versteeg, D.H.G. and Gispen (1990) 4-Aminopyridine stimulates B-50 (GAP-43) phosphorylation and [^{3}H]noradrenaline release in rat hippocampal slices. *J. Neurochem.*, 54: 863–869.

Hemmings, Jr., H.C., Nairn, A.C., McGuinness, T.L., Huganir, R.L. and Greengard, P. (1989) Role of protein phosphorylation in neuronal signal transduction. *FASEB J.*, 3: 1583–1592.

Hidaka, H. and Hagiwara, M. (1987) Pharmacology of the isoquinoline sulfonamide protein kinase C inhibitors. *Trends Pharmacol. Sci.*, 8: 162–164.

Hidaka, H., Inagaki, M., Kawamoto, S. and Sasaki, Y. (1984) Isoquinolinesulfonamides, novel and potent inhibitors of cyclic nucleotide dependent protein kinase and protein kinase C. *Biochemistry*, 23: 5036–5041.

Hidaka, H., Tanaka, T., Onoda, K., Hagiwara, M., Watanabe, M., Ohta H., Ito, Y., Tsurudome, M. and Yoshida, Y. (1988) Cell type-specific expression of protein kinase C isozymes in the rabbit cerebellum. *J. Biol. Chem.*, 263: 4523–4526.

Ho, A.K., Thomas, T.P., Chik, C.L., Anderson, W.B. and Klein, D.C. (1988) Protein kinase C: subcellular redistribution by increased Ca^{2+} influx. *J. Biol. Chem.*, 263: 9292–9297.

Howell, T.W. and Gomperts, B.D. (1987) Rat mast cells permeabilized with streptolysin O secrete histamine in response to Ca^{2+} at concentrations buffered in micromolar range. *Biochim. Biophys. Acta*, 927: 177–183.

Howell, T.W., Cockcroft, S. and Gomperts, B.D. (1987) Essential synergy between Ca^{2+} and guanine nucleotides in exocytotic secretion from permeabilized rat mast cells. *J. Cell Biol.*, 105: 191–197.

Huang, K.P. (1989) The mechanism of protein kinase C activation. *Trends Neurol. Sci.*, 12: 425–432.

Huang, K.P., Nakabayashi, H. and Huang, F.L. (1986) Isozymic forms of rat brain Ca^{2+} activated and phospholipid-dependent protein kinase. *Proc. Natl. Acad. Sci. USA*, 83: 8535–8539.

Huang, F.L., Yoshida, Y., Nakabayashi, H., Knopf, J.L., Young, W.S. and Huang, K.P. (1987) Immunochemical identification of protein kinase C isozymes as products of discrete genes. *Biochem. Biophys. Res. Commun.*, 149: 946–952.

Huang, H.Y., Allgaier, C., Hertting, G. and Jackisch, R. (1988a) Phorbol ester-mediated enhancement of hippocampal noradrenaline release: which ion channels are involved? *Eur. J. Pharmacol.*, 153: 175–184.

Huang, K.P., Huang, F.L., Nakabayashi, H. and Yoshida, Y. (1988b) Biochemical characterization of rat brain protein kinase C isozymes. *J. Biol. Chem.*, 263: 14839–14845.

Inoue, M., Kishimoto, A., Takai, Y., and Nishizuka, Y. (1977) Studies on a cyclic nucleotide-independent protein kinase and its proenzyme in mammalian tissues. II. Proenzyme and its activation by calcium-dependent protease from rat brain. *J. Biol. Chem.*, 252: 2610–7616.

Kaczmarek, L.K. and Levitan, I. (Eds.) (1987) *Neuromodulation, the Biochemical Control of Neuronal Excitability*, Oxford Press, New York.

Kandel, E.R. and Schwartz, J.H. (1982) Molecular biology of learning: modulation of transmitter release. *Science*, 218: 433–442.

Karns, L.R., Ng, S.-C., Freeman, J.A. and Fishman, M.C. (1987) Cloning of complementary DNA for GAP-43, a neuronal growth-related protein. *Science*, 236: 597–600.

Kawamoto, S. and Hidaka, H. (1984) 1-(5-isoquinolinesulfonyl)-2-methylpiperazine (H-7) is a selective inhibitor of protein kinase C in rabbit platelets. *Biochem. Biophys. Res. Commun.*, 125: 258–264.

Kendall, D.A. and Nahorski, S.R. (1984) Inositol phospholipid hydrolysis in rat cerebral cortical slices. II. Calcium requirement *J. Neurochem.*, 42: 1388–1394.

Kendall, D.A. and Nahorski, S.R. (1985) Dihydropyridine calcium channel activators and antagonists influence depolarization-evoked inositol phospholipid hydrolysis in brain. *Eur. J. Pharmacol.*, 115: 31–36.

Kendall, D.A. and Nahorski, S.R. (1987) Depolarization-evoked release of acetylcholine can mediate phosphoinositide hydrolysis in slices of rat cerebral cortex. *Neuropharmacology*, 26: 513–519.

Kikkawa, U., Takai, Y., Minakuchi, R., Inohara, S. and Nishizuka, Y. (1982) Calcium-activated, phospholipid-dependent protein kinase from rat brain. Subcellular distribution, purification and properties. *J. Biol. Chem.*, 257: 13341–13348.

Kikkawa, U. and Nishizuka, Y. (1986) The role of protein kinase C in transmembrane signalling. *Annu. Rev. Cell Biol.*, 2: 149–178.

Kikkawa, U., Ono, Y., Ogita, K., Fujii, T., Asaoka, Y., Sekigushi, K., Kosaka, Y., Igarashi, K. and Nishizuka, Y. (1987) Identification of the structures of multiple subspecies of protein kinase C expressed in rat brain. *FEBS Lett.*, 217: 227–231.

Kishimoto, A., Takai, Y., Mori, T., Kikkawa, U., and Nishizuka, Y. (1980) Activation of calcium and phospholipid-dependent protein kinase by diacylglycerol, its possible relation to phosphatidylinositol turnover. *J. Biol. Chem.*, 255: 2273–2276.

Knopf, J.L., Lee, M.H., Sultzman, L.A., Kriz, R.W., Loomis, C.R., Hewick, R.M. and Bell, R.M. (1986) Cloning and expression of multiple protein kinase C cDNAs. *Cell*, 46: 491–502.

Kose, A., Saito, N., Ito, H., Kikkawa, U., Nishizuka, Y. and Tanaka, C. (1988) Electron microscopic localization of type I protein kinase C in rat purkinje cells. *J. Neurosci.*, 8: 4262–4268.

Kraft, A.S. and Anderson, W.B. (1983) Phorbol esters increase the amount of Ca^{2+}, phospholipid-dependent protein kinase associated with plasma membrane. *Nature*, 301: 621–623.

Kristjansson, G.I., Zwiers, H., Oestreicher, A.B. and Gispen, W.H. (1982) Evidence that the synaptic phosphoprotein

B-50 is localized exclusively in nerve tissue. *J. Neurochem.*, 39: 371–378.

Leach, K.L., James, M.L. and Blumberg, P.M. (1983) Characterization of a specific phorbol ester aporeceptor in mouse brain cytosol. *Proc. Natl. Acad. Sci. USA*, 80: 4208–4212.

Lester, D.S. (1989) High-pressure extraction of membrane-associated protein kinase C from rat brain. *J. Neurochem.*, 52: 1950–1953.

Limberger, N., Späth, L. and Starke, K. (1986) A search for receptors modulating the release of γ-[^{3}H]aminobutyric acid in rabbit caudate nucleus slices. *J. Neurochem.*, 46: 1109–1117.

Llinas, R., McGuiness, T.L., Leonard, C.S., Sugimori, M. and Greengard, P. (1985) Intraterminal injection of synapsin I or calcium/calmodulin-dependent protein kinase II alters neurotransmitter release at the squid giant synapse. *Proc. Natl. Acad. Sci. USA*, 82: 3035–3039.

Lynch, M.A. and Bliss, T.V.P. (1986) Long-term potentiation of synaptic transmission in the hippocampus of the rat; effect of calmodulin and oleoyl-acetyl-glycerol on release of [^{3}H]glutamate. *Neurosci. Lett.*, 65: 171–176.

Malenka, R.C., Madison, D.V., Andrade, R. and Nicoll, R.A. (1986) Phorbol esters mimic some cholinergic actions in hippocampal pyramidal neurons. *J. Neurosci.*, 6: 475–480.

Malenka, R.C., Ayoub, G.S. and Nicoll, R.A. (1987) Phorbol esters enhance transmitter release in rat hippocampal slices. *Brain Res.*, 403: 198–203.

Malhotra, R.K., Bhave, S.V., Wakade, T.D. and Wakade, A.R. (1988) Protein kinase C of sympathetic neuronal membrane is activated by phorbol ester-correlation between transmitter release, ^{45}Ca^{2+} uptake and the enzyme activity. *J. Neurochem.*, 51: 967–974.

Marais, R.M. and Parker, P.J. (1989) Purification and characterization of bovine brain protein kinase C isotypes α, β and γ. *Eur. J. Biochem.*, 182: 129–137.

Masure, H.R., Alexander, K.A., Wakim, B.T. and Storm, D.R. (1986) Physicochemical and hydrodynamic characterization of P-57, a neurospecific, calmodulin binding protein. *Biochemistry*, 25: 7553–7560.

Matthies, H.J.G., Palfrey, H.C., Hirning, L.D. and Miller, R.J. (1987) Down regulation of protein kinase C in neuronal cells: effects on neurotransmitter release. *J. Neurosci.*, 7: 1198–1206.

Mazzei, G., Katoh, N. and Kuo, J.F. (1982) Polymyxin B is a more selective inhibitor for phospholipid-sensitive Ca^{2+}-dependent protein kinase than for calmodulin-sensitive Ca^{2+}-dependent protein kinase. *Biochem. Biophys. Res. Commun.*, 109: 1129–1133.

Messing, R.O., Carpenter, C.L. and Greenberg, D.A. (1986) Inhibition of calcium flux and calcium channel antagonist binding in the PC12 neural cell line by phorbol esters and protein kinase C. *Biochem. Biophys. Res. Commun.*, 136: 1049–1056.

Michell, R.H. (1975) Inositol phospholipids and cell surface receptor function. *Biochim. Biophys. Acta*, 415: 81–147.

Murphy, R.L.W. and Smith, M.E. (1987) Effects of diacylglycerol and phorbol ester on acetylcholine release and action at the neuromuscular junction in mice. *Br. J. Pharmacol.*, 90: 327–334.

Nestler, E.J. and Greengard, P. (1984) *Protein Phosphorylation in the Nervous System*, Wiley, New York.

Nicholls, D.G., Sihra, T.S. and Sanchez-Prieto, J. (1987) Calcium-dependent and -independent release of glutamate from synaptosomes monitored by continuous fluorometry. *J. Neurochem.*, 49: 50–57.

Nichols, R.A., Haycock, J.W., Wang, J.K.T. and Greengard, P. (1987) Phorbol ester enhancement of neurotransmitter release from rat brain synaptosomes. *J. Neurochem.*, 48: 615–621.

Niedel, J.E., Kuhn, L.J., and Vandenbark, G.R. (1983) Phorbol diester receptor copurifies with protein kinase C. *Proc. Natl. Acad. Sci. USA*, 80: 36–40.

Nielander, H.B., Schrama, L.H., Van Rozen, A.J., Kasperaitis, M., Oestreicher, A.B., De Graan, P.N.E., Gispen, W.H. and Schotman, P. (1987) Primary structure of the neuron-specific phosphoprotein B-50 is identical to growth-associated protein GAP-43. *Neurosci. Res. Commun.*, 1: 163–172.

Nielander, H.B., Schrama, L.H., Van Rozen, A.J., Kasperaitis, M., Oestreicher, A.B., Gispen, W.H. and Schotman, P. (1990) Mutation of serine 41 in the neuron-specific protein B-50 (GAP-43) prohibits phosphorylation by protein kinase C. *J. Neurochem.*, 55: 1442–1445.

Nishizuka, Y. (1984) The role of protein kinase C in cell surface signal transduction and tumour promotion. *Nature*, 308: 693–698.

Nishizuka, Y. (1986) Studies and perspectives of protein kinase C. *Science*, 233: 305–312.

Nishizuka, Y. (1988) The molecular heterogeneity of protein kinase C and its implications for cellular regulation. *Nature*, 334: 661–665.

Oestreicher, A.B., Zwiers, H., Schotman, P. and Gispen, W.H. (1981) Immunohistochemical localization of a phosphoprotein (B-50) isolated from rat brain synaptosomal membranes. *Brain Res. Bull.*, 6: 145–153.

Oestreicher, A.B., Van Dongen, C.J., Zwiers, H. and Gispen, W.H. (1983) Affinity-purified anti-B-50 protein antibody: interference with the function of phosphoprotein B-50 in synaptic plasma membranes. *J. Neurochem.*, 41: 331–340.

Oestreicher, A.B., Van Duin, M., Zwiers, H. and Gispen, W.H. (1984) Cross-reaction of anti-rat B-50: characterization and isolation of a "B-50 phosphoprotein" from bovine brain. *J. Neurochem.*, 43: 935–943.

Oestreicher, A.B. and Gispen, W.H. (1986) Comparison of the immunocytochemical distribution of the phosphoprotein B-50 in the cerebellum and hippocampus of immature and adult rat brain. *Brain Res.*, 375: 267–279.

Oestreicher, A.B., Dekker, L.V. and Gispen, W.H. (1986) A radioimmunoassay for the phosphoprotein B-50: distribution in rat brain. *J. Neurochem.*, 46: 1366–1369.

Ohno, S., Kawasaki, H., Imajoh, S., Suzuli, K., Inagaki, M., Yokokura H., Sakoh, T. and Hidaka, H. (1987) Tissue-specific expression of three distinct types of rabbit protein kinase C. *Nature,* 325: 161–166.

Ono, Y., Kikkawa, U., Ogita, K., Fujii, T., Kurokawa, T., Asaoka, Y., Sekigushi, K., Ase, K., Igarashi, K. and Nishizuka, Y. (1987a) Expression and properties of two types of protein kinase C: alternative splicing from a single gene. *Science,* 236: 1116–1120.

Ono, Y., Fujii, T., Ogita, K., Kikkawa, U., Igarashi, K. and Nishizuka, Y. (1987b) Identification of three additional members of rat protein kinase C family: delta, epsilon and zeta subspecies. *FEBS Lett.*, 226: 125–128.

Ono, Y., Fujii, T., Ogita, K., Kikkawa, U., Igarashi, K. and Nishizuka, Y. (1988) The structure, expression and properties of additional members of the protein kinase C family. *J. Biol. Chem.*, 263: 6927–6932.

Ono, Y., Fujii, T., Ogita, K., Kikkawa, U., Igarashi, K. and Nishizuka, Y. (1989) Protein kinase C zeta subspecies from rat brain: its structure, expression and properties. *Proc. Natl. Acad. Sci. USA*, 86: 3099–3103.

Parker, P.J., Coussens, L., Totty, N., Rhee, L., Young, S., Chen, E., Stabel, S., Waterfield, M.D. and Ullrich, A. (1986) The complete primary structure of protein kinase C – the major phorbol ester receptor. *Science,* 233: 853–859.

Parker, P.J., Kour, G., Marais, R., Mitchell, F., Pears, C., Schaap D., Stabel, S. and Webster, C. (1989) Protein kinase C – a family affair. *Mol. Cell. Endocrinol.*, 65: 1–11.

Paupardin-Tritsch, D., Hammond, C., Gerschenfeld, H.M., Nairn, A.C. and Greengard, P. (1986) cGMP-dependent protein kinase enhances Ca^{2+} current and potentiates the serotonin-induced Ca^{2+} current increase in snail neurons. *Nature,* 323: 812–814.

Peterfreund, R.A. and Vale, W.W. (1983) Phorbol diesters stimulate somatostatin secretion from cultured brain cells. *Endocrinology*, 113: 200–208.

Pickard, M.R. and Hawthorne, J.N. (1977) The labelling of nerve ending phospholipids in guinea-pig brain *in vivo* and the effect of electrical stimulation in phosphatidylinositol metabolism in prelabeled synaptosomes. *J. Neurochem.*, 30: 145–155.

Pozzan, T., Gatti, G., Dozio, N., Vicentini, L.M. and Meldolesi, J. (1984) Ca^{++}-dependent and -independent release of neurotransmitters from PC12 cells: a role for protein kinase C activation. *J. Cell Biol.*, 99: 628–638.

Publicover, S.J. (1985) Stimulation of spontaneous transmitter release by the phorbol ester, 12-O-tetradecanoylphorbol-13-acetate, an activator of protein kinase C. *Brain Res.*, 333: 185–187.

Quimet, C.C., Wang, J.K.T., Walaas, S.I., Albert, K.A. and Greengard, P. (1990) Localization of the MARCKS (87 kDa) protein, a major specific substrate for protein kinase C, in rat brain. *J. Neurosci.*, 10: 1683–1698.

Rane, S.G. and Dunlap, K. (1986) Kinase C activator 1,2-oleoylacetylglycerol attenuates voltage-dependent calcium current in sensory neurons. *Proc. Natl. Acad. Sci. USA*, 83: 184–188.

Rodnight, R. and Perrett, C. (1986) Protein phosphorylation and synaptic transmission: receptor mediated modulation of protein kinase C in a rat brain fraction enriched in synaptosomes. *J. Physiol.*, 81: 340–348.

Rooney, T.A. and Nahorski, S.R. (1986) Regional characterization of agonist and depolarization-induced phosphoinositide hydrolysis in rat brain. *J. Pharmacol. Exp. Ther.*, 239: 873–880.

Rosenthal, A., Chan, S.Y., Henzel, W., Haskell, C., Kuang, W.-J., Chen E., Wilcox, J.N., Ullrich, A., Goeddel, D.V. and Routtenberg, A. (1987) Primary structure and mRNA localization of protein F1, a growth-related protein kinase C substrate associated with synaptic plasticity. *EMBO J.*, 6: 3641–3646.

Schaap, D., Parker, P.J., Bristol, A., Kriz, R. and Knopf, J. (1989) Unique substrate specificity and regulatory properties of PKC-epsilon: a rationale for diversity. *FEBS Lett.*, 243: 351–357.

Schaap, D. and Parker, P.J. (1990) Expression, purification and characterization of protein kinase C-epsilon. *J. Biol. Chem.*, 265: 7301–7307.

Schrama, L.H., De Graan, P.N.E., Oestreicher, A.B. and Gispen, W.H. (1987) B-50 phosphorylation, protein kinase C and the induction of excessive grooming behaviour in the rat. In: Ehrlich, Y.H., Lenox, R.H., Kornecki, E. and Berry, W.O. (Eds.), *Molecular Mechanisms of Neural Responsiveness*, pp. 393–408, Plenum Press, New York.

Sekigushi, K., Tsukuda, M., Ogita, K., Kikkawa, U. and Nishizuka, Y. (1987) Three distinct forms of rat brain protein kinase C: differential response to unsaturated fatty acids. *Biochem. Biophys. Res. Commun.*, 145: 797–802.

Shapira, R., Silberberg, S.D., Ginsburg, S. and Rahamimoff, R. (1987) Activation of protein kinase C augments evoked transmitter release. *Nature,* 325: 58–60.

Shearman, M.S., Noar, Z., Sekigushi, K., Kishimoto, A. and Nishizuka, Y. (1989) Selective activation of the γ-subspecies of protein kinase C from bovine cerebellum by arachidonic acid and its lipoxygenase metabolites. *FEBS Lett.*, 243: 177–182.

Shu, C. and Selmanoff, M. (1988) Phorbol esters potentiate rapid dopamine release from median eminence and striatal synaptosomes. *Endocrinology,* 122: 2699–2709.

Shuntoh, H. and Tanaka, C. (1986) Activation of protein kinase C potentiates norepinephrine release from sinus node. *Am. J. Physiol.*, 251: C833-C840.

Shuntoh, H., Taniyama, K., Fukuzaki, H. and Tanaka, C.

232

(1988) Inhibition by cAMP of phorbol ester-potentiated norepinephrine release from guinea pig brain cortical synaptosomes. *J. Neurochem.*, 52: 1565–1572.

Shuntoh, H., Taniyama, K. and Tanaka, C. (1989) Involvement of protein kinase C in the Ca^{2+}-dependent vesicular release of GABA from central and enteric neurons of the guinea pig. *Brain, Res.*, 483: 384–388.

Skene, J.H.P. (1989) Axonal growth-associated proteins. *Annu. Rev. Neurosci.*, 12: 127–156.

Skene, J.H.P. and Virag, I. (1989) Posttranslational membrane attachment and dynamic fatty acylation of a neuronal growth cone protein, GAP-43. *J. Cell Biol.*, 108: 613–624.

Smith, S.J. and Augustine, G.J. (1988) Calcium ions, active zones and synaptic transmitter release. *Trends Neurosci.*, 11: 458–464.

Sörensen, R.G., Kleine, L.P. and Mahler, H.R. (1981) Presynaptic localization of phosphoprotein B-50. *Brain Res. Bull.*, 7: 57–61.

Strittmatter, S.M., Valenzuela, D., Kennedy, T.E., Neer, E.J. and Fishman, M.C. (1990) G_o is a major growth cone protein subject to regulation by GAP-43. *Nature,* 344: 836–841.

Strong, J.A., Fox, A.P., Tsien, R.W. and Kaczmarek, L.K. (1987) Stimulation of protein kinase C recruits covert calcium channels in *Aplysia* bag cell neurons. *Nature,* 325: 714–717.

Stumpo, D.J., Graff, J.M., Albert, K.A., Greengard, P. and Blackshear P.J. (1989) Molecular cloning, characterization, and expression of a cDNA encoding the "80–87 kDa" myristoylated alanine-rich C kinase substrate: A major cellular substrate for protein kinase C. *Proc. Natl. Acad. Sci. USA*, 86: 4012–4016.

Takai, Y., Kishimoto, A., Iwasa, Y., Kawahara, Y., Mori, T. and Nishizuka, Y. (1979) Calcium-dependent activation of a multifunctional protein kinase by membrane phospholipids. *J. Biol. Chem.*, 254: 3692–3695.

Tamaoki, T., Nomoto, N., Takahashi, I., Kato, Y., Morimoto, M. and Tomita, P. (1986) Staurosporine, a potent inhibitor of phospholipid/Ca^{2+}-dependent protein kinase. *Biochem. Biophys. Res. Commun.*, 135: 397–402.

Tanaka, C., Taniyama, K. and Kusunoki, M. (1984) A phorbol ester and A23187 act synergistically to release acetylcholine from the guinea pig ileum. *FEBS Lett.*, 175: 165–169.

Tanaka, C., Fujiwara, H. and Fujii, Y. (1986) Acetylcholine release from guinea pig caudate slices evoked by phorbol esters and calcium. *FEBS Lett.*, 195, 129–134.

Thorn, N.A., Treiman, M. and Petersen, O.H. (Eds.) (1988) *Molecular Mechanisms in Secretion*, Munksgaard, Copenhagen.

Tsujino, T., Kose, A., Saito, N. and Tanaka, C. (1990) Light and electron microscopic localization of β_I-, β_{II}-, and γ-subspecies of protein kinase C in rat cerebral neocortex. *J. Neurosci.*, 10: 870–884.

Van Hooff, C.O.M., De Graan, P.N.E., Oestreicher, A.B. and

Gispen W.H. (1988) B-50 phosphorylation and polyphosphoinositide metabolism in nerve growth cone membranes. *J. Neurosci.*, 8: 1789–1795.

Van Hooff, C.O.M., De Graan, P.N.E., Oestreicher, A.B. and Gispen W.H. (1989a) Muscarinic receptor activation stimulates B-50 (GAP-43) phosphorylation in isolated growth cones. *J. Neurosci.*, 9: 3753–3759.

Van Hooff, C.O.M., Holthuis, J.C.M., Oestreicher, A.B., Boonstra, J., De Graan, P.N.E. and Gispen, W.H. (1989b) Nerve growth factor-induced changes in the intracellular localization of the protein kinase C substrate B-50 in pheochromocytoma PC12 cells. *J. Cell Biol.*, 108: 1115–1125.

Van Lookeren Campagne, M., Oestreicher, A.B., Van Bergen en Henegouwen, P.M.P. and Gispen, W.H. (1989) Ultrastructural immunocytochemical localization of B-50/GAP43, a protein kinase C substrate, in isolated presynaptic nerve terminals and neuronal growth cones. *J. Neurocytol.*, 18: 479–489.

Versteeg, D.H.G. and Florijn, W.J. (1987) Phorbol 12,13-dibutyrate enhances electrically stimulated neuromessenger release from rat dorsal hippocampal slices *in vitro*. *Life Sci.*, 40: 1237–1243.

Versteeg, D.H.G. and Ulenkate, H.J.L.M. (1987) Basal and electrically stimulated release of [^{3}H]noradrenaline and [^{3}H]dopamine from rat amygdala slices *in vitro*: effects of 4β-phorbol 12,13-dibutyrate, 4α-phorbol 12,13-didecanoate and polymyxin B. *Brain Res.*, 416: 343–348.

Wakade, A.R., Malhotra, R.K. and Wakade, T.D. (1986) Phorbol ester, an activator of protein kinase C, enhances calcium-dependent release of sympathetic neurotransmitter. *Naunyn-Schmiedeberg's Arch. Pharmacol.*, 331: 122–124.

Wakade, A.R., Wakade, T.D., Malhotra, R.K. and Bhave, S.V. (1988) Excess K^+ and phorbol ester activate protein kinase C and support the survival of chick sympathetic neurons in culture. *J. Neurochem.*, 51: 975–983.

Wang, H.Y. and Friedman, E. (1987) Protein kinase C: regulation of serotonin release from rat brain cortical slices. *Eur. J. Pharmacol.*, 141: 15–21.

Wang, J.K.T., Walaas, S.I. and Greengard, P. (1988) Protein phosphorylation in nerve terminals: comparison of calcium/calmodulin-dependent and calmodulin/diacylglycerol-dependent systems. *J. Neurosci.*, 8: 281–288.

Wang, J.K.T., Walaas, S.I., Sihra, T.S., Aderem, A., and Greengard, P. (1989) Phosphorylation and associated translocation of the 87-kDa protein, a major protein kinase C substrate, in isolated nerve terminals. *Proc. Natl. Acad. Sci. USA*, 80: 2253–2256.

Weiss, S., Ellis, J., Hendley, D.D. and Lenox, R.H. (1989) Translocation and activation of protein kinase C in striatal neurons in primary culture: relationship to phorbol dibutyrate actions on the inositol phosphate generating system and neurotransmitter release. *J. Neurochem.*, 52: 530–536.

Wolf, M., LeVine, III, H., May, Jr., W.S., Cuatrecasas, P. and

Sahyoun, N. (1985) A model for intracellular translocation of protein kinase C involving synergism between Ca^{2+} and phorbol esters. *Nature,* 317: 546–549.

Wolf, M. and Baggiolini, M. (1988) The protein kinase C inhibitor staurosporine, like phorbol esters, induces the association of protein kinase C with membranes. *Biochem. Biophys. Res. Commun.,* 154: 1273–1279.

Wu, W.C.-S., Walaas, S.I., Nairn, A.C. and Greengard, P. (1982) Calcium/phospholipid regulates phosphorylation of a M_r "87k" substrate protein in brain synaptosomes. *Proc. Natl. Acad. Sci. USA*, 79: 5249–5253.

Yoshida, Y., Huang, F.L., Nakabayashi, H. and Huang, K.P. (1988) Tissue distribution and development expression of protein kinase C isozymes. *J. Biol. Chem.,* 263: 9868–9873.

Zatz, M. (1986) Translocation of protein kinase C in rat hippocampal slices. *Brain Res.,* 385: 174–178.

Zatz, M., Mahan, L.C. and Reisine, T. (1987) Translocation of protein kinase C in anterior pituitary tumor cells. *J. Neurochem.,* 48: 106–110.

Zuber, M.X., Strittmatter, S. and Fishman, M.C. (1989) A membrane-targeting signal in the amino terminus of the neuronal protein GAP-43. *Nature*, 341: 345–348.

Zurgil, N. and Zisapel, N. (1985) Phorbol esters and calcium act synergistically to enhance neurotransmitter release by brain neurons in culture. *FEBS Lett.,* 185: 257–261.

Zurgil, N., Yarom, M. and Zisapel, N. (1986) Concerted enhancement of calcium influx, neurotransmitter release and protein phosphorylation by a phorbol ester in cultured brain neurons. *Neuroscience,* 19: 1255–1264.

Zwiers, H., Schotman, P. and Gispen, W.H. (1980) Purification and some characteristics of an ACTH-sensitive protein kinase and its substrate protein in rat brain membranes. *J. Neurochem.,* 34: 1689–1699.

W.H. Gispen and A. Routtenberg (Eds.)
Progress in Brain Research, Vol. 89
© 1991 Elsevier Science Publishers B.V.

CHAPTER 15

Activation of protein kinase C phosphorylation pathways: a role for storage of associative memory

David S. Lester and Daniel L. Alkon

*Section of Neural Systems, Laboratory of Molecular and Cellular Neurobiology, National Institutes of Health,
Bethesda, MD 20892, U.S.A.*

1. Introduction

Current evidence suggests that the mechanisms underlying learning and memory involve changes at many different levels of cellular function. These include the sensory, integrative and motor command operations of neuronal networks, the flux of ions through channels in neuronal membranes, the molecular regulation of that flux, intraneuronal transport of subcellular organelles, and neuronal architecture. Existing molecular constituents, such as lipids and proteins would be expected to undergo short-term modifications, lasting longer than the relatively rapid mechanisms of signal transduction and the synaptic transmission of these signals to the central nervous system (CNS). Longer-term modifications would be necessary for permanent storage and retrieval. To study neuronal changes which store

Correspondence: Dr. D.L. Alkon, Section of Neural Systems, Laboratory of Molecular and Cellular Neurobiology, NIH, NINDS, Park Building, Room 435, Bethesda, MD 20892, U.S.A.

associative memory, Pavlovian conditioning was demonstrated in a snail preparation where it was possible to precisely trace the flow of information through well-defined neuronal networks which mediate the learning. This animal can be taught to associate a light stimulus (conditioned stimulus) with vestibular stimulation (unconditioned stimulus) (Alkon, 1988). The excitable state in identified photoreceptors of the eye is elevated following conditioning due to persistent changes in the molecular regulation of specific potassium currents for 14 days or longer (Alkon et al., 1985). In a second system, rabbits are conditioned to associate an auditory tone (conditioned stimulus) with a puff of air to the surface of the eye (unconditioned stimulus) causing the nictitating membrane of the eye to extend on subsequent days in response to the tone alone (Alkon, 1989). Memory-specific changes in ion channel properties and their molecular basis, similar to those observed in *Hermissenda*, have been measured in the CA1 pyramidal cells of the rabbit hippocampus. The parallels between mechanisms of memory storage in these and other model preparations suggest conserved molecular mechanisms of memory storage.

236

2. Protein phosphorylation mechanisms

Changes in phosphorylation profiles in eyes of associatively trained *Hermissenda* were first observed in 1981 (Neary et al., 1981). A number of pharmacological agents that produce K$^+$ current changes similar to those found in trained animals also caused changes in phosphorylation patterns of eyes (Neary et al., 1981, 1986). It was of interest to determine which of the second mes-

senger-dependent protein kinases was responsible for phosphorylation of these substrates. The type II Ca^{2+}/calmodulin-dependent protein kinase induced a reduction in K$^+$ currents which could be prevented by addition of calmodulin inhibitors (Sakikibara et al., 1986b) (Fig. 1C). However, these inhibitors are known to also inhibit protein kinase C largely via their effect on the membrane lipid bilayer (Epand and Lester, 1990).

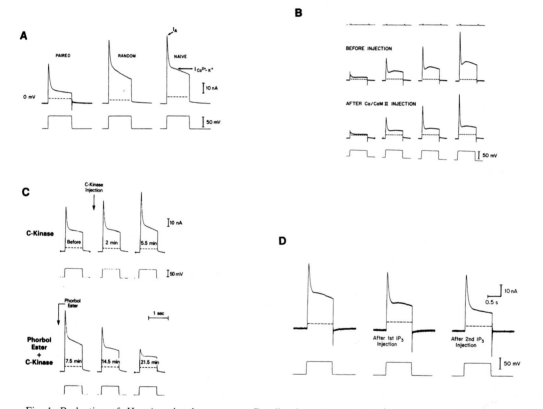

Fig. 1. Reduction of *Hermissenda* photoreceptor B cell voltage-dependent K$^+$ currents (I_A) upon associative learning can be mimicked by iontophoretic injection Ca^{2+}-dependent protein kinases. Voltage clamp conditions were maintained across membranes of *Hermissenda* photoreceptors. (A) Comparison of voltage-dependent K$^+$ currents measured in type B somata isolated from paired, random and naive animals. Photoreceptors from animals that have undergone paired stimuli revealed a substantially reduced I_A current (Alkon et al., 1985). (B) Injection of active Ca^{2+}/calmodulin-dependent protein kinase paired with a light step causes a marked and prolonged reduction in the I_A current (Sakakibara et al., 1986b). (C) Injection of protein kinase C causes a slight increase in the I_A current of the photoreceptor (upper trace). Upon further addition of phorbol ester the same cell shows a progressive and marked decrease of the K$^+$ current (lower trace) (Alkon et al., 1988). (D) Injection of IP$_3$ reduces the outward K$^+$ current which persisted for the duration of the recording. A second injection caused an even further reduction (Sakakibara et al., 1986a).

Protein kinase C (PKC) was detected by biochemical assays in *Hermissenda* (Neary et al., 1986). Its enzymatic properties determined in a crude extract were similar to those observed from numerous other cellular and tissue sources (Kikkawa et al., 1982). The molecular weight of the *Hermissenda* PKC as determined using a polyclonal antibody prepared against rat brain PKC (kindly provided by K.-P. Huang) was 73–75 kDa, slightly less than the 80 kDa forms found in rat brain (D.S. Lester and D.L. Alkon, unpublished observations). The enzyme is predominantly in the cytosolic (or inactive) form. However, upon application of phorbol ester or a diacylglycerol analog, the enzyme is primarily localized in the membrane or particulate fraction (Alkon et al., 1988) (Table I). Also, a principal PKC substrate of *Hermissenda* which is membrane-associated shows a sustained or long-term increase in phosphorylation in response to phorbol ester application (Alkon et al., 1988; Naito et al., 1988). Additionally, PKC activators, such as phorbol dibutyrate and oleoyl acetylglycerol, together with elevated Ca^{2+} reduce K^+ currents and cause changes in phosphoprotein patterns in photoreceptors in a manner similar to that observed in the trained animal (Alkon et al., 1986). Injection of purified cytosolic rat brain PKC into photoreceptors increases K^+ currents, while, upon the addition of phorbol ester there is a decrease in potassium currents resulting in a more excitable cellular state similar to that found in conditioned *Hermissenda* (Alkon et al., 1988) (Fig. 1D). This suggests that exogenous soluble PKC must be converted into a membrane-associated active form in order to exert its influence. Thus, activation of exogenous PKC application or activation of endogenous PKC results in a physiological state of excitation of the photoreceptor equivalent to that measured in the conditioned animal.

With the rabbit eyelid conditioning, there is a long-term increase in the excitability of hippocampal CA1 cells, similar to that observed in the photoreceptor of conditioned *Hermissenda* (Disterhoft et al., 1986). PKC in the rabbit hippocampal CA1 neurons was observed to undergo a change in subcellular localization in conditioned rabbits (Bank et al., 1988). There was an increase in the membrane or particulate fraction of PKC (Table I) which is considered to be indicative of PKC activation (Nishizuka, 1986). The change in distribution was measured 24 h after the conditioning process had taken place suggesting a sustained physiological response (Bank et al., 1988). The change in localization was representative of the alteration in membrane to cytosolic PKC ratio and not due to differences in basal or total (cytosol + membrane) PKC activity. Using an additional technique, autoradiographic visualization of [³H]phorbol dibutyrate binding to intact rabbit hippocampal slices in conditioned and naive animals, an increase in the labelling of the soma and dendritic layers in the CA1 cell region was observed 24 h after conditioning (Olds et al., 1989). This technique developed by James Olds and other members in the laboratory selects for specific phorbol ester binding sites. The dif-

TABLE I

Changes in subcellular localization of protein kinase C in animal systems used for associative learning studies

Animal system	Activity (%)			
	Control		Experimental	
Subcellular localization [c]:	mb	cyt	mb	cyt
Hermissenda [a]	52.3	47.8	90.1	9.9
Rabbit CA1 hippocampus [b]	42.0	58.0	63.3	26.7

[a] From Alkon et al. (1988). Experimental is after the addition of phorbol ester.
[b] From Bank et al. (1988). The experimental results are from animals that have undergone conditioning.
[c] mb = membrane or particulate fraction; cyt = cytosolic or soluble fraction.

ferences in labelling intensity continued to change even 72 h after conditioning, predominantly in the cellular distribution of this PKC probe, i.e. soma vs. dendrite (Olds et al., 1989). Thus, the results obtained using autoradiographic visualization correlate with the biochemical and electrophysiological analyses and support the notion that PKC plays an important role in the conditioning process and the subsequent maintenance of stored information.

A third learning model recently introduced in the laboratory in collaboration with Drs. D. Olton and M. Mishkin, is spatial maze learning in rats. Using the [^{3}H]phorbol dibutyrate autoradiographic visualization procedure described above, changes in the distribution of PKC in the rat hippocampal CA3 region were measured (Olds et al., 1990). In this case of the rat hippocampus, there was a decrease in the amount of [^{3}H]phorbol dibutyrate bound in these specific regions. Although much remains to be understood about these different responses, they all indicate that the one common regulatory element in the associative learning studies performed in this laboratory is the regulatory signal transduction enzyme, protein kinase C.

3. Phosphoprotein substrates

Changes in the phosphorylation state of only a small number of specific phosphoproteins have been found in learning-related processes (Neary et al., 1981; Nelson et al., 1990). Initially, a 20 kDa phosphoprotein was identified in eyes upon associative learning using molecular weight sizing polyacrylamide gels under denaturing conditions (Neary et al., 1981). Using high performance liquid chromatography to improve separation and resolution, significantly more phosphoproteins were identified, including species of M_r = 16, 18, 20 and 27 kDa (Nelson et al., 1990) (Table II). Conditioning-induced increases in the optical

TABLE II

Changes in phosphorylation states of specific phosphoprotein substrates of *Hermissenda* central nervous system or photoreceptors upon learning or pharmacological treatment

Treatment	Learning [a]	Depolarization [b]	Phorbo ester [c]	4-Aminopyridine [d]
M_r (kDa)				
16	+ [e] (s) [f]	n.d.	n.d.	n.d.
18	+ (s)	n.d.	n.d.	n.d.
20	+ (s)	− [e]	− (p) [f]	n.d.
20.5	n.d. [g]	n.d.	+ (p)	n.d.
23	n.d.	n.d.	+ (s)	+
25	n.d.	−	− (p)	−
27, 28	− (s)	n.d.	n.d.	n.d.
56	n.d.	+	+ (p)	n.d.

[a] From Neary et al. (1981) and Nelson et al. (1990).
[b] From Naito et al. (1988).
[c] From Alkon et al. (1988).
[d] From Neary and Alkon (1983).
[e] + = phosphorylation; − = dephosphorylation.
[f] s = soluble fraction; p = particulate fraction.
[g] n.d. = not deteced.

density absorbance of these 4 proteins were perhaps due to changes in their state of phosphorylation. The 20 and 27 kDa were labelled with phosphate at stoichiometries of 0.69 and 0.23 mol ^{32}P/mol of protein, respectively. Analyses of biochemical characteristics of these proteins revealed that the 20 kDa phosphoprotein (cp20) binds GTP and has GTPase activity (Nelson et al., 1990). There is a large family of low molecular weight (20–25 kDa) GTP-binding proteins that have oncogenic properties (Chiarugi et al., 1989). Injection of cp20 into photoreceptors resulted in blocking of K$^+$ channels similar to that observed in conditioned animals (Nelson et al., 1990) (Fig. 2A). Interestingly, the injection of the 21 kDa proto-oncogene v-*ras* modulates photoreceptor K$^+$ channels in a manner similar to cp20 (Collin et al., 1990) (Fig. 2B). Some of the *ras*-related GTP-binding proteins have been shown to alter cellular secretion suggesting a role in intracellular transport processes of secretory vesicles (Santos and Nebrada, 1990). We have recently found that the cp20 may inhibit axonal transport of specific cellular organelles in the crab giant neuron suggesting another correlation with established oncogene products (Moshiach, Nelson and Alkon, in preparation). Some *ras* proto-oncogenes can be phosphorylated by protein kinase C. However, the effect of this posttranslational modification is not known (Magee and Hanley, 1988). *ras* must undergo a series of posttranslational modifications including proteolytic cleavage followed by acylation. This renders the protein more hydrophobic such that it associates with the membrane where it is active (Santos and Nebrada, 1990). It is not yet known at what cellular site cp20 is active, but we have initiated studies to search for acylated proteins of this molecular weight in *Hermissenda*.

It is not yet known with certainty which protein kinase(s) phosphorylates cp20 in the photoreceptor. However, based on previous studies phosphorylation of this protein is Ca^{2+}-dependent with PKC being a likely candidate. The state of phosphorylation of a phosphoprotein of similar molecular weight in photoreceptors changes when treated with pharmacological agents that affect Ca^{2+}-dependent protein kinases, such as phorbol esters and calmodulin antagonists, and induce a state of cellular electrical activity similar to the eye removed from a trained animal (Alkon et al., 1988; Sakakibara et al., 1986b) (Table II). In vitro analyses of potential protein kinase C substrates in *Hermissenda* identified a number of potential

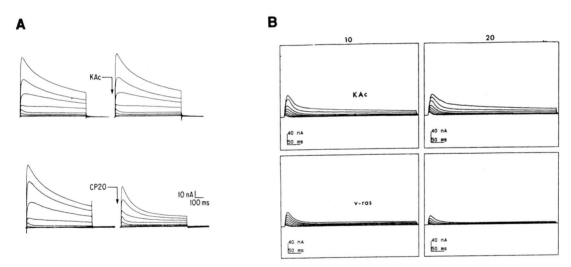

Fig. 2. Injection of low molecular weight GTP-binding proteins into *Hermissenda* type B photoreceptors causes a reduction of voltage-dependent K^+ currents (I_A) similar to that observed in photoreceptors of animals that have undergone associative learning. (A) Cells were held at -60 mV for 10 min and the cp20 then injected inside (lower trace). The first V-I relationship was determined with 500-ms steps from -60 mV at 10 mV increments to $+20$ mV, with one pulse occurring every 60 s. Immediately after the first V-I relationship determination, -2.0 nA was applied through the voltage electrode for 2 min, and then a second V-I relationship was measured (Nelson et al., 1990). Potassium acetate (KAc) was used as a control. (B) Upon impalement of the cell with the v-*ras*-filled electrode, there was an immediate reduction in the outward K_+ currents which was further enhanced after an additional period of time (20′) (lower trace). This was in contrast to the cp20 effect which had a slower onset. A hyperpolarizing 500 ms conditioning step to -90 mV from the holding potential of -60 mV preceded a series of 500 ms depolarizing steps of -40 to -5 mV (Collin et al., 1990). KAc was used as a control (upper trace). c-*ras* had no effect (not shown).

species. The predominant species was a protein of $M_r = 56$ kDa, while other proteins including species of 20 and 28 kDa were also identified (Alkon et al., 1986; Neary et al., 1986). Application of the protein kinase C activator phorbol dibutyrate to photoreceptors or central nervous systems stimulates phosphorylation of 5 phosphoproteins ($M_r = 20.5, 23, 56, 62$ and 165 kDa) and a decrease in two proteins (20 and 25 kDa) (Alkon et al., 1988) (Table II). The 20.5, 28 and 56 kDa proteins were associated with the particulate fraction upon addition of phorbol ester, consistent with the active protein kinase C phosphorylating membrane-associated substrates. Depolarization of the CNS also elevated the state of phosphorylation of the 56 kDa while reducing the 20 and 25 kDa indicating some dephosphorylation process (Naito et al., 1988) (Table II). The elevated phosphorylation was Ca^{2+}-dependent suggesting either protein kinase C or Ca^{2+}/calmodulin protein kinase involvement. The dephosphorylation of the 25 kDa phosphoprotein was independent of Ca^{2+} and long-lasting (Naito et al., 1988). It was previously shown that pharmacologically-induced reduction of K4 currents in the eye and/or ganglia reduces the level of ^{32}P incorporation into a 25 kDa protein (Neary and Alkon, 1983). Dephosphorylation of the 20 kDa phosphoprotein was enhanced upon the removal of Ca^{2+} and recovery from the depolarization treatment (high K^+) was quicker than for the 25 kDa protein.

Thus, the 20, 28 and 56 kDa proteins are good candidates for protein kinase C substrates in the *Hermissenda* eye and central nervous system. The nature of the 28 and 56 kDa protein has not yet been determined. The difficulty in determining whether these proteins are unequivocally protein kinase C substrates is largely due to the lack of specific protein kinase C inhibitors (Epand and Lester, 1990) and the inability to purify large quantities of protein. Most of the inhibitors also interact with other second messenger-dependent protein kinases. Also, the majority of inhibitors have been designed according to their action in in vitro analyses, thus, their action in the cell could be expected to be different. It is not yet known whether reduction in phosphorylated substrate upon high K^+ exposure is due to the activation of a phosphatase or due to inactivation of an active protein kinase. The dephosphorylation of the 25 kDa phosphoprotein is not dependent on Ca^{2+} nor is it inhibited by sodium fluoride, an established inhibitor of many protein phosphatases. In contrast, the dephosphorylation of the 20 kDa species is enhanced by the removal of Ca^{2+} which may be indicative of an upset in the equilibrium of the active phosphorylation/dephosphorylation of this protein, rather than an activation of a Ca^{2+}-inhibited phosphatase. Many of the details in the role of protein phosphorylation/dephosphorylation in cellular memory processes remain unclear largely due to our lack of understanding of the phosphorylation/dephosphorylation mechanisms. In spite of this serious shortcoming, the available data suggest a model for the potential role(s) of protein phosphorylation/dephosphorylation in the associative memory process in *Hermissenda*, some of which is applicable to the rabbit hippocampus.

4. A model for associative memory in *Hermissenda*

The immediate transduction and transmission processes underlying associative memory would be expected to involve changes or modifications of existing membrane proteins and lipids, hence, the analyses of signal transduction and electrical activity. In *Hermissenda* changes in potassium currents of photoreceptors correlate with different stages of the learning process. These ion fluxes can modulate or be modulated by signal transduction processes common to numerous biological systems. For example, an influx of Ca^{2+} via the calcium channel results in a rise in intracellular Ca^{2+}. A rise in intracellular Ca^{2+} is associated with activation of protein kinase C which

can in turn phosphorylate and modify K^+ currents (Kaczmarek, 1987). In the proposed model we are correlating data collected in our laboratory to the events during associative learning and

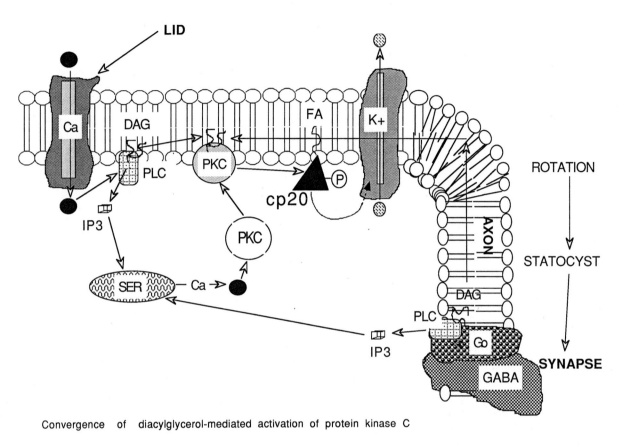

Convergence of diacylglycerol-mediated activation of protein kinase C

Fig. 3. A model for diacylglycerol-mediated activation of protein kinase C and its subsequent action in the photoreceptor upon associative learning in *Hermissenda*. Open arrows indicated an inhibitory action while closed or filled is activating. Dashed lines are correlative evidence while the solid lines are correlated to data collected in the laboratory. Light (conditioned stimulus) acts upon rhodopsin in the rhabdomere inducing depolarization (LID) of the cell body of the photoreceptor. The rise in intracellular Ca^{2+} aids in the activation of the phosphatidylinositol-specific phospholipase C (PLC). The release of diacylglycerol (DAG) by the action of PLC would activate protein kinase C (PKC) while the soluble PLC product inositol triphosphate (IP3) would act on intracellular Ca^{2+} storage organelles, such as the endoplasmic reticulum (SER), to release Ca^{2+}. The elevated intracellular Ca^{2+} would associate with PKC, inducing its translocation from the cytosol (○) to the membrane (●) where it could interact with the DAG and become activated. The potential PKC substrate (cp20) phosphorylated by membrane-active PKC is proposed to be acylated (FA tail). The cp20 either in a cytosolic or membrane-associated form reduces outward K^+ currents. Meanwhile, while the influx of Ca^{2+} and efflux of K^+ has caused the cell to depolarize, the unconditioned stimulus, rotation, causes release by the statocyst of GABA at a synapse on the axonal tract which activates a $GABA_B$ receptor resulting in activation of PLC via a different G-protein (G_o). The release of DAG and IP3 via this mechanism could interact with and/or enhance the light depolarizing effect in the cell body. This suggests that there is a convergence of the intracellular signals between the soma and the axon, induced by light and rotation.

adding potential biochemical events that may link these observations (see Fig. 3). The model is for *Hermissenda*. However, some of the observations have also been made in the rabbit hippocampus. So, portions of the pathways proposed in this model may be applicable to associative learning processes in more complex animals.

The associated learning events in *Hermissenda* are a light stimulus and a vestibular stimulus. Light energy is transduced into a biochemical cellular response in invertebrate and vertebrate eyes via the visual receptor transmembrane protein, rhodopsin. This light-induced conformational change in the protein promotes its interaction with a specific GTPase which hydrolyses GTP. The rhodopsin-coupled G-protein is called transducin (G_t) and has been shown to act exclusively via a cyclic GMP phosphodiesterase (Hingorani et al., 1988). The cyclic GMP phosphodiesterase hydrolyses cyclic GMP which has been shown to be a substrate for the cyclic GMP-dependent protein kinase (Edelman et al., 1987) and can also directly activate the cation channel in rod outer segments (Cote et al., 1989). A consequence of light stimulation in *Hermissenda* is that the photoreceptor is depolarized. Depolarization by light activates a sustained voltage-dependent Ca^{2+} current. The resulting inward flux will elevate intracellular Ca^{2+} levels. An influx of Ca^{2+} would be expected to mediate a variety of Ca^{2+}-dependent processes including protein kinase C and phospholipase A_2 activation, and the initiation of Ca^{2+}/calmodulin-dependent processes. The cooperative effect between protein kinase C and Ca^{2+} (Fig. 3) is well established (Nishizuka, 1986) as is the Ca^{2+}-dependence of phospholipase A_2 (Fig. 5) (Waite, 1987). Calmodulin is an activator of numerous regulatory enzymes including the calmodulin-dependent protein kinase and the calmodulin-dependent phosphatase (calcineurin) (Kennedy, 1989). It also binds to GAP-43 (B50, Fl, SNAP) in

a Ca^{2+}-independent manner (Strittmater et al., 1990). Depolarization has also been linked to hydrolysis of phosphatidylinositol via activation of the phosphatidylinositol-dependent phospholipase C (Miller, 1986). In addition, we have recently shown that stimulating the hair cells causes presynaptic release of GABA onto postsynaptic type B photoreceptors (Matzel and Alkon, submitted). This hyperpolarizes the photoreceptor by acting on a $GABA_A$-type receptor to open Cl^- channels (Alkon, Anderson, Matzel and Nelson, submitted). GABA also causes a slow prolonged depolarization of the type B cell presumably by closing K^+ channels. This slow depolarization is more apparent in the presence of Cl^- channel blockers and is greatly enhanced by depolarization of the type B cell. When this GABA-induced depolarization is paired with depolarization of the photoreceptor, there is a protein kinase-mediated increase in neuronal resistance of the photoreceptor (Matzel and Alkon, submitted). The $GABA_B$ receptor has been proposed to act via a G_0 transducer which activates phosphatidylinositol-specific phospholipase C to release diacylglycerol and inositol triphosphate (Goh and Pennefather, 1989). The $GABA_B$-mediated depolarization of the type B cell was abolished if Ca^{2+} was removed from the bath, or if the PKC inhibitor, H7, was added suggesting the involvement of PKC in this process (Matzel and Alkon, submitted). The GABA effect is consistent with other experiments which demonstrated that application of both diacylglycerol analogs or inositol triphosphate modulate K^+ currents.

Activation of protein kinase C is considered to have occurred when there is an increase in the membrane to cytosol ratio (Nishizuka, 1986). The membrane or particulate form has a different secondary structure than the cytosolic or soluble form even after solubilization and subsequent extraction from the membrane (Lester et al.,

1990). Thus, this changed structure is maintained or remembered upon extraction. This could explain the physical nature of the persistent translocated protein kinase C activity found in animals that have undergone associative learning processes. The physiological consequence would be an increased basal level of protein kinase C activity. A key mediating factor(s) in protein kinase C activation is considered to be the lipid environment (Epand and Lester, 1990). Diacylglycerol production could contribute to a suitable lipid support for PKC activity. Arachidonic acid, which is released by the action of phospholipase A_2 on phospholipids, is an additional lipid component that could be affected by these depolarizing conditions. The release of arachidonic acid would modulate the localized bilayer properties and could also activate protein kinase C in a manner distinct from diacylglycerol as has been shown for another unsaturated fatty acid, linoleic acid (Lester, 1990). Arachidonic acid is a known in vitro activator of protein kinase C (Nishizuka, 1988). These types of changes in lipid composition could induce a persistent protein kinase C translocation which would eventually be reversed upon reorganization or restoration of the lipid matrix. Thus, these long-term changes would not require actions at the level of the gene.

The action of protein kinases is the phosphorylation of specific substrates. The unique characteristic of protein kinase C as a second messenger-dependent protein kinase is that it is active when associated with the membrane. Therefore, it would be expected that the substrates be in close proximity or associated with the membrane. Transmembrane proteins such as ion channels are potential targets for membrane-associated PKC (Kaczmarek, 1987). Soluble proteins could also become suitable targets if they were somehow brought to the membrane surface. The principal protein kinase C substrate in synaptosomes undergoes a posttranslational modification whereby a fatty acid is attached (acylation) rendering the protein more hydrophobic. Upon association with the membrane it can then be phosphorylated (Wang et al., 1989). The GAP-43 (SNAP) protein is a protein kinase C substrate that is found in both a soluble and particulate form and also has a potential acylation site (Van Lookeren Campagne et al., 1989). The 56 kDa protein kinase C substrate in *Hermissenda* is found in both soluble (Alkon et al., 1986) and particulate (Alkon et al., 1988) fractions. Another protein that must undergo acylation before it is active is *ras* (Santos and Nebrada, 1990). *ras* acts on the intracellular membrane surface resulting in the modification of a number of cellular processes including neurite outgrowth in PC12 cells (Satoh et al., 1987). Morphological changes in the *Hermissenda* photoreceptor have been observed in response to stimuli related to learning (Alkon et al., 1990; Lederhendler et al., 1990). Other G-proteins of molecular weight similar to *ras* are found in both the cytosol and the membrane (Burgoyne et al., 1989b). One such protein was found to be translocated from the membrane to the cytosol upon phosphorylation (Magee and Hanley, 1988). These proteins are also considered to play an important role in cell organelle trafficking (Santos and Nebrada, 1990). Cp20 is a good candidate for such a protein as it is phosphorylated, it alters movement of axonal organelles (Moshiach, Nelson and Alkon, in preparation), and modulates K^+ currents in photoreceptors in a manner similar to *ras* (Collin et al., 1990; Nelson et al., 1990). The pathway by which cp20 could modulate its cellular responses remains as much a mystery as it does for *ras* and similar *ras*-like proteins. However, we have combined the above low molecular weight G protein-associated observations to propose a mechanism for action of cp20 (see Fig. 3). This may also be applicable to other *ras*-like G proteins. It is proposed that these types of G-proteins act upon a "cluster" of regulatory proteins such that numerous cellular activities could be modified by a single action.

This model indicates that the conditioned and unconditioned stimuli have distinct pathways, yet some of the biochemical changes that occur involve common components. PKC has been shown to be activated by the second messengers, diacylglycerol and arachidonic acid (Nishizuka, 1988). The enzymes which produce these two lipid products, phospholipase C and phospholipase A_2, re-

spectively, can be activated both by G proteins and Ca^{2+} (Axelrod et al., 1988). Thus, the subcellular localization of the G-proteins could play a major role in the action of their products. In *Hermissenda* light causes depolarization in the cell body while the hair cell input via a postsynaptic GABA receptor is located at synapses on the terminal branches. The specific G protein that activates these phospholipase(s) would be expected to be in close proximity to these enzymes in order for activation to take place.

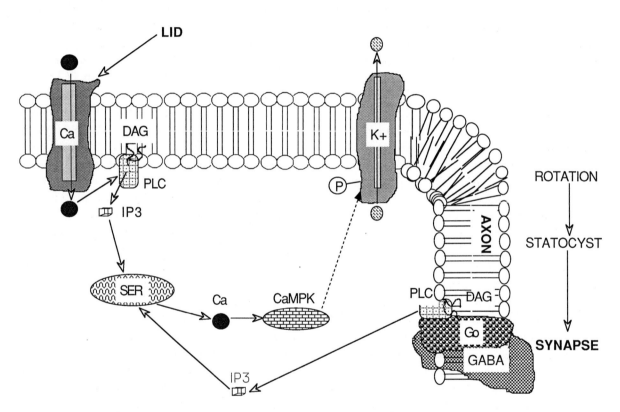

Regulation of K⁺ channels by Ca/calmodulin-dependent protein kinase

Fig. 4. Proposed model for the activation of Ca^{2+}/calmodulin-dependent protein kinase and its subsequent action in the *Hermissenda* photoreceptor upon associative learning. The lines, arrows and symbols are used as described in Fig. 3. Upon depolarization of the photoreceptor soma, PLC is activated and DAG and IP3 are released. IP3 acts on SER to release Ca^{2+} which would bind to cytosolic calmodulin and this complex would interact with and activate the Ca^{2+}/calmodulin-dependent protein kinase (CaMPK) which is proposed to act directly on the K⁺ channel and phosphorylate it. It could also phosphorylate cp20 and have a similar effect as described in Fig. 3. The elevation in intracellular Ca^{2+} could also be obtained via the GABA mediated response at the synapse. Thus, there is a potential convergence of the activation pathways of CaMPK also.

Long-term potentiation (LTP) is a synaptic condition that has been suggested to underlie certain forms of learning and memory. Upon tetanic or repetitive stimulation upstream of specific synapses in the hippocampus, there is a long-lasting use-dependent increase in the efficacy of excitatory transmission of the stimulated synapse. This is termed LTP. This condition has also been demonstrated in dissociated hippocampal neuronal cultures (Bekkers and Stevens, 1990). In contrast to the generation and induction mechanisms of LTP localized specifically at a

particular single synapse, associative learning processes involve intricate neural networks. Some of the biochemical events that are associated with LTP processes in the hippocampus share certain characteristics with the associative memory processes in *Hermissenda*. However, the mechanisms resulting in these responses are unique for each system. In order to obtain LTP, two subtypes of postsynaptic glutamate receptors are activated, the NMDA and the quisqualate receptors. Upon induction of LTP, in order to maintain this state there is also a lasting change in excitability due to

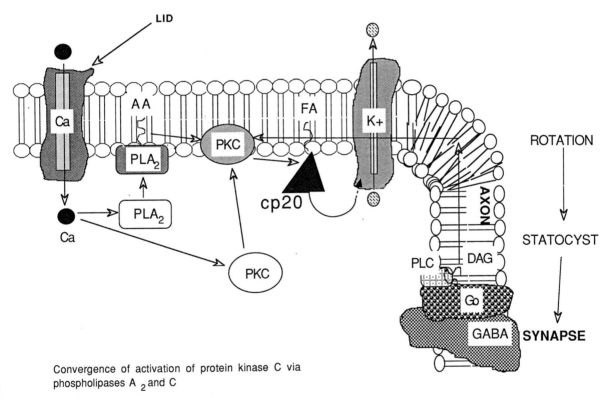

Convergence of activation of protein kinase C via phospholipases A$_2$ and C

Fig. 5. Proposed model for the phospholipase-mediated activation of protein kinase C in the *Hermissenda* photoreceptor upon associative learning. Lines, symbols and arrows are as described in Fig. 3. Upon depolarization of the photoreceptor cell body, the elevated intracellular Ca^{2+} would activate phospholipase A$_2$ (PLA$_2$) which would release fatty acids, including arachadonic acid (AA). This could also activate PKC in a different way such as was observed for PKC activation by linoleic acid (Lester, 1990). The unconditioned stimulus, rotation, would activate PLC to release DAG which may migrate to the cell body and interact with the PKC-AA complex. This may also result in a synergistic activation of PKC as was seen for linoleic acid and DAG (Lester, 1990). The active membrane-associated PKC would act to decrease K$^+$ efflux via a similar mechanism as proposed in Fig. 3. This suggests a convergence of two different second messengers, AA and DAG, resulting in modulation of a similar pathway.

increased activation of the postsynaptic quisqualate receptors. This receptor directly mediates postsynaptic channel activity (Lynch, 1989). This glutamate receptor-mediated process involved in LTP production may be compared to the $GABA_B$ receptors in *Hermissenda* (Fig. 6). However, the lasting change in photoreceptor excitability upon associative learning is due to modulation of voltage-gated K^+ channels by Ca^{2+}-dependent protein kinases. K^+ currents have not been shown to change with LTP. In LTP, enzymes such as protein kinase C, the Ca^{2+}/calmodulin-dependent protein kinase, and phospholipase A_2 are activated (Lynch, 1989), all of which are involved in associative learning in *Hermissenda* (Figs. 3, 4 and 5). In LTP there is a persistent protein kinase activity (for about 1 h) which may be due to the proteolytic release of the catalytic subunit of PKC (Oliver et al., 1989) or persistent protein kinase C activity (Burgoyne, 1989a). In associative conditioning of the rabbit, activation of protein kinase C is due to long-term translocation of the

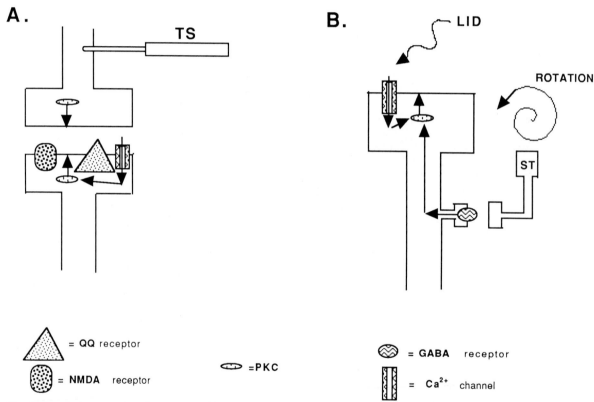

Fig. 6. Simplistic comparison between specific cellular events in long-term potentiation and associative learning. (A) Tetanic stimulation (TS) induces release of glutamate from the presynapse which associates with two postsynaptic glutamate receptors, the *N*-methyl-D-aspartate (NMDA) and the quisqualate (QQ) receptors. There is an accompanying activation of protein kinase C (PKC) or Ca^{2+}-dependent protein kinase C. PKC is also activated in the presynapse as is observed by the phosphorylation of the F1 or B50 protein. (B) Light induces depolarization (LIGHT) of the cell body of the *Hermissenda* photoreceptor and rotation causes depolarization of the statocyst (ST) resulting in the release of GABA from a synapse localized on the axonal tract of the photoreceptor. The resulting activation of protein kinase C by both of these pathways causes K^+ efflux to be reduced. This suggests that the intracellular signals induced by the two conditioning stimuli converge to act at a similar target in the cell body of the eye.

enzyme to the membrane. This membrane-active form can then modulate the voltage-gated K^+ channel. Another difference between the two systems is at the spatial or organizational level of the stimulatory inputs for LTP and associative learning. The receptors which are responsible for LTP induction are exclusively at the synapse, while both the cell body and dendritic locales play a role in *Hermissenda* type B cell conditioning (Fig. 6). These specific details are not yet established for the associative learning process in the hippocampus upon rabbit eye conditioning, but it should be noted that if GABA is also involved in this process as for *Hermissenda,* the $GABA_B$ receptors are located in the dendrites of the CA1 region (Newbury and Nicoll, 1985) in proximity to the the principle excitatory input.

A striking feature of many of the signal transduction pathways is that they use similar components even though different stimuli initiate them. Those features or components responsive to different stimuli may provide for an event such as learning. An intriguing question is to what degree are these processes or their components shared in different animal learning systems, e.g. the rabbit and *Hermissenda*? Such questions provoke development of new technologies and lines of thinking, and promise that the field of learning and memory will be most exciting in the years to come.

References

Alkon, D.L. (1988) Memory Traces in the Brain, Cambridge University Press, Cambridge.

Alkon, D.L. (1989) Memory storage and neural systems. *Sci. Amer.*, 261: 42–50.

Alkon, D.L., Sakakibara, M., Forman, R., Harrigan, J., Lederhendler, I. and Farley, J. (1985) Reduction of two voltage-dependent K_+ currents mediates retention of a learned association. *Behav. Neural Biol.*, 44: 278–300.

Alkon, D.L., Kubota, M., Neary, J.T., Naito, S., Coulter, D. and Rasmussen, H. (1986) C-kinase activation prolongs

Ca^{2+}-dependent inactivation of K^+ currents. *Biochem. Biophys. Res. Commun.*, 143: 1245–1253.

Alkon, D.L., Naito, S., Kubota, M., Chen, C., Bank, B., Smallwood, J., Gallant, P. and Rasmussen, H. (1988) Regulation of *Hermissenda* K^+ channels by cytoplasmic and membrane-associated C-kinase. *J. Neurochem.*, 51: 903–917.

Alkon, D.L., Ikeno, H., Dworkin, J., McPhie, D.L., Olds, J.L., Lederhndler, I., Matzel, L., Schreurs, B.G., Kuzirian, A, Collin, C. and Yamoah, E. (1990) Contraction of neuronal branching volume: an anatomic correlate of Pavlovian conditioning. *Proc. Natl. Acad. Sci. USA*, 87: 1611–1614.

Axelrod, J., Burch, R.M. and Jelsema, C.L. (1988) Receptor-mediated activation of phospholipase A_2 via GTP-binding proteins: arachidonic acid and its metabolites as second messengers. *Trends Neurol. Sci.*, 11: 117–123.

Bank, B., DeWeer, A., Kuzirian, A.M., Rasmussen, H. and Alkon, D.L. (1988) Classical conditioning induces long-term translocation of protein kinase C in rabbit hippocampal CA1 cells. *Proc. Natl. Acad. Sci. USA*, 85: 1988–1992.

Bekker, J.M. and Stevens, C.F. (1990) Presynaptic mechanism for long-term potentiation in the hippocampus. *Nature*, 346: 724–729.

Burgoyne, R.D. (1989a) A role for membrane-inserted protein kinase C in cellular memory? *Trends Biol. Sci.*, 14: 87–88.

Burgoyne, R.D. (1989b) Small GTP-binding proteins. *Trends Biol. Sci.*, 14: 394–396.

Chiarugi, V.P., Ruggiero, M. and Corradetti, R. (1989) Oncogenes, protein kinase C, neuronal differentiation and memory. *Neurochem. Int.*, 14: 1–9.

Collin, C., Papageorge, A.G., Sakakibara, M., Huddie, P.L., Lowy, D.R. and Alkon, D.L. (1990) Early regulation of membrane excitability by *ras* oncogene proteins. *Biophys. J.*, in press.

Cote, R.H., Nicol, G.D., Burke, S.A. and Bownds, M.D. (1989) Cyclic GMP levels and membrane current during onset, recovery, and light adaptation of the photoresponse of detached frog photoreceptors. *J. Biol. Chem.*, 264: 15384–15391.

Disterhoft, J.F., Coulter, D.A. and Alkon, D.L. (1986) Conditioning-specific membrane changes of rabbit hippocampal neurons measured in vitro. *Proc. Natl. Acad. Sci. USA*, 83: 2733–2737.

Edelman, AM., Blumenthal, D.K and Krebs, E.G. (1987) Protein serine/threonine kinases. *Annu. Rev. Biochem.*, 56: 567–613.

Epand, R.M. and Lester, D.S. (1990) The role of membrane biophysical properties in the regulation of protein kinase C activity. *Trends Pharmacol. Sci.*, 11: 317–320.

Goh, J.W. and Pennefather, P.S. (1989) A pertussis toxin-sensitive G protein in hippocampal long-term potentiation. *Science*, 244: 980–983.

Hingorani, V.N., Tobias, D.T., Henderson, J.T. and Ho, Y.-K. (1988) Chemical crosslinking of bovine retinal transducin and cGMP phosphodiesterase. *J. Biol. Chem.*, 263: 6919–6926.

Kaczmarek, L. (1987) The role of protein kinase C in the regulation of ion channels and neurotransmitter release. *Trends Neurol. Sci.*, 10: 30–34.

Kennedy, M.B. (1989) Regulation of neuronal function by calcium. *Trends Neurol. Sci.*, 12: 417–419.

Kikkawa, U., Takai, Y., Minakuchi, R., Inohara, S., and Nishizuka, Y. (1982) Calcium-activated, phospholipid-dependent protein kinase from rat brain: subcellular distribution, purification and properties. *J. Biol. Chem.*, 257: 13341–13348.

Lederhendler, I., Etcheberrigay, R., Yamoah, E., Matzel, L., and Alkon, D.L. (1990) Outgrowths from *Hermissenda* photoreceptors are associated with activation of protein kinase C. *Brain Res.*, 534: 195–200.

Lester, D.S. (1990) In vitro linoleic acid activation of protein kinase C. *Biochim. Biophys. Acta*, 1054: 297–303.

Lester, D.S., Orr, N. and Brumfeld, V. (1990) Structural distinction between soluble and particulate protein kinase C species. *J. Protein Chem.*, 9: 209–220.

Lynch, M.A (1989) Mechanisms underlying induction and maintenance of long-term potentiation in the hippocampus. *BioEssays*, 10: 85–90.

Magee, T. and Hanley, M. (1988) Protein modification. Sticky fingers and CAAX boxes. *Nature*, 335: 114–115,

Miller, R.J. (1986) Protein kinase C: a key regulator of neuronal excitability? *Trends Neurol. Sci.*, 9: 538–541.

Naito, S., Bank, B., and Alkon, D.L. (1988) Transient and persistent depolarizationinduced changes of protein phosphorylation in a molluscan nervous system. *J. Neurochem.*, 50: 704–711.

Neary, J.T. and Alkon, D.L. (1983) Protein phosphorylation/dephosphorylation and the transient, voltage-dependent potassium conductance in *Hermissenda crassicornis*. *J. Biol. Chem.*, 258: 8979–8983.

Neary, T.J., Crow, T. and Alkon, D.L. (1981) Change in a specific phosphoprotein band following associative learning in *Hermissenda*. *Nature*, 293: 658–660.

Neary, J.T., Naito, S., De Weer, A., and Alkon, D.L. (1986) Ca^{2+}/diacylglycerol-activated, phospholipid-dependent protein kinase in the *Hermissenda* CNS. *J. Neurochem.*, 47: 1405–1411.

Nelson, T.J., Collin, C. and Alkon, D.L. (1990) Isolation of a G protein that is modified by learning and reduces potassium currents in *Hermissenda*. *Science*, 247: 1479–1483.

Newbury, N.R. and Nicoll, R.A (1985) Comparison of the action of baclofen with gamma-aminobutyric acid on rat hippocampal pyramidal cells in vitro. *J. Physiol.*, 360: 161–185.

Nishizuka, Y. (1986) Studies and perspectives of protein kinase C. *Science*, 233: 305–311.

Nishizuka, Y. (1988) The molecular heterogeneity of protein kinase C and its implications for cellular regulation. *Nature*, 334: 661–665.

Olds, J.L., Anderson, M.L., McPhie, D.L., Staten, L.D. and Alkon, D.L. (1989) Imaging of memory-specific changes in the distribution of protein kinase C in the hippocampus. *Science,* 245: 866–869.

Olds, J.L., Golski, S., McPhie, D.L., Olton, D., Mishkin, M. and Alkon, D.L. (1990) Discrimination learning alters the distribution of protein kinase C in the hippocampus of rats. *J. Neurosci.*, 10: 3707–3713.

Oliver, M.W., Baudry, M. and Lynch, G. (1989) The protease inhibitor leupeptin interferes with the development of LTP in hippocampal slices. *Brain Res.*, 505: 233–238.

Sakakibara, M., Alkon, D.L., Neary, J.T., Heldman, E., and Gould, R. (1986a) Inositol triphosphate regulation of photoreceptor membrane currents. *Biophys. J.*, 50: 797–803.

Sakakibara, M., Alkon, D.L., DeLorenzo, R., Goldenring, J.R., Neary, J.T., and Heldman, E. (1986b) Modulation of calcium-mediated inactivation of ionic currents by Ca^{2+}/calmodulin-dependent protein kinase II. *Biophys. J.*, 50: 319–327.

Santos, E., and Nebrada, A (1990) Structural and functional properties of ras proteins. *FASEB J.*, 3: 2151–2163.

Satoh,T., Nakamura, S. and Kaziro, Y. (1987) Induction of neurite formation in PCl2 cells by microinjection of protooncogenic Ha-ras protein preincubated with guanosine-5′-O-(3-thiotriphosphate). *Mol. Cell. Biol.*, 7: 4553–4556.

Strittmater, S.M., Valenzuela, D., Kennedy, T.E., Neer, E.J., and Fishman, M.C. (1990) G_0 is a major growth cone protein subject to regulation by GAP-43. *Nature*, 344: 836–841.

Van Lookeren Campagne, M., Oestreicher, A.B., Van Bergen en Henegowen, P.M. and Gispen, W.H. (1989) Ultrastructural immunocytochemical localization of B-50/GAP43, a protein kinase C substrate, in isolated presynaptic nerve terminals and neuronal growth cones. *J. Neurocytol.*, 18: 479–489.

Waite, M. (1987) The phospholipases. In: Hanahan, D.J. (Ed.), *Handbook of Lipid Research,* Vol.5, Plenum Press, London.

Wang, J.K, Walaas, S.I., Sihra, T.S., Aderem, A, and Greengard, P. (1989) Phosphorylation and associated translocation of the 87-kDa protein, a major protein kinase C substrate, in isolated nerve terminals. *Proc. Natl. Acad. Sci. USA*, 86: 2253–2256.

W.H. Gispen and A. Routtenberg (Eds.)
Progress in Brain Research, Vol. 89
© 1991 Elsevier Science Publishers B.V.

CHAPTER 16

A tale of two contingent protein kinase C activators: both neutral and acidic lipids regulate synaptic plasticity and information storage

Aryeh Routtenberg

Northwestern University, Cresap Neuroscience Laboratory, 2021 Sheridan Road, Evanston, IL 60208, U.S.A.

There is now considerable evidence that signal transduction and transmembrane signalling mechanisms pivot around the activation of a lipid-dependent enzyme, protein kinase C (PKC; (Nishizuka, 1984; Berridge and Irvine, 1984). PKC as well as other protein molecules involved in this process have been identified and their primary structure has been described. Since our initial proposal (Routtenberg, 1984a), it has become clear that this signalling mechanism has a critical role in the information storage process in the central nervous system.

PKC is now known to exist as a family of subtypes (Coussens et al., 1986; see Nishizuka, 1988, for review). As originally described (see Takai et al., 1982, for review) phospholipase C cleaves the diacylglycerol (DAG) moiety from the parent phospholipid producing a signal in the plane of the membrane to activate PKC (Fig. 1). This is a calcium and phospholipid dependent activation mechanism, here dubbed the DAG pathway, which had been considered the sole mechanism for PKC activation.

A second mechanism was proposed by Murakami et al. (1985, 1986) based on the discovery that PKC could be fully activated by cis unsaturated fatty acids (CUFAs) in the absence of calcium or phospholipid. It was suggested that calcium activated phospholipase A_2 (PLA2) released CUFAs from the 2-acyl position of phospholipids which in their monomeric form activated PKC (Murakami et al., 1986; Fig. 2). I shall review here the available information from different levels of analysis – molecular, biochemical, physiological, cellular and behavioral – that converge in their support of the view that the CUFA/PKC mechanism has an important function in synaptic plasticity.

Initial biochemical evidence on the CUFA pathway

Based on initial evidence demonstrating increased protein F1 phosphorylation after LTP (Routtenberg et al., 1983, 1985), it was proposed

---

Correspondence: Dr. A. Routtenberg, Northwestern University, Cresap Neuroscience Laboratory, 2021 Sheridan Road, Evanston, IL 60208, U.S.A.

that PKC, which phosphorylates F1, played an important role in LTP (Routtenberg, 1984). Subsequent reports from this and other laboratories have provided different lines of evidence support-

ing this proposal. In a nutshell, (a) LTP activated PKC, (b) PKC activation facilitated or induced LTP, and (c) PKC inhibition blocked its long-term expression (see Linden and Routtenberg, 1989,

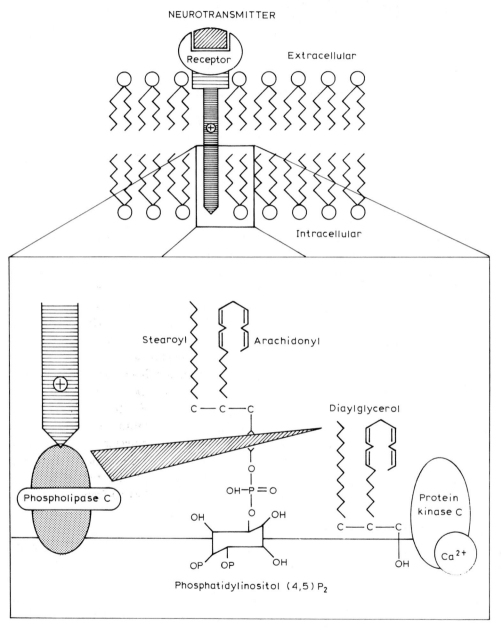

Fig. 1. The DAG pathway.

for review). One activator of PKC, phorbol esters, which facilitated or induced LTP (Routtenberg et al., 1985; Malenka et al., 1986; Colley et al., 1989) raised the issue as to the mechanism by which PKC activation was brought about. The obvious conclusion was through the transmembrane signalling pathway, with DAG as the trigger (Nishizuka, 1986).

We wondered, however, whether the entire DAG molecule would be necessary for activating PKC. In early reports (Kishimoto et al., 1980) cholesterol and free fatty acid were found not to activate PKC and it was stated that "neutral lipids such as triacylglycerol, monoacyl-glycerol and free fatty acid were totally inactive" (Nishizuka et al., 1984, p. 305). In the same year,

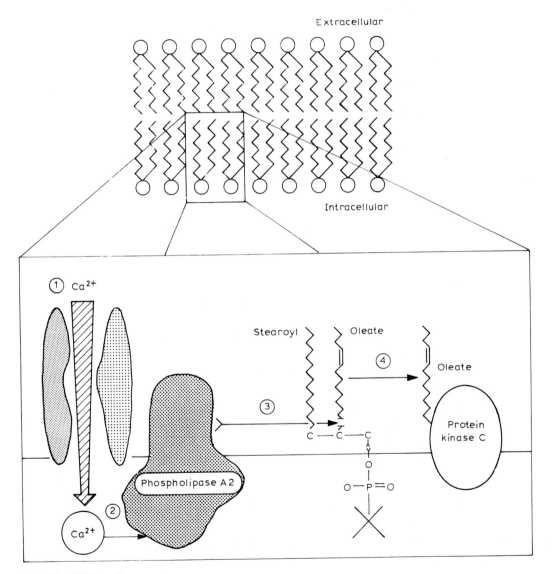

Fig. 2. The CUFA pathway.

in apparent contrast to this statement, McPhail et al. (1984) reported that unsaturated free fatty acid activated protein kinase C in a calcium dependent fashion. But this study used a crude PKC preparation hence the results could have been related to a phosphorylation reaction not involving PKC.

We found that unsaturated, but not saturated, fatty acids were able to activate PKC purified to homogeneity, defined as a single band on one-dimensional gels. Of critical significance was the new finding that purified PKC enzyme could be activated in the absence of calcium and phospholipid (Murakami and Routtenberg, 1985). To ensure that the fatty acid mixture was well suspended in the tris buffer, we strongly vortexed and extensively sonicated the various lipid solutions employed. Because PKC was fully activated by CUFAs in the absence of calcium with no further activation by added PS and calcium, we suggested that there existed two separate pathways for PKC activation: a DAG pathway and a CUFA pathway. The first was mediated via the activation of a phosphatidylinositol specific phospholipase C (PLC) and the second by a calcium-dependent PLA2 (Murakami et al., 1986).

These two activation pathways were distinguished by their behavior after micelle formation. Since increased ionic strength lowers the critical micelle concentration by lowering repulsive forces among polar head groups, Murakami et al. (1986) studied these two mechanisms after altering ionic strength. While the DAG pathway was little affected by the high ionic strength of 130 mM KCl, the CUFA pathway for PKC activation was reduced to near zero (Murakami et al., 1987). This indicated that CUFA-activated PKC was unable to do so when the CUFAs were in micelle form. Of related interest was the finding that increasing the concentration of oleate had no effect on DAG type activation, indicating that oleate was not activating PKC via a detergent action.

A second distinction was observed in the subsequent study of cooperative influence of zinc and calcium on PKC activity (Murakami et al. 1987). Increasing concentrations of zinc only inhibited PKC activation when activated by the DAG pathway, but zinc had no influence whatsoever on the CUFA pathway. Interestingly, in the presence of calcium, CUFA activation could be inhibited by zinc to some extent. This suggests a cooperativity between zinc and calcium, possibly indicating two binding sites on the PKC molecule in the activation/inhibition of the enzyme.

A third distinction between the DAG and CUFA pathways was that the former had an absolute requirement for calcium while the CUFA mechanism did not. When the PS used in activating the DAG pathway was dioleoyl-PS, i.e., oleic acid as the acyl moieties in both the 1 and 2 position, PKC was stimulated, but only in the presence of calcium (Murakami et al., 1986). In order for PS to be active it required the presence of CUFAs as acyl moieties since if dimyristoyl-PS was used, no activation of PKC was observed, even in the presence of calcium. CUFAs are therefore important in both esterified and free fatty acid forms of activation, but as the monomer they are calcium independent; as part of the phospholipid they are calcium dependent.

Do CUFAs activate a particular PKC subtype?

Given the dual PKC activation mechanisms, it was logical to determine whether these were acting on the same kinase or on different PKC subtypes first described by Coussens et al. (1986). Four subtypes from one family have been described (α, β1 and β2, and γ); subsequently 3 others have been demonstrated (δ, ϵ and ζ) by Ono et al. (1987). Although the terminology has varied, that employed by Nishizuka (1988) will be used. As the biochemistry and molecular biology of these subtypes will be discussed extensively elsewhere in this volume (see chapter 9 by Nishizuka and chapter 10 by Huang), it need not be reviewed here.

As one may have anticipated, given the existence of variable regions in the regulatory domain of PKC in the different subtypes, Sekiguchi et al. (1987, 1988) demonstrated differential responses of subtypes to CUFAs. Before reviewing their findings it is important to know that in whole brain β PKC is about twice that of α or γ: considering the α, β and γ subtypes the relative activity is 26, 49 and 25%, respectively (Sekiguchi et al., 1987). The membrane/cytosol distribution of each subtype is not identical and this distribution difference varies as a function of brain location (Shearman et al., 1987). For example, in cerebellum there is more γ than β in the cytosolic fraction; in cerebral cortex the reverse is observed. In cerebellum, PKC is primarily associated with the cytoplasm in all three subtypes. This however varies according to subtype: the γ is 8–10-fold more abundant in cytoplasm than in membrane, the β 3–5-fold greater and the α perhaps twice as great in cytoplasm. In cerebral cortex, in contrast, the distribution in membrane/cytosol of all subtypes is roughly equal (Shearman et al., 1987).

Both Sekiguchi et al. (1988) and Naor et al. (1988) described the preferential activation of γ PKC by low μM arachidonic acid (AA) at 300 μM calcium concentrations. They also describe an inhibition with increasing concentrations of AA. This effect was more pronounced in the Naor et al. (1988) report which used γ PKC from hypothalamus, rather than whole brain PKC as did Sekiguchi et al. (1988). Note, however, that the selective activation of γ PKC by arachidonate is not unique as lipoxin also appears to be a selective γ activator (Hansson et al., 1986).

With increasing concentrations of CUFA there is, in contrast to γ, an increase in β and α PKC activity. This is also the pattern of activation that

we observed with our purified whole PKC preparation suggesting that α and β activation predominates over the decrease in γ. This conclusion is consistent with the fact that β PKC comprises nearly 50% of the PKC in brain.

The studies just reviewed indicate that CUFAs activate PKC and do so in a manner distinct from the DAG pathway. Moreover, it appears that CUFAs preferentially activate different PKC subtypes in vitro as a function of a dosage level. To understand their role in intact cells and intact whole animal preparations, we explored the electrophysiological effects of CUFAs on LTP in the intact hippocampus and on sodium and calcium currents of a defined cell line studied with patch pipettes.

CUFA role in synaptic plasticity: electrophysiology

Earlier studies, from this and other laboratories (reviewed elsewhere, Linden and Routtenberg, 1989) had shown that LTP activates PKC (Akers et al., 1986; Colley et al., 1989) and phorbol ester PKC activators enhance LTP (Colley et al., 1989; Malenka et al., 1986; Routtenberg et al., 1985) while PKC inhibitors block LTP (see Colley et al., 1990, for review). If CUFAs were to activate PKC in vivo, one would predict that application of cis fatty acids should enhance the synaptic plasticity of LTP. To assess this prediction we applied oleic acid, a monounsaturated (18:1) fatty acid and arachidonic acid, a polyunsaturated (20:4) fatty acid using micropresssure injection techniques through glass pipettes in the hippocampal dentate gyrus.

We found that both oleate and arachidonate· enhanced LTP. Elaidic acid the *trans*-isomer of oleic acid was without effect. Moreover, there was a direct relation between the ability of the CUFA to activate PKC in a test tube and the extent to which it facilitated LTP in the intact hippocampus (Linden et al., 1986). To our knowledge, this was the first report demonstrating that CUFAs could influence LTP.

It was proposed by Murakami et al. (1986) that external signals elevating intracellular calcium would activate PLA2 and release CUFAs from the 2-position of membrane phospholipids (see Fig. 1). CUFAs in their monomeric form would activate PKC increase substrate phosphorylation and thereby enhance synaptic communication. This predicts that the most effective ejections should be at the synaptic zone; ejections at the cell body layer should be less effective. This also predicts that blockade of PLA2 activity should prevent expression of LTP or attenuate its longevity. Finally, this blockade should be reversed by exogenous application of CUFAs.

These predictions were confirmed in the Linden et al. (1987) study. Ejection of oleate into the molecular layer facilitated abbreviated LTP to a considerably greater extent than did ejections into the hilus and granule cell layer. For example, when 100 pmol was ejected 10 min prior to abbreviated LTP, a nearly 170% increase was observed in the persistence of the enhanced response measured 240 min after initial facilitation. In contrast, there was no effect of this dosage 350 μm away in the hilar region.

With regard to PLA2 inhibition, pretreatment with mepacrine blocked the persistence of LTP but not its initial component lasting 30 min. This means that the CUFA pathway is not important for the initiation of LTP. When oleate was applied immediately before mepacrine the effects were reversed, suggesting that the inhibitory action of mepacrine was compensated by exogenous oleate.

Measurement of PKC activity indicated a translocation of the kinase to the membrane in oleate-enhanced LTP. The extent of translocation was related to the magnitude of the physio-logical response, specifically the population EPSP slope ($r = +0.760$). An important insight into the mechanism of action of PKC was demonstrated by the fact that oleic acid alone was unable to increase PKC activity nor was it able to enhance synaptic communication. Only when an abbreviated LTP paradigm was used in conjunction with CUFA injection was it possible to observe effects of CUFA. This finding is identical to the effect of phorbol ester when used at appropriate doses (Colley et al., 1989)

This illustrates an important principle of operation of the PKC molecule: that under normal circumstances PKC activators do not activate by themselves. If the activator or second messenger is present when the cell is "primed" or stimulated beyond its equilibrium position, the activator is then in a position to turn on PKC activity which then modulates, either negatively or positively, physiological responses of the cell. Hence the term "contingent" activator. This view is similar to but can be distinguished from the proposal (Nishizuka, 1984) that PKC acts by synergism between a calcium mediated event and a second messenger event. The view here is that the calcium-mediated event or some other priming stimulus is a precondition for and must precede a lipid second messenger event; they do not occur in parallel as suggested in the synergism model since the absence of priming precludes the operation of the lipid second messenger. Indeed, the absence of any response to phorbol ester alone was first demonstrated by Castagna et al. (1982). Further discussion of this will be presented in the coda section of this chapter.

If PKC activation only occurs when deviations from cellular equilibrium occur and persist via a priming stimulus, then facilitators of PKC should only be effective when such cellular disequilibration processes are put into effect. We have been able to support this prediction by comparing in the same animal at the same time the effect of

PKC activators alone vs. PKC activators applied along with priming, here an abbreviated LTP paradigm. We studied PKC effectors on inputs to the same cell population: the medial and lateral perforant path input to the hippocampal granule cells (Lovinger and Routtenberg, 1987).

The term "contingent" activator is used here to emphasize the fact that neither phorbol ester nor oleate by themselves activate PKC. It may be predicted that inputs to the same cell would be differentially influenced by PKC regulators if only one of the inputs was primed. To evaluate this prediction, stimulating electrodes were placed in each of the two pathways converging on the granule cells from the entorhinal cortex: the medial and lateral perforant path. Stimulation producing abbreviated LTP (an enhanced response lasting 15–60 min) was applied to one pathway, while low frequency stimulation (which evokes a stable baseline response but not an enhanced epsp and spike activity) was applied to the other. Application of CUFAs which spread to the synapses of both pathways influenced only the pathway that had high frequency stimulation. Thus, in the absence of CUFA, abbreviated LTP decayed to baseline; in its presence, an LTP-like persistent enhancement was observed as decay was essentially prevented. This same CUFA application had no influence on baseline activation of the other low frequency activated pathway. That the CUFA was influencing both pathways equally was demonstrated by using the same injection site but reversing the pathway which received high frequency and low frequency stimulation in a separate experiment, with parallel results. These results along with preceding (Routtenberg et al., 1986) and following ones (Colley et al., 1989) clearly demonstrate that oleate and phorbol esters at appropriate doses enhance synaptic communication and activate PKC only when the contingent activator is preceded with a priming stimulus.

Localization of PKC manipulations to the synaptic zone

The results of the study just reviewed (Lovinger and Routtenberg, 1987) in addition to demonstrating the nature of contingent PKC activators, also demonstrated that the PKC site of action is the synaptic region. This synapse-specific effect strongly argues for an alteration in the metabolism of PKC at the synaptic junction, rather than at the cell body layer. It suggests too that there are local domains of action, which could be presynaptic or post-synaptic, but restricted to particular locations along the dendrite. In our view, the likelihood is that both sides of the synapse are altered. We have proposed that the activation of presynaptic PKC occurs as a joint result of a retrograde message triggered by elevated post-synaptic calcium and PLA2 triggered by elevated presynaptic calcium (Linden and Routtenberg, 1989). This proposal is compatible with the observation of the delayed effects of mepacrine, a PLA2 inhibiter, on LTP (Linden et al., 1987).

A related, but clearly different, view has been expressed by Bliss and his colleagues who have observed an LTP-dependent release of arachidonic acid (M. Lynch et al., 1989). Since LTP is blocked by NDGA, a lipoxygenase and PLA2 inhibitor (but is also a potent Ca^{2+}-channel blocker), they proposed that arachidonate is released from the post-synaptic cell and itself or a lipoxygenase product then activates the presynaptic terminal to produce LTP by enhancing transmitter release. Although they do detect an elevation of arachidonate in their "push-pullate" it is not necessary to infer that that extracellular CUFA is coming from the post-synaptic cell. Nor is it established that CUFAs are exerting their physiological effect on the presynaptic terminal. Moreover, it should be recalled that there is a plasma membrane barrier to oleate (Linden and Routtenberg, 1989) and presumably to arachidonate as well since it is longer and kinkier than its monounsaturated sister. Until it is determined whether oleate is also released after LTP and whether the effect of non-arachidonate CUFAs are equivalent to arachidonate, it is not necessary to conclude that this particular arachidonate feedback regulation occurs. Moreover, if a feedback mechanism should prove to be important it is not necessary to conclude that it is a particular CUFA like arachidonate that is especially important.

One experiment that would be useful to perform involves the influence of PKC inhibitors on CUFA-enhanced LTP. Our prediction, clearly, is that it would block persistent long-term changes in synaptic efficacy. If the change in LTP persisted in such a case, then a non-PKC mechanism would be implied. If it were blocked, then a CUFA-mediated activation of PKC would be suggested. It would leave unanswered the site of action of the CUFA, whether PKC is pre- or post-synaptic. As discussed elsewhere (Routtenberg, 1985, 1989a,b), the evidence to date supports a presynaptic locus of action. It is clear, irrespective of the site of CUFA action, that these studies lend credence to our proposal that CUFAs have an important physiological role in regulating synaptic plasticity.

How might it carry out such a regulatory process? Could CUFAs by activating PKC then alter ion channel function to change ion movement? There is a growing body of evidence that PKC can regulate ion channels (e.g., Doerner et al., 1988; for review, Shearman et al., 1989). PKC can regulate calcium currents and calcium dependent potassium channels in hippocampus. Until recently, little evidence existed for PKC regulation of the sodium channel.

To determine the effects of CUFAs on ion channel function via PKC activation we first studied their action on a homogenous cell population: the neuroblastoma cell line, N1E-115. We found

that Na^+ currents were attenuated by 5 μM oleate applied externally to N1E cells recorded in a whole cell patch configuration (Linden and Routtenberg, 1988). At 1 μM there was no response after external application, but when the cell was internally perfused with a patch pipette filled with oleate at a 1 μM concentration, a greater than 20% reduction in Na^+ current was observed. This indicates that the plasma membrane does pose a barrier to external CUFAs.

Was this effect related to PKC activation? Several lines of evidence indicated an affirmative answer. PKC inhibitors blocked the effect; down-regulation by phorbol esters of PKC also attenuated the oleate mediated reduction in sodium current. Finally, PKC inhibitors applied internally via whole cell perfusion, were significantly more effective than when applied externally indicating an intracellular site of action. It will now be important to study the effects of CUFAs on hippocampal cells, using acute dissociation of tissue from Ammon's horn or dentate gyrus.

In support of a physiological role of CUFAs in N1E cells it is important to emphasize that oleate effects were observed in the presumed physiological range of fatty acids, low micromolar. In Murakami's studies using purified PKC and histone as substrate, the oleate effects were observed at no less than 50–100 μM. It should be noted, however, that this was using histone, an exogenous, artificial substrate. What would happen if an endogenous substrate were used? Chan et al. (Chan, S.Y., Nelson, R.B., Murakami, K. and Routtenberg, A., in preparation) studied the effects of oleate activation of PKC using protein F1 as substrate. They found that activation of PKC could be detected when oleate concentrated was as low as 5 μM in the presence of 200 μM calcium; without calcium this level of activation was 100 μM. This interaction with calcium suggests a potential mechanism for the "priming" effects observed in the LTP study of Linden et al. (1987). Essentially, the high frequency stimulation could provide increased intracellular calcium above a certain threshold level so that oleate would then be effective in activating PKC. The synergism model suggested that DAG lowered the calcium requirement thereby activating PKC. An alternative model is that calcium elevation lowers the requirement for the activator so that it can be effective at a lower concentration

The study of sodium and calcium currents regulated by PKC revealed that the two pathways of activation, DAG and CUFA, did not have similar consequences on ion channel function (Linden and Routtenberg, 1989). We observed that while oleate reduced sodium currents, phorbol esters were without effect. Two distinct interpretations may be made of this result. First, there is a single PKC subtype within the N1E cell. CUFAs, but not phorbol esters, activate PKC and produce a conformational change that is compatible with reducing sodium channel conductance, possibly directly, by phosphorylating the channel itself (Costa and Cattarall, 1984). With phorbol esters or OAG the kinase is activated (as shown by their effect on calcium channels) but in such a fashion that it does not alter the sodium channel. Another interpretation is that there are two subtypes, one of which is activated solely by CUFAs and modulates only certain channels including Na^+. Another subtype is activated by both DAG and CUFAs and regulates calcium channel function.

To begin to evaluate these alternatives, Murakami et al. (1991) have studied the two-dimensional protein phosphorylation patterns of N1E cells after activation both by CUFAs and by phorbol esters. There is a 40kD phosphoprotein that is phosphorylated by CUFA and to a much lesser extent by the DAG mechanism, lending support to the view that there are distinct PKC systems for activating different pathways in the cell. This study does not reveal, however, whether this uses one PKC subtype which in turn phosphorylates different substrates depending on the contingent activator, or whether there are two subtypes preferentially activated by the different activators.

A distinction between CUFAs and DAG is observed during LTP when the longer term consequences of high dose application are monitored. With phorbol esters, there is a subsequent growth of the response, even though initial potentiation was blocked (Colley et al., 1989). To our knowledge this is the first demonstration of long-term enhancement when the response to high frequency stimulation is blocked so there is an absence of an initial enhanced response. The reason for the oleate/phorbol distinction may reside in the ability of phorbol ester by itself to translocate PKC to the membrane, or that phorbol ester is not as rapidly cleared as CUFAs.

Coda: synergism and contingent activator models of PKC function

The proposal of contingent activation is based on the finding that phorbol esters at low dosages, CUFAs, and PKC inhibitors have no effect by themselves on synaptic transmission, as indexed by baseline activation of the perforant path. When a "priming" stimulus is used, however, then phorbol ester and CUFA effects are observed to enhance transmission in an LTP-like manner. Similarly, PKC inhibitors have no effect on synaptic transmission but if present during LTP production will prevent its long-term expression. Thus, the role of PKC is not to induce function, but rather to regulate its occurrence; amplification in some instances, attenuation in others.

An analogy may be helpful here. Consider a university administration chart. There are a variety of high level administrators (enzymes) that ignore the day-to-day activities of faculty and staff. As long as they go about their business –

use the appropriate channels, transmit messages – these enzymes are quiescent. But now a change in communication is required and a persistent activation occurs from the staff (the squeeky wheel in the bureaucracy, priming here). Will the administrator swing into action? Only if the squeeking goes on for a long enough period of time and only if it is loud enough. If not, the enzyme will take care of another wheel that is making a louder and more persistent noise. In this sense a well reasoned memo with no impact is much like a PKC activator without priming. From the view of the enzyme it has no intention of "micromanaging." If it did so then when important changes had to be made it would be overloaded with too many relatively insignificant items. In this sense then the PKC molecule is not involved with synaptic transmission; it is involved in the change.

The distinction with the original Nishizuka formulation is in the use of the term synergism. The implication of the term is that each mechanism provides some consequence, and the net outcome is greater than the sum of its parts. Here I wish to suggest that the PKC molecule is only called into action after the other process is initiated. Part of the confusion stems from the use of levels of activator or inhibitor that are well beyond the range which can inhibit or activate not only PKC but a variety of other enzyme systems as well. In particular, phorbol esters have been used at high nanomolar and even micromolar levels. This produces abnormally high levels of PKC activation as well as other actions mediated by the other enzyme systems activated. The cellular activities observed are then mistakenly attributed to PKC activation.

In the Lovinger and Routtenberg (1987) study of synapse-specific effects of CUFAs the PKC activator (contingent) by itself has no discernible effect on the physiology of the cell. When the synapse is activated and there is prompt return to baseline no distinction is observed in the physiology of the nonactivated and the PKC-treated synapse. However, if by using high frequency stimulation baseline levels are not returned to, then PKC and the applied activator will act on the synapse and enhance the persistent increase in synaptic communication.

Consistent with this proposal is the finding that after low frequency stimulation PKC activity or its membrane/cytosol distribution are not altered (Linden et al., 1987). This suggests, first, that if PKC is altered at all it is altered for only a brief period of time. Second, since we know that PKC inhibitors do not influence baseline activity, it may be possible to conclude that PKC activity is not necessary for baseline synaptic transmission. But it is critical for synaptic plasticity. It prolongs change (see below) when preceded by the appropriate priming state. PKC ignores the status quo.

Because of the extensive negative and positive feedback systems with which PKC is involved (ion channels, PI turnover, G proteins, cytoskeletal elements), PKC is unlikely to provoke unending enhancement or inhibition of function. At some point these various factors will come to a new equilibrium, that will represent the altered state of enhanced communication of that synapse.

This new equilibrium then represents, one could suppose, the substrate for the storage of a more or less permanent memory, a change in state that the nervous system requires to achieve storage of information. The evidence from LTP studies makes it quite clear that transmission per se, the low frequency activation of synapses, does not change synaptic strength. Storage of information appears to have a threshold and unless that threshold is reached the nervous system will resist the change.

How is that threshold reached? I suggest that it may be necessary to keep the altered state, the change in equilibrium or set point for a prolonged period of time. This could possibly be achieved by the activation of more than one system. For example, there is some evidence that the biogenic amines, particularly noradrenaline, may facilitate LTP (see Bliss and Lynch, 1988, for review). As discussed several years ago (Routtenberg, 1984) intrinsic pathways carrying information, such as the perforant path from cortical-entorhinal origins, and extrinsic pathways, such as the noradrenaline containing from brainstem, which modulate activity, would together determine altered state. How they achieve this action in a cooperative way would be speculative at this moment.

One line of inquiry that may shed light on this issue comes from a recent study by Meberg et al. (1990). The initial impetus for this study was to determine why granule cells showed so little protein F1 mRNA in the adult rat (Rosenthal et al., 1987). As part of the study we surveyed message expression in the adult central nervous system. We found that cells containing biogenic amines serotonin (raphe dorsalis and medialis), norepinephrine (locus ceruleus) and dopamine (substantia nigra, pars compacta and ventral tegmental area) had high levels of expression of protein F1 mRNA. So also did the cells of origin of the perforant path in layer II of the entorhinal cortex. Curiously, structures known to be cholinergic (e.g. medial habenula) showed almost no protein F1 mRNA expression. What this might mean is that two inputs to a CA3 cell, one from entorhinal cortex and one from locus ceruleus, would both contain F1 at their terminals. Only when the two were active, however, would plasticity occur. If protein F1 played a role in prolonging the duration of action of a physiological response, by increasing the duration when enhanced transmitter release occurred, then these two inputs together would have a rather long-lasting effect on the hippocampal cell. One would then have the prolonged disequilibrium necessary to activate

PKC, phosphorylating substrate proteins that change the synaptic state thereby enhancing communication necessary for memory storage.

Note added in proof

Since the writing of this review, Murakami et al. (*Biochem. J.*, (1991) in press) have shown a synergism between these two activators discussed herein. Shinomura et al. (*PNAS*, (1991) 88 (5): 149) have obtained similar results. This suggests that the two pathways of activation may work synergistically to produce a prolongation of PKC activation. Thus under physiological conditions the elevation of CUFAs and DAG along with calcium may be the required "contingency" for triggering synaptic plasticity. This might partially explain why CUFAs, which can fully activate purified PKC, have little influence by themselves, on LTP. Both Murakami and Linden, whom I thank for their comments at the proof stage of this manuscript, independently pointed out the importance of this particular issue.

References

Akers, R., Lovinger, D., Colley, P., Linden, D. and Routtenberg, A. (1986) Translocation of protein kinase C activity after long-term potentiation may mediate synaptic plasticity. *Science,* 231: 587–589.

Berridge, M.J. and Irvine, R.F. (1984) Inositol trisphosphate, a novel second messenger in cellular signal transduction. *Nature,* 312: 315–321.

Bliss, T.V.P. and Lynch, M.A. (1988) Long-term potentiation of synaptic transmission in the hippocampus; properties and mechanisms. In P.W. Landfield and S.A. Deadwyler (Eds.), *Long-Term Potentiation: From Biophysics to Behavior,* Alan R. Liss Inc., New York, pp. 3–72.

Castagna, M., Takai, K., Sano, K., Kikkawa, U. and Nishizuka, Y. (1982a) Direct activation of a calcium-activated phospholipid dependent protein kinase by tumor-promoting phorbol esters. *J. Biol. Chem.,* 257: pp. 7551–7847.

Castagna, M., Takai, Y., Kaibuchi, K., Sano, K., Kikkawa, U. and Nishizuka, Y. (1982b) Distribution activation of calcium-activated, phospholipid-dependent protein kinase by tumor-promoting phorbol esters. *J. Biol. Chem.,* 257: 7847–7851.

260

Colley, P.A. and Routtenberg, A. (1989) Dose-dependent phorbol ester facilitation or blockade of hippocampal long-term potentiation: relation to membrane/cytosol distribution of protein kinase C activity. *Brain Res.*, 495: 205–216.

Colley, P.A., Sheu, F.-S. and Routtenberg, A. (1990) Inhibition of protein kinase C blocks two components of LTP persistence leaving initial potentiation intact. *J. Neurosci.*, in press.

Costa, M.R.C. and Catterall, W.A. (1984) Phosphorylation of the subunit of the sodium channel by protein kinase C. *Cell. Mol. Neurobiol.*, 4: 291–297.

Coussens, L., Parker, P.J., Rhee, L., Yang-Feng, T.L., Chen, E., Waterfield, M.D., Francke, U. and Ullrich, A. (1986) Multiple, distinct forms of bovine and human protein kinase C suggest diversity in cellular signalling pathways. *Science*, 233: 859–866.

Doerner, D., Pitler, T.A. and Alger, B.E. (1988) Protein kinase C activators block specific calcium and potassium current components in isolated hippocampal neurons. *J. Neurosci.*, 8: 4069–4078.

Hansson, A., Serhan, C.N., Haeggstrom, J., Ingelman-Sundberg, M. and Samuelsson, B. (1986) Activation of protein kinase C by lipoxin A and other eicosanoids. Intracellular action of oxygenation products of arachidonic acid. *Biochem. Biophys. Res. Commun.*, 134: 1215–1222.

Kishimoto, A., Takai, Y., Mori, T., Kikkawa, U. and Nishizuka, Y. (1980) Activation of calcium and phospholipid-dependent protein kinase by diacylglycerol, its possible relation to phosphotidylinositol turnover. *J. Biol. Chem.*, 255: 2273–2276.

Linden, D.J. and Routtenberg, A. (1989a) The role of protein kinase C in long-term potentiation: a testable model. *Brain Res. Rev.*, 14: 279–296.

Linden, D.J. and Routtenberg, A. (1989b) cis-Fatty acids, which activate protein kinase C, attenuate voltage-dependent Na^+ current in mouse neuroblastoma cells. *J. Physiol.*, 419: 95–119.

Linden, D., Murakami, K. and Routtenberg, A. (1986) A newly discovered protein kinase C activator (oleic acid) preserves synaptic plasticity and promotes the growth of long-term potentiation. *Brain Res.*, 379: 358–363.

Linden, D., Sheu, F.-S., Murakami, K. and Routtenberg, A. (1987) cis-Fatty acid regulation of synaptic potentiation: relation to phospholipase A_2 and protein kinase C activation. *J. Neurosci.*, 7: 3783–3792.

Linden, D.J., Wong, K.L., Sheu, F.-S. and Routtenberg, A. (1988) NMDA receptor blockade prevents the increase in protein kinase C substrate (protein F1) phosphorylation produced by long-term potentiation. *Brain Res.*, 458: 142–146.

Lovinger, D.M. and Routtenberg, A. (1989) Synapse specific protein kinase C activation enhances maintenance of long-term potentiation in rat hippocampus. *J. Physiol. (Lond.)*, 400: 321–333.

Lynch, M.A., Errington, M.L. and Bliss, T.V.P. (1989) Norihydroguaiaretic acid blocks the synaptic component of long-term potentiation and the associated increases in release of glutamate and arachidonate: in vivo study in the dentate gyrus of the rat. *Neuroscience*, 30: 693–701.

Malenka, R.C., Madison, D.V. and Nicoll, R.A. (1986) Potentiation of synaptic transmission in the hippocampus by phorbol esters. *Nature*, 321: 175–177.

McPhail, L., Clayton, C.C., Snyderman, R. (1984) A potential second messenger role for unsaturated fatty acids: Activation of Ca^{2+}-dependent protein kinase C. *Science*, 224: 622–625.

Meberg, P.J. and Routtenberg, A. (1991) Selective expression of F1/GAP43 mRNA in pyramidal but not granule cells of the hippocampus, *Neuroscience*, in press.

Murakami, K. and Routtenberg, A. (1985) Direct activation of purified protein kinase C by unsaturated fatty acids (oleic acid and arachidonic acid) in the absence of phospholipids and Ca^{2+}. *FEBS Lett.*, 192: 189–193.

Murakami, K. and Routtenberg, A. (1990) Second messenger specific subsets of protein kinase C substrates in N1E-115 cell, submitted.

Murakami, K., Chan, S.Y. and Routtenberg, A. (1986) Protein kinase C activation by cis-fatty acid in the absence of Ca^{2+} and phospholipids. *J. Biol. Chem.*, 261: 15424–15429.

Murakami, K., Whiteley, M.K. and Routtenberg, A. (1987) Regulation of protein kinase C activity by cooperative interaction of Zn^{2+} and Ca^{2+}. *J. Biol. Chem.*, 262: 13902–13906.

Naor, Z., Shearman, M.S., Kishimoto, A. and Nishizuka, Y. (1988) Calcium-independent activation of hypothalamic type 1 protein kinase C by unsaturated fatty acids. *Mol. Endocrinol.*, 2: 1044–1048.

Nishizuka, Y. (1986) Studies and perspectives of protein kinase C. *Nature*, 233: 305–312.

Nishizuka, Y. (1988) The molecular heterogeneity of protein kinase C and its implications for cellular regulation. *Nature*, 334: 661–665.

Nishizuka, Y., Takai, Y., Kishimoto, A., Kikkawa, U. and Kaibuchi, K. (1984) Phospholipid turnover in hormone action. *Recent Progr. Hormone Res.*, 40: 301–345.

Ono, Y., Fujii, T., Ogita, K., Kikkawa, U., Igarashi, K. and Nishizuka, Y. (1987) Identification of three additional members of rat protein kinase C family: delta, epsilon and zeta subspecies. *FEBS Lett.*, 226: 125–128.

Rosenthal, A., Chan, S.Y., Henzel, W., Haskell, C., Kuang, W.-J., Chen, E., Wilcox, J.N., Ullrich, A., Goeddel, D.V. and Routtenberg, A. (1987) Primary structure and mRNA localization of protein F1, a growth-related protein kinase C substrate, associated with synaptic plasticity. *EMBO J.*, 6: 3641–3646.

Routtenberg, A. (1984a) Brain phosphoproteins kinase C and protein F1: Protagonists of plasticity in particular pathways. In: G. Lynch, J. McGaugh and N. Weinberger (Eds.), *Neurobiology of Learning and Memory*, The Guilford Press, New York, pp. 479–490.

Routtenberg, A. (1984b) The CA3 pyramidal cell in the hippocampus: Site of intrinsic expression and extrinsic control of memory formation. In: L. Squire and N. Butters (Eds.), *Neuropsychology of Memory*, The Guilford Press, New York, pp. 536–546.

Routtenberg, A. (1989a) Role of protein kinase C and protein F1 in presynaptic terminal growth leading to information storage. In: H. Rahmann (Ed.), *Fundamentals of Memory Formation, Progress in Zoology*, Vol. 37, pp. 283–295.

Routtenberg, A. (1989b) Molecular basis of Hebb Synapse: Preserved mechanisms of axonal growth. In: M. Ito and Y. Nishizuka (Eds.), *Brain Signal Transduction and Memory*, Academic Press, New York, pp. 213–227.

Routtenberg, A., Lovinger, D., Cain, S., Akers, R. and Steward, O. (1983) Effects of long-term potentiation of perforant path synapses in the intact hippocampus on in vitro phosphorylation of a 47 kD protein (F-1). *Fed. Proc.*, 42: 755.

Routtenberg, A., Lovinger, D.M. and Steward, O. (1985) Selective increase in phosphorylation state of a 47kD

protein (F1) directly related to long-term potentiation. *Behav. Neural Biol.*, 43: 3–11.

Routtenberg, A., Colley, P., Linden, D., Lovinger, D. and Murakami, K. (1986) Phorbol ester promotes growth of synaptic plasticity. *Brain Res.*, 378: 374–378.

Sekiguchi, K., Tsukuda, M., Ogita, K., Kikkawa, U. and Nishizuka, Y. (1987) Three distinct forms of protein kinase C: differential response to unsaturated fatty acids. *Biochem. Biophys. Res. Commun.*, 145: 797–802.

Sekiguchi, K., Tsukuda, M., Ase, K., Kikkawa, U. and Nishizuka, Y. (1988), Mode of activation and kinetic properties of three distinct forms of protein kinase C from rat brain. *J. Biochem. (Tokyo)*, 103: 759–765.

Shearman, M.S., Naor, Z., Kikkawa, U. and Nishizuka, Y. (1987) Differential expression of multiple protein kinase C subspecies in rat central nervous tissue. *Biochem. Biophys. Res. Commun.*, 147: 911–919.

Shearman, M., Sekiguchi, K. and Nishizuka, Y. (1989) Moduation of ion channel activity: a key function of the protein kinase C enzyme family. *Pharmacol. Rev.*, 41: 211–237.

Takai, Y., Minakuchi, R., Kikkawa, U., Sano, K., Kaibuchi, K., Yu, B., Matsubara, T. and Nishizuka, Y. (1982) Membrane phospholipid turnover, receptor function and protein phosphorylation. *Progr. Brain Res.*, 56: 287–301.

W.H. Gispen and A. Routtenberg (Eds.)
Progress in Brain Research, Vol. 89
© 1991 Elsevier Science Publishers B.V.

CHAPTER 17

The neuropil and GAP-43/B-50 in normally aging and Alzheimer's disease human brain

Paul D. Coleman [1], Kathryn E. Rogers [1] and Dorothy G. Flood [1,2]

[1] *Department of Neurobiology and Anatomy,* [2] *Department of Neurology, School of Medicine and Dentistry, University of Rochester, 601 Elmwood Avenue, Rochester, NY 14642, U.S.A.*

Studies of dendritic extent

Morphological data from our laboratory, as well as from other laboratories, have indicated that in some regions of the normally aging brain there is an age-related increase of dendritic extent of single neurons. Our first computer-aided, quantitative study of dendritic extent of single randomly sampled Golgi-Cox stained layer II pyramidal neurons in the parahippocampal gyrus of normally aging human brain showed an age-related increase in dendritic extent (Buell and Coleman, 1979) (Fig. 1). We interpreted these data as suggesting a compensatory, plastic response of surviving neurons to the death of their neighbors. Shortly thereafter, Cupp and Uemura (1980) reported an age-related increase in dendritic extent of neurons in frontal association cortex of macaques. Their results, however, differed from our

earlier results in showing decreased dendritic extent in their oldest monkeys. This inverted U-shaped curve for dendritic extent as a function of age had also been described with regard to the mitral/tufted cells of the rodent olfactory bulb, a region where age-related neuron loss also is found (Hinds and McNelly, 1977).

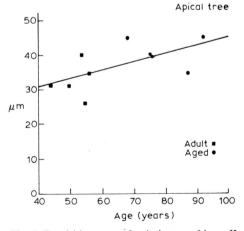

Fig. 1. Dendritic extent of apical trees of layer II pyramidal neurons of parahippocampal gyrus as a function of age in control brains. From Coleman and Buell (1983).

Correspondence: Dr. P.D. Coleman, Department of Neurobiology and Anatomy, University of Rochester, 601 Elmwood Avenue, Rochester, NY 14642, U.S.A. Tel.: (716) 275-2581; Fax: (716) 442-8766.

Studies in our laboratories of dendritic extent in other regions of the human brain revealed regional heterogeneity of age-related changes in dendritic extent. The granule cells of the dentate gyrus showed an inverted U-shaped curve (Fig. 2)

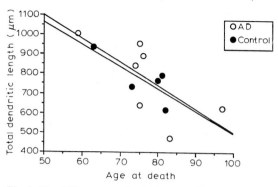

Fig. 3. Dendritic extent of apical trees of layer II pyramidal neurons from area 9 as a function of age. Least squares linear fit to control and AD cases show similar trends.

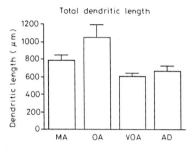

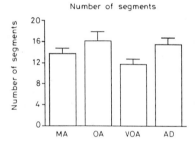

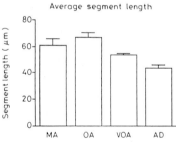

Fig. 2. Total dendritic length, numbers of segments, and average segment length for the apical dendritic trees of dentate gyrus granule cells. Age groups for neurologically and psychiatrically normal subjects are: middle-aged subjects (MA) with a mean age in the early fifties, old-aged subjects (OA) with a mean age in the early seventies, and very old-aged subjects (VOA) with a mean age in the early nineties. Subjects with clinically and neuropathologically verified Alzheimer's disease (AD) averaged in their mid-seventies. Error bars represent standard error of the mean. Redrawn from Flood and Coleman (1986).

similar to that described by Cupp and Uemura in monkey cortex and Hinds and McNelly in rodent olfactory bulb (Flood et al., 1985).

On the other hand, layer II pyramidal neurons of frontal association cortex (middle frontal gyrus, areas 9 and 46) showed a continual age-related decrease in dendritic extent (Flood, unpublished results) (Fig. 3). In human neocortex it is difficult to determine whether such age-related decreases in dendritic extent are a consequence of partial deafferentation of the dendritic tree, or are antecedent to neuron death. However, the rat hypothalamic supraoptic nucleus does not lose neurons with increasing age (Hsu and Peng, 1978), but the dendritic zone of the magnocellular neurons is partially deafferented by an age-related loss and dorsal shift of catecholaminergic axonal afferents (Sladek et al., 1980). We have found in these magnocellular neurons an age-related 34% decrease in dendritic extent that appears to be coordinated with the partial loss of afferent axons (Flood and Coleman, unpublished results). Consideration of these data, as well as the data of others, has led us to hypothesize that in the normally aging brain neighbor neuron death serves as a dendrite proliferative force, while partial loss of afferent axonal supply is a regressive force. We propose that when both proliferative and regressive phenomena are present, the effect on the dendritic tree under consideration is a function of the balance between these two influences.

In Alzheimer's disease (AD) our consistent finding has been an absence of age-related increases in dendritic extent in regions which have shown such increases in normal aging. This has been so in parahippocampal gyrus (Buell and Coleman, 1979) and in dentate gyrus (Flood et al., 1987) (see Fig. 2). In frontal association cortex, where dendritic extent declined with normal aging it also declined, approximately equally, in AD (see Fig. 3). In CA1 (Hanks and Flood, 1991) and subiculum (Flood, 1991), there is no change in net dendritic extent during normal aging, but dendritic extent is reduced in AD cases. These data have led us to hypothesize that one of the more significant neurobiological defects in AD is a loss of neuronal plasticity. This suggestion of a loss of neuronal plasticity in AD may seem to be at variance with other reports suggesting plasticity in the AD brain (e.g. Geddes et al., 1985, who suggest plasticity in AD of axons afferent to the dentate gyrus). However, our data on dendrites of dentate gyrus granule cells indicate a form of aborted plasticity in AD cases in that dendrites do show numbers of branches consistent with the increased branching seen in normally aging brain, but in AD this branching does not result in an increased dendritic extent because the dendritic segments are abnormally short compared to normal (Flood et al., 1987). This "abortive" branching is unique to the dentate gyrus of all regions we have examined. In addition, we are aware of recent immunohistochemical studies utilizing tau or MAP2 antibodies that reported sprouting and regeneration in the AD brain (e.g., Ihara, 1988; McKee et al., 1989). Although it is unclear from these qualitative reports how prevalent the sprouting and regeneration may be, such reports

may not be inconsistent with our Golgi-based evidence, since we have seen similar phenomena, but base our reports on statistical averages. Thus, in AD sprouting or regeneration are, in our hands, much less frequent than degenerative phenomena. Nevertheless, the variety of results emphasizes the need for study of the prevalence of growth of neuronal processes in the AD brain through other means such as quantification of the growth-associated protein, GAP-43/B-50/F1, and its message.

Studies of GAP-43/B-50/F1

Our study of GAP-43/B-50/F1 in the normally aging and AD brain is not intended to be a direct reflection of events in the dendritic compartment of the neuropil. We do argue, however, that the dendritic and axonal compartments of the neuropil mirror each other under most circumstances. Dendrites regress in response to lesions of afferent axons in mature (e.g., Caceres and Steward, 1983) as well as in developing organisms (e.g., Jones and Thomas, 1962). Dendrites also expand in response to increased availability of afferent supply (Perry and Linden, 1982). In our laboratory, a Golgi-serial reconstruction electron microscopic study of the tips of dendrites in rodent olfactory bulb (which show age-related proliferation and regression (Hinds and McNelly, 1977)) found that as dendritic extent changes, density of axonal contacts (synapses) in the tip region remains stable from 12 to 30 months, implying sprouting and loss of axon terminals as dendrites proliferate and regress (Carboni et al., 1985). These data indicate that dendritic extent is related to afferent axons and that changes of dendritic extent are a reflection of the axonal supply, among other factors.

In this morphological context we have initiated studies of GAP-43/B-50/F1 in normally aging and AD human brain. Our initial efforts have concentrated on quantitative studies of message levels rather than on the protein. A number of laboratories have reported relative stability of message in postmortem human brain, and this is so in our hands also, as indicated by the appearance of RNA in denaturing gels and on Northern blots for a variety of mRNA species. Initial studies established ranges for parameters to ensure that determinations were within linear ranges with regard to the quantity of mRNA bound to the nylon membrane, the response of the x-ray film and the laser densitometer. Conditions were chosen such that hybridizations were done under saturating conditions in the presence of excess probe. For reasons of tissue availability, our most complete data set on GAP-43/B-50/F1 message levels is in frontal association cortex (superior frontal gyrus, area 9), a region in which dendrites show continual regression over the age range studied in both normal aging and AD (see above). The amount of total mRNA extracted did not differ between control and AD cases (Fig. 4). In this region, in normal aging there is a 63% decline in level of GAP-43 message from middle age (early fifties) to old age (seventies to nineties). In AD the level of GAP-43/B-50/F1 message declines 32% compared to control cases of similar age, but note that there are no AD cases to

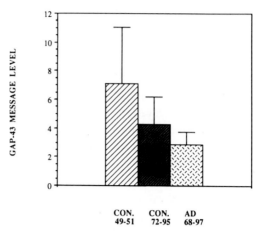

Fig. 5. GAP-43 message levels in area 9 as a function of age and AD.

compare with our youngest control cases. If the small decrease is valid it can probably be explained as a consequence of the excess neuronal loss in AD (Fig. 5).

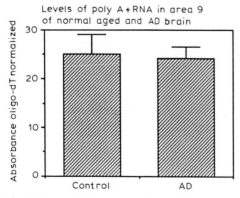

Fig. 4. Total mRNA extracted from AD and control brain, area 9. Data adjusted for weight of tissue.

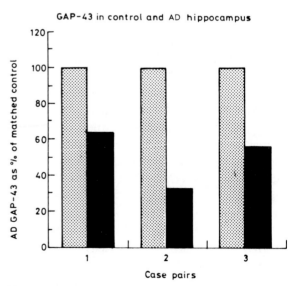

Fig. 6. GAP-43 protein levels in hippocampus determined by ELISA. Levels of control cases are set as 100% for comparison with matched AD cases.

In neocortical area 11 the prevalence of GAP-43 message as a function of age presents a rather different picture. In area 11 there is no decline in message level in normal aging within the age range studied to date. In AD there is about 20% reduction in GAP-43 message level compared to control cases. This reduction in message level can be entirely accounted for by a similar reduction in neuronal density in this region in AD (see Coleman and Flood, 1987, for review).

Very preliminary data in the hippocampus indicate a greater decline in message level in AD, which is paralleled by an approximately 50% decline in GAP-43/B-50/F1 protein as determined by ELISA (Fig. 6). However, some zones of the hippocampus appear to be among the regions showing the most severe excess neuron loss in AD (e.g., Ball, 1977; Doebler et al., 1987), and a 50% decrease in protein or message level in this region may be entirely accounted for by excess neuron loss. In collaboration with G. Higgins we have initiated in situ hybridization studies to clarify this issue.

Preliminary studies with two-dimensional gels followed by Western blotting for GAP-43/B-50/F1 protein in control human hippocampal samples show three major isoforms differing in pI but not in molecular weight. In AD samples the most acidic of these major isoforms appears to be appreciably reduced (Fig. 7). These preliminary data suggest a defect in posttranslational processing of GAP-43/B-50/F1 in AD, which is consistent with a report of decreased protein kinase C activity in AD brain (Cole et al., 1988). Data presented elsewhere in this volume are consistent with a hypothesis that such a defect in posttranslational processing may be related to decreased neuronal plasticity in AD, consistent with the morphological data reviewed above.

Conclusions

Morphological studies of dendritic extent in normal aging and Alzheimer's disease have suggested the hypotheses that normally aging neurons are able to mount plastic, compensatory responses to the death of their neighbors. There are regional differences in the balance between proliferative and regressive influences on dendritic extent that lead to regional differences in age-related changes in dendritic extent. In AD dendritic extent has been found to not increase

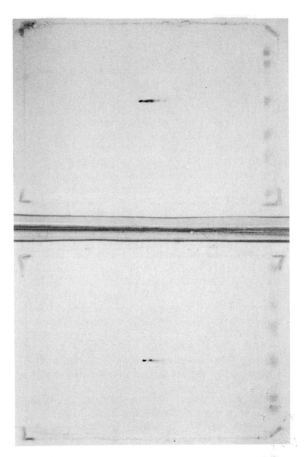

Fig. 7. Two-dimensional gel of butanol-extracted protein from AD and control hippocampus blotted with antibody to GAP-43. Top: control brain; bottom: AD brain. The decreased isoform in the AD brain is the most acidic isoform.

with increasing age, even in regions where there is normally an age-related increase in dendritic extent, suggesting the hypothesis that one of the neurobiological deficits of AD is reduced neuronal plasticity. We have examined GAP-43/B50/F1 message levels as a potential marker of neuronal plasticity. Prevalence of this message as a function of normal aging also shows regional differences, with area 9 showing a decrease in GAP message level in normal aging, while data obtained from area 11 suggest no significant change of message level *per cell* in normal aging. Any decreases in GAP-43 message level found in AD brain may be accounted for on the basis of excess neuron loss in AD. However, preliminary data suggest that there may be defective post-translational processing of GAP-43 in AD.

Acknowledgements

The authors received support from National Institute on Aging grants AG 03644 and AG 01121 and Alzheimer's Disease and Related Disorders grants IIRG-87-053 and PRG-89-120. P.D.C. is a LEAD awardee of the National Institute on Aging. The authors thank J.H.P. Skene for generous provision of GAP-43 probe and antibody.

References

Ball, M.J. (1977) Neuronal loss, neurofibrillary tangles and granulovacuolar degeneration in the hippocampus with ageing and dementia. *Acta Neuropathol. (Berl.)*, 37: 111–118.

Buell, S.J. and Coleman, P.D. (1979) Dendritic growth in the aged human brain and failure of growth in senile dementia. *Science*, 206: 854–856.

Caceres, A. and Steward, O. (1983) Dendritic reorganization in the denervated dentate gyrus of the rat following entorhinal cortical lesions: a Golgi and electron microscopic analysis. *J. Comp. Neurol.*, 214: 387–403.

Carboni, A.A., Jr., del Cerro, M. and Coleman, P.D. (1985) A combined Golgi high voltage-e.m. study of dendrite tips in olfactory bulb in aging Sprague-Dawley rat. *Soc. Neurosci. Abstr.*, 11: 896.

Cole, G., Dobkins, K.R., Hansen, L.A., Terry, R.D. and Saitoh, T. (1988) Decreased levels of protein kinase C in Alzheimer brain. *Brain Res.*, 452: 165–174.

Coleman, P.D. and Buell, S.J. (1983) Dendritic growth in aging brain? In W.H. Gispen and J. Traber (Eds.), *Aging of the Brain*, Elsevier Science Publishers, Amsterdam, pp. 3–8.

Coleman, P.D. and Flood, D.G. (1987) Neuron numbers and dendritic extent in normal aging and Alzheimer's disease. *Neurobiol. Aging*, 8: 521–545.

Cupp, C.J. and Uemura, E. (1980) Age-related changes in prefrontal cortex of *Macaca mulatta*: quantitative analysis of dendritic branching patterns. *Exp. Neurol.*, 69: 143–163.

Doebler, J.A., Markesbery, W.R., Anthony, A. and Rhoads, R.E. (1987) Neuronal RNA in relation to neuronal loss and neurofibrillary pathology in the hippocampus in Alzheimer's disease. *J. Neuropathol. Exp. Neurol.*, 46: 28–39.

Flood, D.G. (1991) Region-specific stability of dendritic extent in normal human aging and regression in Alzheimer's disease. II. Subiculum. *Brain Res.*, 540: 83–95.

Flood, D.G., Buell, S.J., DeFiore, C.H., Horwitz, G.J. and Coleman, P.D. (1985) Age-related dendritic growth in dentate gyrus of human brain is followed by regression in the 'oldest old'. *Brain Res.*, 345: 366–368.

Flood, D.G., Buell, S.J., Horwitz, G.J. and Coleman, P.D. (1987) Dendritic extent in human dentate gyrus granule cells in normal aging and senile dementia. *Brain Res.*, 402: 205–216.

Flood, D.G. and Coleman, P.D. (1986) Failed compensatory dendritic growth as a pathophysiological process in Alzheimer's disease. *Can. J. Neurol. Sci.*, 13: 475–479.

Geddes, J.W., Monaghan, D.T., Cotman, C.W., Lott, I.T., Kim, R.C. and Chui, H.C. (1985) Plasticity of hippocampal circuitry in Alzheimer's disease. *Science*, 230: 1179–1181.

Hanks, S.D. and Flood, D.G. (1991) Region-specific stability of dendritic extent in normal human aging and regression in Alzheimer's disease. I. CA1 of hippocampus. *Brain Res.*, 540: 63–82.

Hinds, J.W. and McNelly, N.A. (1977) Aging of the rat olfactory bulb: growth and atrophy of constituent layers and changes in size and number of mitral cells. *J. Comp. Neurol.*, 171: 345–368.

Hsu, H.K. and Peng, M.T. (1978) Hypothalamic neuron number of old female rats. *Gerontology*, 24: 434–440.

Ihara, Y. (1988) Massive somatodendritic sprouting of cortical neurons in Alzheimer's disease. *Brain Res.*, 459: 138–144.

Jones, W.H. and Thomas, D.B. (1962) Changes in the dendritic organization of neurons in the cerebral cortex following deafferentation. *J. Anat.*, 96: 375–381.

McKee, A.C., Kowall, N.W. and Kosik, K.S. (1989) Microtubular reorganization and dendritic growth response in Alzheimer's disease. *Ann. Neurol.,* 26: 652–659.

Perry, V.H. and Linden, R. (1982) Evidence for dendritic competition in the developing retina. *Nature*, 297: 683–685.

Sladek, J.R., Jr., Khachaturian, H., Hoffman, G.E. and Scholer, J. (1980) Aging of central endocrine neurons and their aminergic afferents. *Peptides,* 1 (Suppl. 1): 141–157.

W.H. Gispen and A. Routtenberg (Eds.)
Progress in Brain Research, Vol. 89
© 1991 Elsevier Science Publishers B.V.

CHAPTER 18

Long-term potentiation: postsynaptic activation of Ca^{2+}-dependent protein kinases with subsequent presynaptic enhancement

Roberto Malinow [1] and Richard W. Tsien [2]

[1] *Department of Physiology and Biophysics, University of Iowa, Iowa City, IA 52242, U.S.A., and* [2] *Department of Molecular and Cellular Physiology, Beckman Center, Stanford University Medical Center, Stanford, CA 94305, U.S.A.*

It is generally believed that persistent alterations in synaptic function provide the cellular basis for learning and memory in both vertebrates and invertebrates (Hebb, 1949; Eccles, 1953; Kandel and Schwartz, 1985; Alkon and Nelson, 1990). The most thoroughly studied example of such synaptic plasticity in the mammalian nervous system is long-term potentiation (LTP). The remarkable feature of LTP is that a short burst of synaptic activity can trigger persistent enhancement of synaptic transmission lasting at least several hours, and possibly weeks or longer (Bliss and Gardner-Medwin, 1973). First found in the hippocampus (Lomo, 1966; Bliss and Lomo, 1973), this phenomenon has recently been described in areas of the mammalian CNS including visual cortex (Artola and Singer, 1988) and motor cortex (Iriki et al., 1989). Many properties of LTP have been described over the past years, such as

threshold (McNaughton et al., 1978), associativity (Barrionuevo and Brown, 1983), and persistence (Bliss and Gardner-Medwin, 1973). These properties make LTP a very likely player in the generation and persistence of memories, a hypothesis that has gained experimental support (McNaughton and Morris, 1987). Thus, there is intense interest in understanding the cellular and molecular basis for this form of synaptic plasticity (Lynch and Baudry, 1984; Collingridge and Bliss, 1987; Bliss and Lynch, 1988; Nicoll et al., 1988; Stevens, 1988; Brown et al., 1988; Cotman et al., 1989; Kennedy, 1989).

Most studies on mechanisms of LTP have focused on synaptic transmission in the CA1 field of the hippocampus, at synapses between Schaffer collaterals and CA1 pyramidal cells, an area rich in NMDA receptors (Cotman et al., 1989) and Ca^{2+}-dependent protein kinases (Nishizuka, 1988; Kennedy, 1989). Induction of LTP requires a temporal conjunction of presynaptic transmitter release and postsynaptic depolarization (Collingridge and Bliss, 1987; Brown et al., 1988), a combination of events resembling that envisioned

Correspondence: Dr. R. Malinow, Department of Physiology and Biophysics, University of Iowa, Iowa City, IA 52242, U.S.A.

by Hebb (1949). It is generally agreed that these factors work together: glutamate to activate the NMDA receptor channel and postsynaptic depolarization to free the channel from block by extracellular Mg^{2+} (Ascher and Nowak, 1988; Mayer and Westbrook, 1988). The NMDA receptor allows a significant Ca^{2+} influx (MacDermott et al., 1986) that increases $[Ca^{2+}]_i$ in postsynaptic dendrites (Regehr and Tank, 1990) and this rise in $[Ca^{2+}]_i$ is necessary for LTP (Lynch et al., 1983; Malenka et al., 1988).

Many questions remain about subsequent cellular signals. What is the mechanism of the long-lasting synaptic modification and its relationship to Ca^{2+} signalling? Our work has been directed toward understanding the role of Ca^{2+}-dependent protein kinases and elucidating the nature of the persistent change. A key question is whether synaptic transmission is strengthened by increased transmitter release or enhanced postsynaptic receptivity. There is considerable controversy about this issue. Evidence in favor of a strictly postsynaptic mechanism for the expression of LTP has come from the groups of Gary Lynch (Muller et al., 1988) and Roger Nicoll (Kauer et al., 1988). Their results indicated that the expression of LTP was associated with a selective enhancement of the response of AMPA receptors, but not NMDA receptors, although both of these glutamate-receptor types are on individual postsynaptic synapses (Bekkers and Stevens, 1989). Thus, they concluded that the locus of expression must be postsynaptic. This conclusion is appealingly simple: if induction and expression of LTP are both localized within the postsynaptic spine, they could be linked by purely intracellular signalling mechanisms.

On the other hand, early evidence from Tim Bliss and colleagues and others has provided support for enhancement of presynaptic release. Using a push-pull cannula system in vivo, Dolphin and colleagues (1982) found evidence for increased release of [^{3}H]glutamate in the dentate gyrus in association with LTP, and proposed that presynaptic release was elevated (see also Skrede and Malthe-Sorenssen, 1981). However, Bliss et al. (1986) were careful to point out that some questions remain unanswered. Which type of cell is the source of the increase? Could the increased concentration of glutamate in the perfusate be due to decreased uptake? If presynaptic glutamate release is truly increased, is this evoked by action potentials or merely an increase in nonquantal leakage? What is the nature of the retrograde communication across the synaptic cleft that must be postulated to link postsynaptic induction and presynaptic expression? Having discussed these questions, Bliss and Lynch (1987) concluded that the available evidence provided "plausible though not conclusive reasons for attributing the enhanced release to an increase in the amount of transmitter released per action potential from potentiated terminals; it is probable that only a rigorous quantal analysis, not presently available, will finally settle the issue".

This chapter describes progress toward such analysis. A major obstacle to statistical analysis of synaptic variability has been the poor signal-to-noise ratio of conventional intracellular recordings (see Redman, 1990 for review). We have overcome this problem by applying the whole-cell voltage clamp technique to study synaptic transmission in conventional hippocampal slices (Malinow and Tsien, 1990a). We find that robust LTP can be recorded with much improved signal resolution and biochemical access to the postsynaptic cell. Prolonged dialysis of the postsynaptic cell blocks the triggering of LTP, with no effect on expression of LTP. The improved signal resolution unmasks a large trial-to-trial variability, reflecting the probabilistic nature of transmitter release. Changes in the synaptic variability, and a decrease in the proportion of synaptic failures during LTP provide two lines of evidence to demonstrate that transmitter release is significantly enhanced (Malinow and Tsien, 1990a). These biophysical approaches are complemented by studies in which we explored mechanisms and localization of LTP induction and expression by direct intracellular injection of protein kinase inhibitors into individual postsynaptic cells (Malinow et al., 1989).

Experimental strategy and methods

Transverse hippocampal slices (400–500 μm) were obtained from 3–5-week-old rats by standard methods (Alger and Nicoll, 1982). Slices were submerged and superfused continuously with a modified Earle's solution containing (in mM): NaCl, 119; KCl, 2.5; MgCl$_2$, 1.3; CaCl$_2$, 2.5; NaH$_2$PO$_4$, 1.0; NaHCO$_3$, 26.2; glucose, 11; and gassed with 95% O$_2$/5% CO$_2$ (pH 7.4) at room temperature (22°C). Picrotoxin (20 μM) was used to block inhibitory transmission; a cut was made between regions CA3 and CA1 to prevent epileptiform activity. Synaptic transmission was elicited with bipolar stainless steel stimulating electrodes at two locations; each electrode delivered a stimulus every 4 s; stimuli from the two electrodes were separated by 2 s. Epscs were recorded with a patch electrode (3–7 MΩ tip resistance, no fire polishing or Sylgard coating) in the whole-cell mode (Axopatch 1D). Pipette solution contained (in mM): Cs-gluconate, 100; EGTA, 0.6; MgCl$_2$, 5; ATP, 2; GTP, 0.3; Hepes, 40 (pH 7.2 with CsOH). Holding potential was kept constant at a level between -60 and -70 mV. Epscs were amplified 50–500-fold, filtered at 1 kHz, digitized at 10 kHz and stored. For a given experiment, epsc amplitudes were determined by averaging the current over a 10–20 ms window at the peak response and subtracting a baseline estimate from the same record. A similar procedure was used to measure the background noise before the synaptic stimulation (Sayer et al., 1989).

Results

Whole cell recordings from hippocampal slices

We were able to make whole-cell voltage clamp recordings from CA1 neurons in rat hippocampal slices without special procedures for exposing or cleaning the neurons (see also Barnes and Werblin, 1987; Coleman and Miller, 1989; Blanton et al., 1990). Positive pressure was applied to the recording pipette during penetration of the slice; high-resistance (1–5 GΩ) seals on cell bodies 2–3 cell diameters below the surface were then obtained by suction. After breaking into the cell (input resistance 80–190 MΩ), synaptic currents were elicited with bipolar stimulation electrodes placed in the stratum radiatum of CA1. Stable synaptic recordings could be maintained for as long as 10 h.

Whole cell current recording offers better resolution than even the most favorable microelectrode recordings (Sayer et al., 1989; Redman, 1990; Foster and McNaughton, 1991). The low background noise of whole cell recordings facilitated the study of small synaptic responses. In most experiments, we used minimal stimulation, only slightly stronger than the highest stimulus that gave only failures. Under these conditions, activation is thought to be restricted to a single synapse onto the monitored neuron (McNaughton et al., 1981; Foster and McNaughton, 1991). This hypothesis is supported by experiments in which a single CA3 neuron was directly stimulated by an intracellular microelectrode while simultaneously recording from a CA1 neuron (Malinow and Tsien, 1990b). The resulting synaptic currents showed a large inter-trial variability that was much greater than the background noise, but had similar time courses (Fig. 1a,b), supporting the view that they originated from the same synapse. There were occasional clear failures (Figs. 1a, 4a,b), and sporadic spontaneous events that resembled the elicited response, ranging in amplitude from < 1 pA up to ~ 10 pA (Fig. 1b, inset).

The average synaptic current for minimal stimulation was 6.7 pA in a representative series of 17 experiments. This is in reasonable agreement with estimates of synaptic current from minimal epsp's recorded with high resistance microelectrodes if one allows for the measured input resistance in those experiments (Sayer et al., 1989; Foster and McNaughton, 1991).

Whole cell recordings of LTP

To optimize the chances of obtaining LTP under whole cell recording conditions, we used internal solutions with minimal Ca^{2+} buffering to avoid block of LTP, and with Cs^+ as the main

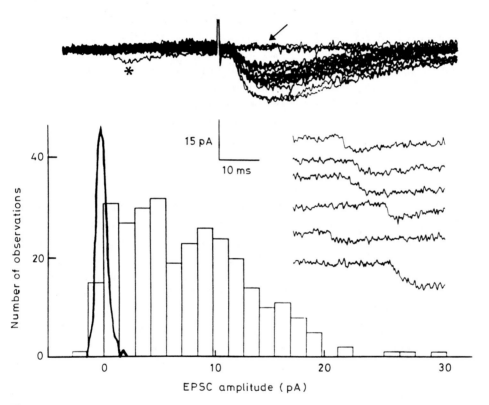

Fig. 1. Whole-cell recordings of synaptic transmission in conventional hippocampal slices show large inter-trial variability, synapse-specific LTP and loss of induction but not expression with prolonged dialysis. (a) 16 consecutive records of synaptic currents, superimposed. Note spontaneous event in baseline period (*) and synaptic failures (arrow). (b) Amplitude distribution histograms for 200 consecutive synaptic currents (bars) and baseline noise (smooth curve). Bin sizes were 1.4 pA for epsc, and 0.14 pA for baseline noise. Inset: representative spontaneous synaptic currents. Note the variability in amplitude and time course but general similarity to evoked responses. (c,d) Plots of synaptic responses against time in a postsynaptic cell receiving two independent inputs. (c) LTP was selectively induced in one pathway by a pairing procedure 15 min following break-in (arrow); in the other pathway (d), the pairing procedure was applied 25 min later, but failed to induce LTP, in contrast to the maintenance of LTP in c. In the pairing procedure, the cell was depolarized from − 70 mV to ~ 0 mV while the paired (conditioned) pathway was stimulated 40 times at 2 Hz with no change in stimulus strength. Non-conditioned pathway received no stimuli during the postsynaptic depolarization. Inset: families of consecutive current records showing epscs of conditioned pathway, collected before (left) and 30 min after pairing (right). From Malinow and Tsien (1990a).

cation to enhance voltage control of the postsynaptic membrane. We also included ATP and GTP in the internal solution to allow activity of protein kinases or GTP-binding proteins. Following a stable baseline period of 10–15 min, we were able to induce LTP by pairing a steady

postsynaptic depolarization to ~ 0 mV with continued activation of the test pathway (40 stimuli at 2 Hz). This pairing procedure resulted in LTP in 14/18 experiments (Figs. 1c, 3a, 4c). In contrast, no potentiation was found in transmission through a simultaneously monitored control path-

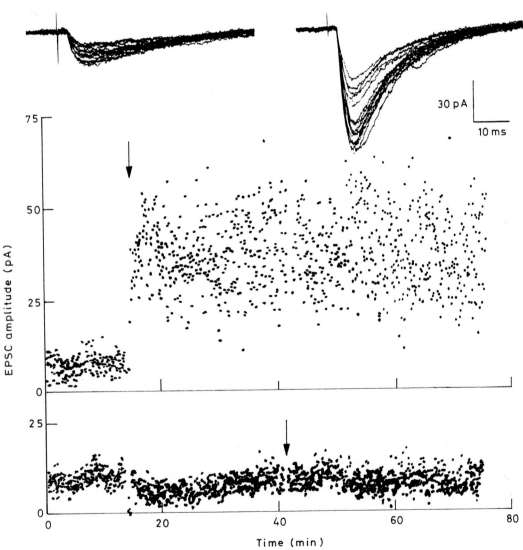

Fig. 1. (continued).

276

way that did not receive presynaptic stimuli during the postsynaptic depolarization (Figs. 1d, 3b). Furthermore, synaptic stimulation at 2 Hz with-out postsynaptic depolarization did not give synaptic enhancement ($n = 5$). Thus, the potentiation was synapse-specific and required a combi-

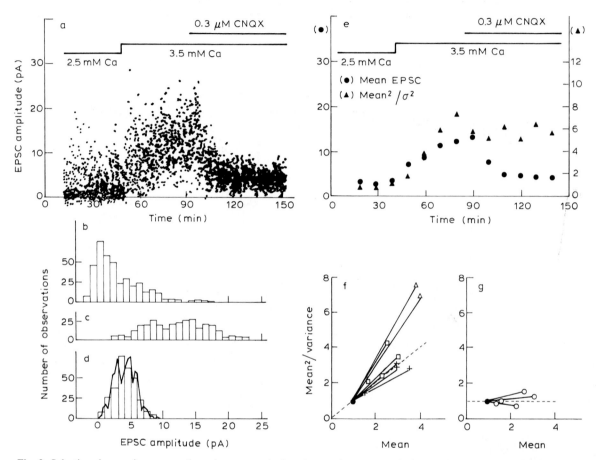

Fig. 2. Selective changes in presynaptic and postsynaptic function produce expected changes in mean and variance of synaptic currents. Agents acting through presynaptic mechanisms affect shape of distribution histograms and M^2/σ^2 whereas inhibition of postsynaptic receptors does not. (a) Epsc amplitude plotted against time. Bathing calcium was elevated from 2.5 to 3.5 mM and 0.3 μM CNQX was applied where shown. (b–d) Amplitude distribution histograms for 300 epscs before manipulation (b), after elevation of calcium (c), and after subsequent addition of CNQX (d). Continuous line in (d) shows data from (c) normalized to mean epsc amplitude in (d). Note close agreement of normalized histograms. (e) Plot of mean epsc amplitude, M ($\circ$, $\bullet$) and M^2/σ^2 ($\triangle$, $\blacktriangle$) for the experiment shown in (a) with $\tau = 10$ min (150 trials). (f) Plot of M^2/σ^2 against M for several presynaptic manipulations: Ca increased from 2.5 to 3.5 mM (triangles); Mg decreased from 1.3 to 0.65 mM in the presence of 50 μM APV (squares); 10 μM 4-aminopyridine (diamonds); increased stimulus strength to recruit more fibers ($\times$). (g) Plot of M^2/σ^2 against M for experiments in which CNQX (0.2–0.3 μM) was bath applied. For each experiment in (f) and (g) lines connect points obtained before and after manipulation; M and M^2/σ^2 were computed for 300 trials prior to the manipulation and 300 trials at the peak of the effect and normalized by the values corresponding to the smaller M for that experiment. Note that presynaptic manipulations that are thought to increase p affect M^2/σ^2 more than the mean, whereas increasing stimulus strength, which increases N, affects M^2/σ^2 as much as M. These findings are expected for a binomial release process (see text). From Malinow and Tsien (1990a).

nation of presynaptic activity and postsynaptic depolarization, just as seen in conventional recordings of LTP.

To investigate whether diffusible cytoplasmic factors are involved in potentiation, we attempted to trigger LTP at different times after gaining whole-cell access (Fig. 1c,d). Pairing soon after beginning whole-cell recording (~ 20 min or less) consistently resulted in LTP lasting > 1 h (Figs. 1, 3, 4). However, no potentiation was found with pairing > 30 min after whole-cell access (6/7 experiments). One possibility is that some diffusible postsynaptic component is needed to trigger LTP, but not to maintain the potentiation.

Analysis of presynaptic and postsynaptic factors

To understand the basis of the synaptic variability and its possible relation to fluctuations in transmitter release, we modified presynaptic or postsynaptic functions by changing the bathing medium (Fig. 2). Elevating $[Ca^{2+}]_o$ to increase presynaptic release enhanced the synaptic currents (Fig. 2a). The amplitude histogram changed from a highly skewed distribution (Fig. 2b) to a nearly symmetrical bell-shape (Fig. 2c). In contrast, when we added low concentrations of the glutamate receptor antagonist 6-cyano-7-nitroquinoxaline-2,3-dione (CNQX) to modify postsynaptic responsiveness, the average synaptic current was dramatically reduced but the shape of the distribution remained unchanged (Fig. 2a,e). Thus, the distributions before and after CNQX matched closely when normalized by their means (Fig. 2d). Similar results were obtained with CNQX at lower $[Ca^{2+}]_o$ (not shown).

To obtain a simple and revealing index of synaptic variability, we computed $CV^{-2} = M^2/\sigma^2$ where CV is the coefficient of variation, M is the mean synaptic current and σ^2 is the variance about M, for a given epoch of consecutive responses from t to $t + \tau$. This kind of analysis has been applied to synaptic transmission in other systems (e.g., del Castillo and Katz, 1954; Martin, 1977). We can compute the expected behaviour of M^2/σ^2 with few assumptions regarding the mechanisms underlying synaptic transmission. In a general case, transmitter release, x, is allowed to vary over all possible values of transmitter release. The postsynaptic response is assumed to be zx, where z is a constant postsynaptic response factor. From elementary probability theory, the mean synaptic response is given by $\langle zx \rangle$ and the variance about the mean is $\langle (zx - \langle zx \rangle)^2 \rangle$, where $\langle \rangle$ denotes the expectation for trials within a given epoch. Then

$$M^2/\sigma^2 = \langle zx \rangle^2 / \langle (zx - \langle zx \rangle)^2 \rangle$$

$$= z^2 \langle x \rangle^2 / \left(\langle (zx)^2 \rangle - \langle zx \rangle^2 \right)$$

$$= z^2 \langle x \rangle^2 / z^2 \left(\langle x^2 \rangle - \langle x \rangle^2 \right)$$

$$= \langle x \rangle^2 / \left(\langle x^2 \rangle - \langle x \rangle^2 \right).$$

Thus, M^2/σ^2 depends only on characteristics of the release process for that epoch, and not on z. This formalism assumes that postsynaptic responsivity is linear and does not vary from trial to trial (i.e., z is constant) within the epoch τ, as found at the neuromuscular junction (del Castillo and Katz, 1954; Martin, 1977). It ignores the possibility of rapid and concerted changes in the receptivity of ensembles of receptors or a spatial heterogeneity in receptivity that responds differentially to independent sites of transmitter release.

Assuming z is constant for an epoch, the theory predicts that changes in M^2/σ^2 between epochs will result from changes in release characteristics but not postsynaptic modifications. This was tested by examining changes in M^2/σ^2 following interventions known to affect presynaptic or postsynaptic mechanisms. As illustrated in Fig. 2e, raising extracellular calcium dramatically in-

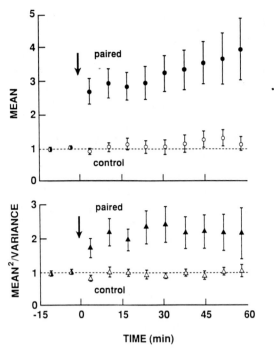

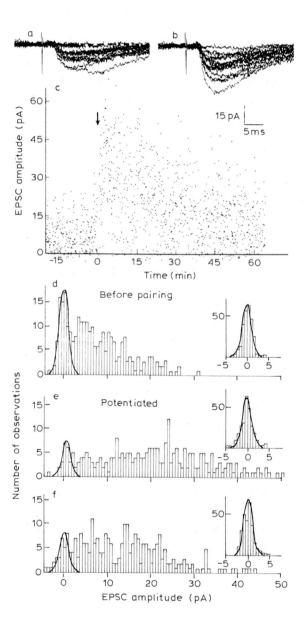

creased M^2/σ^2. This was found for all experimental maneuvers expected to affect presynaptic release: lowering magnesium, application of 4-aminopyridine (not shown) or increasing the stimulus strength to recruit more afferent fibres (Fig. 2f, $n = 10$). In general, M^2/σ^2 increased at least

Fig. 3. A synapse-specific increase in M^2/σ^2 associated with LTP. Results from the 14 of 18 experiments that showed LTP with pairing (in the other 4 experiments with no LTP, M^2/σ^2 remained unchanged, not shown). (a) Ensemble averages and SEM of M for the paired pathway (filled circles) and the control pathway ($\bigcirc$, $n = 10$; in 4 experiments this pathway was not monitored). Arrow marks time of pairing procedure (as in Figs. 1, 2). (b) Ensemble averages of M^2/σ^2 for paired ($\blacktriangle$) and non-paired ($\triangle$) pathways for the same experiments as in (a). From Malinow and Tsien (1990a).

Fig. 4. Decreases in the proportion of synaptic failures associated with LTP. a,b, groups of 16 consecutive epscs taken before (a) and 40 min after (b) a pairing procedure to induce LTP. (c) Amplitude of epsc's elicited by minimal stimulation plotted against time. Arrow marks pairing procedure (see Fig. 1). (d,e) Amplitude distribution histograms for 17 min epochs prior to pairing (d, from $t = -17$ to $t = 0$) and after pairing (e, from $t = 5$ to $t = 22$ min; (f) from $t = 40$ to $t = 57$ min). Smooth curves are Gaussians whose height was adjusted to fit the first peak in the amplitude distribution. The half-width of the gaussians, 2.5 pA, was obtained by fitting amplitude histograms of the baseline noise during the same epochs (insets). The proportion of failures was estimated as the area under the gaussian curve. From Malinow and Tsien (1990a).

as much as the mean synaptic current. In contrast, addition of CNQX produced no significant change in M^2/σ^2, despite its large effect on synaptic transmission ($n = 5$; see Fig. 2e,g).

Changes during LTP

Does M^2/σ^2 change with LTP? Fig. 3b compares M^2/σ^2 of the test pathway with that of the control (unpaired) pathway, for the 14 experiments showing potentiation of mean synaptic current (Fig. 3a). Following pairing, M^2/σ^2 increased significantly in the test pathway, with no change in the control pathway. The increase in

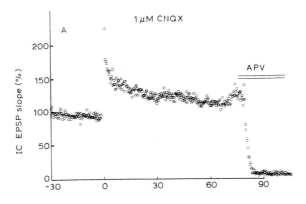

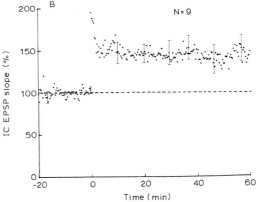

Fig. 6. Enhancement of NMDA-receptor response during LTP. (A) Representative experiment in the presence of 1 μM CNQX to block K/Q receptors. Relative amplitude of epsp, recorded with an intracellular microelectrode plotted against time. Tetanic stimulation was applied at the arrow. Abolition of epsps by 50 μM APV demonstrates that epsp was mediated by NMDA receptors. For further experimental details, see Malinow et al. (1989). (B) Collected results from similar experiments in a total of 9 slices. Epsps amplitudes were normalized to control values before tetanic stimulation, then ensemble averaged. Bars indicate SEM. Note significant and long-lasting potentiation of the NMDA-receptor-mediated response.

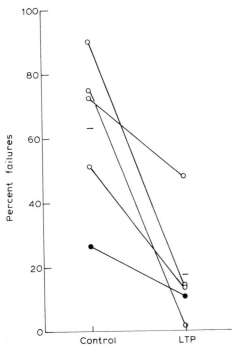

Fig. 5. Consistent decrease in proportion of synaptic failures during LTP. Collected results from five experiments. Horizontal lines show mean percentage of failures ($63 \pm 11\%$ in control before pairing, $17 \pm 8\%$ during LTP after pairing). Filled symbols show results from experiment illustrated in Fig. 4. The 2 experiments with the highest percentage of failures in control were recordings of one-to-one transmission from presynaptic CA3 neurons, stimulated with an intracellular microelectrode. Data from Malinow and Tsien (1990a).

M^2/σ^2 with LTP could arise from various presynaptic mechanisms: an increase in the number of available vesicles; an increase in the probability of release of some or all vesicles; or changes in non-vesicular release. It would not be expected from uniform changes in the number, sensitivity or conductance of glutamate receptor channels, or in the effectiveness of charge transfer from spines to dendrites.

If the expression of LTP involves an increased probability of release or a greater number of available quanta, one would expect fewer failures of transmission during LTP. Fig. 4 illustrates an experiment where failures and responses were clearly resolvable. The amplitude distribution histogram displayed a distinct peak at zero amplitude which matched the amplitude distribution of the noise (inset). The area under the zero amplitude peak was used to estimate the percentage of failures. In this experiment, the proportion of failures was 27% prior to pairing (Fig. 4d), falling to 11% between 5 and 22 min after pairing (Fig. 4e). This change accompanied a 2.8-fold increase in mean synaptic current from d to e. During a later epoch (from $t = 40$ to $t = 57$ min), the synaptic enhancement was 2.1-fold relative to control, and the proportion of failures was 15% (Fig. 4f).

This decrease in the proportion of failures was a consistent finding. Fig. 5 shows collected results from 5 experiments where failures could be clearly resolved. Three of the recordings were obtained with minimal stimulation and two with intracellu-

lar microelectrode stimulation of a CA3 neuron (Malinow and Tsien, 1990b). The collected data from all 5 experiments showed a mean ($\pm$ SEM) decrease in failures from $63 \pm 11\%$ in control to $17 \pm 8\%$ after pairing, consistent with a greater likelihood of transmitter release during LTP. This conclusion is supported by the overall amplitude distribution, which changes from a skewed shape in control (Fig. 4d) to a bell-shape during LTP (Fig. 4e), as with increasing extracellular Ca (Fig 2b,c).

Changes in NMDA receptor-mediated transmission can be detected

If presynaptic transmitter release is enhanced, one might expect to find an increase in both the kainate/quisqualate-sensitive and NMDA-sensitive components of the epsp. The available evidence is not consistent. On one hand, Kauer et al. (1988) and Muller and Lynch (1988) found no significant increase in the NMDA component under conditions that normally produce LTP; thus, they concluded that the expression of LTP must be purely postsynaptic. On the other hand, Bashir and Collingridge (1990) have reported a significant potentiation of NMDA-mediated transmission under very similar conditions.

Fig. 7. Selective postsynaptic block of PKC or CaMKII prevents LTP. Diagram shows stimulation and recording conditions (see text). (A) Extracellular recordings of transmission show persistent potentiation following a tetanus (arrow). (B) Simultaneous monitoring of synaptic potentials using an intracellular microelectrode whose tip is filled with 3 mM PKC(19–31) ($n = 8$). Following the tetanus there is no persistent potentiation. (C) Transmission in a non-tetanized pathway, monitored through the same PKC(19–31)-containing electrode, is constant throughout the experiment, indicating no non-specific depressant effect on basal synaptic transmission. (D) Transmission monitored in a different set of slices with 3 mM [Glu^{27}]PKC(19–31) in the intracellular electrode shows LTP after a conditioning tetanus (n = 6 pathways from three slices). (E–H) Recordings of transmission in experiments testing the involvement of CaMKII. (E) Extracellular monitoring shows LTP after tetanic stimulation. (F) Simultaneous monitoring of synaptic potentials with intracellular electrode containing 1.1 mM CaMKII(273–302) shows no persistent potentiation after tetanic conditioning. (G) Transmission in a nontetanized pathway, monitored with the CaMKII(273–302) containing electrode, is constant throughout the experiment. (H) Transmission monitored in a different set of slices, using 1.1 mM CaMKII(284–302) in the intracellular electrode shows LTP after a tetanus ($n = 5$ pathways from three slices). Error bars indicate SEMs for representative time points. Insets: average of 10 consecutive potentials obtained at the times designated on time axis. Scale bars for A and E: 0.33 mV, 12.5 ms; for B, D, F and H: 5 mV and 12.5 ms. Peptides PKC(19–31), PKC(19–36) were generously provided by Dr. John J. Nestor, Jr., Dr. Bruce Kemp and Dr. Tim Mietzner, respectively. From Malinow et al. (1989).

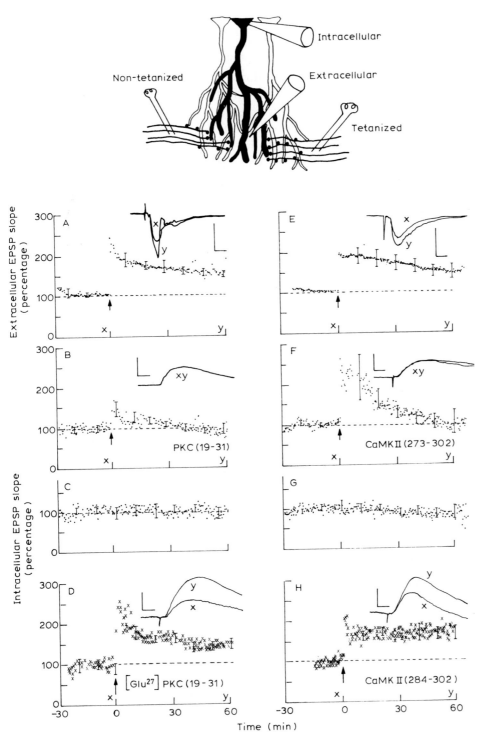

Non-tetanized

Intracellular

Extracellular

Tetanized

Time (min)

We have looked for changes in the NMDA-sensitive component under conditions similar to those in our other experiments. To reveal the NMDA receptor component, the external $[Mg^{2+}]_o$ was lowered to 0.65 mM, the membrane potential was held near -60 mV, and K/Q receptors were blocked by including 1–10 μM CNQX in the superfusing solutions. Fig. 6 shows a representative experiment (A) and collected results from 9 recordings, obtained in the presence of 1 μM CNQX (B). The results show a clear and significant increase in the NMDA component during LTP in the test pathway (B), but not in the control pathways (not shown). The potentiated epsp's were completely blocked by the NMDA receptor blocker APV. Thus, our findings show that an enhancement of the NMDA-receptor component can be detected.

Further work will be needed to understand how our results and those of Bashir and Collingridge (1989) may be reconciled with the results of Kauer et al. (1988) and Muller and Lynch (1988). We find that the NMDA response is quite sensitive to membrane potential – even a few millivolts hyperpolarization (which can occur after a tetanus) can reduce the NMDA response by 50%. Thus in all our experiments, the NMDA-mediated transmission in the control pathway was monitored, and found to be unchanged. Similarly, ambient glycine concentration can affect NMDA responses (Johnson and Ascher, 1987), and this action can be modulated by CNQX (Lester et al., 1989). To allow for such complications, we looked for LTP of the NMDA component under a variety of conditions: with 1 μM CNQX (Fig. 6), with 10 μM CNQX, and with 10 μM CNQX plus exogenous glycine. In all cases, there was clear and significant potentiation of an APV-sensitive epsp. If anything, the potentiation was more pronounced with 1 μM CNQX or with 10 μM CNQX + 300 μM glycine than with 10 μM CNQX alone, but the differences were not dramatic enough to be statistically significant.

In addition to the experimental considerations mentioned above, another possibility is that the NMDA receptor was closer to saturation in the earlier experiments. In our opinion, the possibility of NMDA saturation was not excluded by previous arguments based on the demonstration of synaptic potentiation with paired pulses (Muller and Lynch, 1988) or after tetani (Kauer et al., 1988); these forms of potentiation can take place by recruitment of silent boutons as well as previously active boutons, so some enhancement would be seen regardless of NMDA receptor saturation at previously active boutons. In contrast, during LTP, enhancement of transmission is thought to be restricted to previously active synapses since local transmitter release must occur to induce LTP. NMDA receptor saturation would thus impose a more stringent limitation during LTP than with paired pulse facilitation or PTP.

Unravelling signal transduction between postsynaptic induction and presynaptic expression: role of Ca^{2+}-dependent protein kinases

A central issue at this point is how a rise in postsynaptic $[Ca^{2+}]_i$ can lead to a long-lasting enhancement of presynaptic function. A generic working hypothesis, favored by many investigators, is that postsynaptic Ca^{2+} acts through a signal transduction pathway that involves a Ca^{2+}-dependent protein kinase such as protein kinase C (PKC) or the multifunctional Ca/calmodulin-dependent protein kinase (CaMKII) (for reviews, see Cotman et al., 1989; Kennedy, 1989; Malenka et al., 1989). We have tested these ideas by postsynaptic intracellular injection of peptides that are potent and selective inhibitors of either PKC or CaMKII (Malinow et al., 1989). PKC(19–31) or PKC (19–36) are peptide fragments forming the pseudosubstrate region of the PKC regulatory domain (House and Kemp, 1987). They are > 600-fold more potent as blockers of PKC (IC_{50} ~ 0.1 μM) than as blockers of CaMK. In contrast, the Glu^{27} derivative of PKC(19–31) is relatively inactive against either kinase and serves as a useful control against nonspecific effects.

The results of postsynaptic injection of peptides are illustrated in Fig. 7 (left). PKC(19–31) and PKC(19–36) blocked LTP when delivered to the postsynaptic cell with the recording intracellular microelectrode (LTP should have been expected as all surrounding cells showed enhanced transmission as monitored with an extracellular electrode); [Glu^{27}]PKC(19–31), the control peptide, failed to block LTP. These results are compatible with effects of injecting PKC (Hu et al., 1987), or PKC inhibitory peptides (Andersen et al., 1990).

To investigate the role of postsynaptic CaMKII, we used the peptide fragment CaMKII (273–302), which inhibits CaMKII at much lower concentrations than PKC. A control was provided by CaMKII(284–302), a shorter peptide that is much less effective in blocking CaMKII. When delivered to postsynaptic cells through the recording microelectrode, CaMKII(273–302) blocks LTP (Fig. 7, right), while CaMKII(284–302) does not. We obtained similar results with a calmodulin-blocking peptide (not shown; see also Malenka et al., 1989). The effects of the CaMKII fragments are almost certainly not explained by block of PKC, since CaMKII(284–302) is more effective than CaMKII(273–302) in blocking PKC but was ineffective in preventing LTP (Malinow et al., 1989). We thus conclude that postsynaptic PKC and CaMKII are both required for the establishment of LTP.

These experiments do not test directly if postsynaptic activity of these kinases is required during tetanic stimulation, or if they are generally active at some basal level and that this basal activity is necessary. It is possible, for instance,

that continual phosphorylation of NMDA receptors (MacDonald et al., 1989) is required for their function, and that in the absence of kinase activity, a phosphotase could render NMDA receptors inactive and thereby prevent induction of LTP. The exact temporal role and interaction between postsynaptic PKC and CaMKII in the generation of LTP will require more investigation.

To determine if the postsynaptic kinases are involved in the *expression* of LTP, we impaled

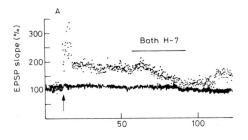

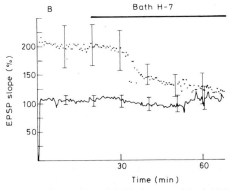

Fig. 8. Expression of LTP is reversibly inhibited by bath applied H-7. A, representative experiment. Arrow marks tetanic stimulation. For experimental details, see Malinow et al., 1988. B, ensemble average of epsp amplitudes in 20 experiments, plotted against time. Transmission from two different pathways is monitored with an extracellular electrode. One pathway (unconnected dots) was previously potentiated by delivery of a high-frequency stimulus, whereas the other pathway (connected dots) was not conditioned. H-7 (50–300 μM) was applied to the bath as indicated. Note that drug acts relatively selectively in suppressing the potentiation in the conditioned pathway while leaving transmission in the control pathway essentially unaffected.

cells with microelectrodes containing protein kinase inhibitors *after* the high frequency stimulation (arrow) had already established potentiated transmission (insets, Fig. 9). Consistent with the idea that the expression of LTP involves a presynaptic modification, we found that established LTP was *not* suppressed by intracellular postsynaptic H-7. Delayed introduction of a combination of PKC(19–31) and CaMKII(273–302) was similarly ineffective (Malinow and Tsien, 1990c). Control procedures showed that the agents were successfully delivered (D) and that LTP had in-

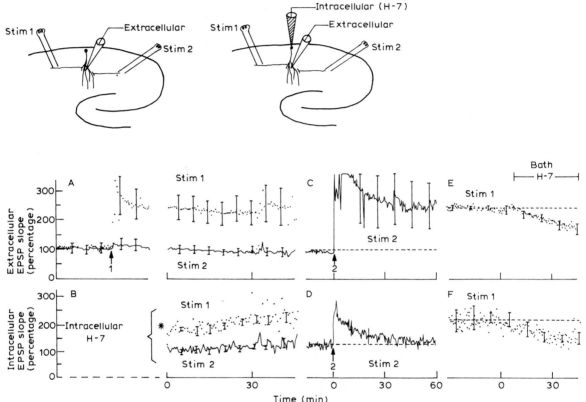

Fig. 9. Expression of LTP is insensitive to postsynaptic H-7 application. Insets indicate recording configurations before and after delayed impalement with H-7 containing microelectrode. (A) Synaptic potentials are monitored extracellularly (no microelectrode) in response to alternate stimulation of two independent pathways (unconnected points: stim 1; and connected points: stim 2). Tetanic conditioning is delivered to stim 1 (arrow, 1). After the establishment of stable LTP an intracellular recording with an H-7-containing electrode is obtained (B) and monitoring of synaptic potentials begins within 2 minutes of penetration (at $t = 0$ on lower axis). As monitored by the H-7-containing intracellular electrode, synaptic transmission from the potentiated pathway (stim 1, *) does not decay during the observation period and parallels transmission from the unpotentiated pathway (stim 2) ($n = 13$). In this panel, synaptic strength in each pathway is normalized relative to average data for the first five minutes of transmission in the untetanized pathway. (C, D) To determine if H-7 from the intracellular electrode has reached the synaptic zone, the previously unpotentiated pathway (stim 2) is tested for the ability to generate LTP (7 slices). Delivery of a tetanus to stim 2 results in slowly decaying potentiation but no LTP as monitored with the intracellular electrode (D), despite a large persistent potentiation seen with the extracellular electrode (C). (E,F) In 8 slices, H-7 is subsequently bath applied and a comparable synaptic diminution of pathway 1 is seen with both extracellular monitoring and with the H-7-filled intracellular microelectrode. Occlusion or reduction of the effect of externally applied H-7 would have been expected if postsynaptic H-7 had already inhibited the potentiated transmission. From Malinow et al. (1989).

deed been established before the impalement (E,F). Thus, once established, the persistent signal is inaccessible to postsynaptic injection of H-7 (or to the kinase-blocking peptides). Interestingly, established LTP remains sensitive to bath application of H-7 (Fig. 9E), and we find it to be relatively selective to potentiated transmission (Fig. 8). Thus, we conclude that the maintenance of LTP depends upon H-7-sensitive persistent signalling somewhere outside the postsynaptic cell, possibly in the presynaptic terminal.

Discussion and Conclusions

The main conclusions of our studies are as follows:

(1) whole cell recordings can be used to study synaptic transmission in neurons well below the surface of conventional brain slices over many hours. This method allows access to deeper cells that are more likely to retain intact dendritic structures and functional properties. The approach is an alternative to methods that require cleaning of the surface of slices (Edwards et al., 1989).

(2) Whole cell recordings offer much better signal-to-noise than conventional intracellular recordings, facilitating statistical analysis of small synaptic signals. Under favorable circumstances, quantal responses can be distinguished from synaptic failures (Figs. 1 and 4).

(3) Experiments conform with theory in supporting M^2/σ^2 as a simple measure of changes in presynaptic function in hippocampal slices. This approach might be useful in characterizing neuroactive drugs whose locus of action is unknown.

(4) Synapses can display LTP even after the postsynaptic cell has been accessed by a whole-cell recording pipette. This opens up the possibility of probing the postsynaptic mechanisms of LTP with a wide range of biochemical compounds which could not be reliably delivered with intracellular

microelectrodes. Some obvious possibilities include large proteins or even nucleic acids.

(5) Interestingly, the ability to undergo LTP was lost following longer periods of whole cell recording, as if some key cytoplasmic constituent were washed out, even though LTP established earlier during the recording did not decay. This contrast supports the idea that induction and persistence of LTP involve different molecular events, possibly in different locations (Malinow et al., 1989; Goh and Pennefather, 1989; Segal and Patchornik, 1989).

(6) When LTP was induced relatively soon after break-in, the degree of potentiation was large, ranging up to tenfold and averaging more than threefold. The magnitude of the enhancement is considerably greater than seen with other recording methods. A large degree of persistent synaptic plasticity is of obvious interest for neural modelling (e.g., McNaughton and Morris, 1987).

(7) Long-term potentiation of up to 10-fold was seen with intracellular stimulation of single CA3 neurons (Malinow and Tsien, 1990b). This provides the first direct evidence that LTP is a property of one-to-one connections between presynaptic and postsynaptic neurons, rather than an emergent property of a large number of converging presynaptic inputs (cf. Friedlander et al., 1990).

(8) Biophysical analysis and biochemical approaches provide three complementary lines of evidence for an enhancement of presynaptic function during LTP, as suggested previously by other approaches (Skrede and Malthe-Sorenssen, 1981; Dolphin et al., 1982; Malenka et al. 1987; Nelson et al., 1989; Malinow et al., 1989; Desmond and Levy, 1986).

(9) The analysis of M^2/σ^2 is relatively model independent, and does not require failures. M^2/σ^2 increased significantly and remained elevated during LTP; it was unchanged in pathways not undergoing potentiation (Fig. 3b). LTP was

associated with a clear change in M^2/σ^2 and the shape of the amplitude histogram was similar to that seen with elevated $[Ca^{2+}]_o$.

(10) An analysis based on failures gives a more direct view of presynaptic function. This method does not require that a single synapse be activated, an issue for analyses based on changes in M^2/σ^2. In all cases where unitary events were large relative to the background noise so that failures were clearly resolved, the occurrence of failures was greatly diminished; the proportion of failures fell by about 5-fold during LTP (e.g. Fig. 5). In some experiments, failures were common during the control run and completely disappeared with LTP.

(11) While all the biophysical results support an increase in the likelihood of transmitter release, they do not exclude some change in postsynaptic responsivity. We found that M increased more than M^2/σ^2, particularly 30–60 min after pairing; the late change in M might reflect a delayed increase in postsynaptic responsivity as suggested by iontophoretic application of AMPA (Davies et al., 1989).

(12) A third and rather different line of evidence is based on postsynaptic injection of inhibitors of protein kinases, H-7 or peptide inhibitors of Ca^{2+}-dependent kinases. After its establishment, LTP appears unresponsive to postsynaptic H-7 or to delivery of both peptides in combination, although it remains sensitive to externally applied H-7 (Figs. 8 and 9). These results suggest that LTP is maintained by a signalling pathway not entirely contained within the postsynaptic cell. One possibility is that the pathway includes a presynaptic protein kinase such as PKC.

(13) Our evidence supporting presynaptic changes with LTP is striking in light of previous studies showing that induction is postsynaptic (Wigstrom et al., 1986; Malinow and Miller, 1986; Sastry et al., 1986; Kelso et al., 1986; Lynch et al., 1983; Malenka et al., 1988; Malinow et al., 1989). The

combination of results makes it difficult to escape the conclusion that a retrograde message must travel from the conditioned postsynaptic cell to modify presynaptic function (see Bliss and Lynch, 1988). Since the proportion of failures decreases and M^2/σ^2 increases soon after pairing, the retrograde signalling must occur promptly.

(14) A retrograde message might act jointly with a presynaptic signal (e.g., elevated $[Ca^{2+}]_i$). This would constitute a presynaptic AND function, somewhat similar to the AND operation of glutamate + depolarization at the postsynaptic NMDA receptor. In principle, the retrograde signal could spread laterally along the presynaptic afferent to modify synaptic boutons onto other postsynaptic cells (Bonhoeffer et al., 1989).

(15) An enhancement of presynaptic function effectively increases the signal-to-noise ratio of evoked synaptic release relative to "background noise" produced by the ambient level of glutamate within brain tissue. For signal processing, this may be a more robust and efficient mechanism than an elevated postsynaptic receptivity, which would increase both signal and noise together.

Acknowledgments

We are grateful to Dr. Daniel Madison for contributing some of the experiments illustrated in Fig. 8. We thank Drs. D.D. Friel, J.A. Kauer, D.V. Madison and R.S. Zucker for helpful discussion. This work was supported by Javits Investigator Award NS24067 to R.W.T.

References

Andersen, P., Godfraind, J.M., Greengard, P., Hvalby, O., Nairn, A., Raastad, M. and Storm, J.F. (1990) Injection of a peptide inhibitor of protein kinase C blocks the induction of long-term potentiation in rat hippocampal cells in *vitro. J. Physiol.*, in press.

Alger, B.E. and Nicoll, R.A. (1982) Feed-forward dendritic

inhibition of rat hippocampal pyramidal cells studied *in vitro*. *J. Physiol*. 328: 105–123.

Alkon, D.L. and Nelson, T.J. (1990) Specificity of molecular changes in neurons involved in memory storage. *FASEB J.*, 4: 1567–1576.

Artola, A. and Singer, W. (1987) Long-term potentiation and NMDA receptors in rat visual cortex. *Nature*, 330: 649–652.

Ascher, P. and Nowak, L. (1988) Electrophysiological studies of NMDA receptors. *Trends Neurosci.*, 10: 284–288.

Barnes, S. and Werblin, F. (1987) Gated currents generate single spike activity in amacrine cells of the tiger salamander retina. *Proc. Natl. Acad. Sci. USA*, 83: 1509–1512.

Bashir, Z.I. and Collingridge, G.L. (1990) Potentiation of an NMDA receptor-mediated EPSP in rat hippocampal slices *in vitro*. *J. Physiol.*, 425: 23P.

Bekkers, J.M. and Stevens, C.F. (1989) NMDA and non-NMDA receptors are co-localized at individual excitatory synapses in cultured rat hippocampus. *Nature*, 241: 230–233.

Blanton, M.G., Lo Turco, J.J. and Kriegstein, A.R. (1989) Whole cell recording from neurons in slices of reptilian and mammalian cerebral cortex. *J. Neurosci. Methods*, 30: 203–210.

Bliss, T.V.P. and Gardner-Medwin, A. (1973) Long-lasting potentiation of synaptic transmission in the dentate area of the unanaesthetized rabbit following stimulation of the perforant path. *J. Physiol.*, 232: 357–74.

Bliss, T.V.P. and Lomo, T. (1973) Long-lasting potentiation of synaptic transmission in the dentate area of the anaesthetized rabbit following stimulation of the perforant path. *J. Physiol. (Lond.)*, 232: 331–56.

Bliss, T.V.P. and Lynch, M. (1988) Long-term potentiation of synaptic transmission in the hippocampus: properties and mechanisms. In: P.W. Landfield and S.A. Deadwyler (Eds.), *Long-Term Potentiation: Mechanisms and Key Issues, From Biophysics to Behavior*, Alan R. Liss, New York, pp. 3–72.

Bliss, T.V.P., Douglas, R.M., Errington, M.L. and Lynch, M.A. (1986) Correlation between long-term potentiation and release of endogenous amino acids from dentate gyrus of anaesthetized rats. *J. Physiol.*, 377: 391–408.

Bonhoeffer, T., Staiger, V. and Aertsen, A. (1989) Synaptic plasticity in rat hippocampal slice cultures: local "Hebbian" conjunction of pre- and postsynaptic stimulation leads to distributed synaptic enhancement. *Proc. Natl. Acad. Sci. USA*, 86: 8113–8117.

Brown, T.H., Chapman, P.F., Kairiss, E.W. and Keenan, C.L. (1988) Long-term synaptic potentiation. *Science*, 242: 724–728.

Chang, F. and Greenough, W.T. (1984) Transient and enduring morphological correlates of synaptic activity and efficacy change in the rat hippocampal slice. *Brain Res.*, 309: 35–46.

Coleman, P.A. and Miller, R.F. (1989) Measurement of passive membrane parameters with whole-cell recordings from neurons in the intact amphibian retina. *J. Neurophysiol.*, 61: 218–230.

Cotman, C.W., Bridges, R.J., Taube, J.S., Clark, A.S., Geddes, J.W. and Monaghan, D.T. (1989) The role of the NMDA receptor in central nervous system plasticity and pathology. *J. NIH Res.*, 1(2): 65–74.

Davies, S.N., Lester, R.A.J., Reymann, K.G. and Collingridge, G.L. (1989) Temporally distinct pre- and post-synaptic mechanisms maintain long-term potentiation. *Nature*, 330: 500.

Del Castillo, J. and Katz, B. (1954) Quantal components of the end-plate potential. *J. Physiol.*, 124: 560–573.

Dolphin, A.C., Errington, M.L. and Bliss, T.V.P. (1982) Long-term potentiation of the perforant path *in vivo* is associated with increased glutamate release. *Nature*, 297: 496.

Eccles, J.C. (1953) *The Neurophysiological Basis of Mind, The Principles of Neurophysiology*, Clarendon Press, Oxford.

Edwards, F.A., Konnerth, A. and Sakmann, T. (1989) A thin slice preparation for patch clamp recordings from synaptically connected neurones of the mammalian central nervous system. *Pfluegers Arch.*, 414: 600.

Foster, T.C. and McNaughton, B.L. (1991) Long-term synaptic enhancement in CA1 is due to increased quantal size, not quantal content. *Hippocampus*, in press.

Friedlander, J.J., Sayer, R.J. and Redman, S.J. (1990) Evaluation of long-term potentiation of small compound and unitary EPSPs at the hippocampal CA3-CA1 synapse. *J. Neurosci.*, 10: 814–825.

Goelet, P., V.F. Castellucci, S. Schacher and E.R. Kandel. (1985) The long and short of long-term memory – a molecular framework. *Nature*, 322: 419–422.

Goh, J.W. and Pennefather, P.A. (1989) Pertussis toxin-sensitive G protein in hippocampal long-term potentiation. *Science*, 244: 980–983.

Gustafsson, B., Wigstrom, H., Abraham, W.C. and Huang, Y.Y. (1987) Long-term potentiation in the hippocampus using depolarizing current pulses as the conditioning stimulus to single volley synaptic potentials. *J. Neurosci.*, 7: 774–780.

Hebb, D.O.(1949) *Organization of Behavior*, Wiley, New York.

House, C. and Kemp, B.E. (1987) Protein kinase C contains a psendosubstrate prototype in its regulatory domain. *Science*, 238: 1726–1728.

Hu, G.Y., Hvalby, O., Walaas, S.I., Albert, K.A., Skjelfo, P., Andersen, P. and Greengard, P. (1987) Protein kinase C injection into hippocampal pyramidal cells elicits features of long-term potentation. *Nature*, 328: 426–429.

Iriki, A., Pavlides, C., Keller, A. and Asanuma, H. (1989) Long-term potentiation in the motor cortex. *Science*, 246: 1385–1387.

Johnson, J.W. and Ascher, P. (1987) Glycine potentiates the

288

NMDA response in cultured mouse brain neurons. *Nature*, 325: 529–531.

Kandel, E.R. and Schwartz, J.H. (1982) Molecular biology of learning: Modulation of transmitter release. *Science*, 218: 433–443.

Kauer, J.A., Malenka, R.C. and Nicoll, R.A. (1988) A persistent postsynaptic modification mediates long-term potentiation in the hippocampus. *Neuron*, 1: 911–917.

Kelso, S.R., Ganong, A.H. and Brown, T.H. (1986) *Proc. Natl. Acad. Sci. USA*, 83: 5326–5330.

Kennedy, M.B. (1989) Regulation of synaptic transmission in the central nervous system: long-term potentiation. *Cell*, 59: 777–787.

Lester, A.J.R, Quarum, M.L., Parker, J.D., Weber, E. and Jahr, C.E. (1989) Interaction of 6-cyano-7-nitroquinoxaline-2,3-dione with the *N*-methyl-D-aspartate receptor-associated glycine binding site. *Mol. Pharmacol.*, 35: 565–570.

Levy, W.B. and Steward, O. (1983) Temporal contiguity requirements for long-term associative potentiation/depression in the hippocampus. *Neuroscience*, 8: 791–797.

Lomo, T. (1966) Frequency potentation of excitatory synaptic activity in the dentate area of the hippocampal formation. *Acta Physiol. Scand.*, Suppl. 277: 128.

Lynch, G. and Baudry, M. (1984) The biochemistry of memory: a new and specific hypothesis. *Science*, 224: 1057–1063.

Lynch, G., Larson, J., Kelso, S., Barrionuevo, G. and Schottler, F. (1983) Intracellular injections of EGTA block induction of hippocampal long-term potentiation. *Nature*, 305: 719–721.

MacDermott, A.B., Mayer, M.L., Westbrook, G.L., Smith, S.J. and Barker, J.L. (1986) NMDA-receptor activation increases cytoplasmic calcium concentration in cultured spinal cord neurones. *Nature*, 321: 519–522.

McNaughton, B.L. and R.G.M. Morris. (1987) Hippocampal synaptic enhancement and information storage within a distributed memory system. *Trends Neurosci.*, 10: 408–415.

McNaughton, B.L., Barnes, C.A. and Andersen, P. (1981) Synaptic efficiacy and EPSP summation in granule cells of rat fascia dentate studied in vitro. *J. Neurophysiol.*, 46: 952–966.

Malenka, R.C., Kauer, J.A., Zucker, R.S. and Nicoll, R.A. (1988) Postsynaptic calcium is sufficient for potentiation of hippocampal synaptic transmission. *Science*, 242: 81–84.

Malenka, R.C., Kauer, J.A., Perkel, D.J. and Nicoll, R.A. (1989a) *Trends Neurosci.*, 12: 444–450.

Malenka, R.C., Kauer, J.A., Perkel, D.J., Mauk, M.D., Kelly, P.T., Nicoll, R.A. and Waxham, M.N. (1989b) *Nature*, 340: 554–557.

Malinow, R. and Miller J.P. (1986) Postsynaptic hyperpolarization during conditioning reversibly blocks induction of long-term potentiaton. *Nature*, 321: 529–530.

Malinow, R. and Tsien, R.W. (1990a) Presynaptic enhancement revealed by whole cell recordings of long-term potentiation in rat hippocampal slices. *Nature*, in press.

Malinow, R. and Tsien, R.W. (1990b) Long-term potentiation of synaptic transmission between individual CA3 and CA1 neurons in rat hippocampal slices. *Soc. Neurosci. Abs.*, in press.

Malinow, R., Madison, D.V. and Tsien R.W. (1988) Persistent protein kinase activity underlying long-term potentiation. *Nature*, 335: 820–824.

Malinow, R., Schulman, H. and Tsien, R.W. (1989) Inhibition of postsynaptic PKC or CaMKII blocks induction but not expression of LTP. *Science*, 245: 862–866.

Martin, A.R. (1977) Junctional transmission II. Presynaptic mechanisms. In: *Handbook of Physiology: The Nervous System*, pp. 329–355, American Physiological Society.

Mayer, M.L. and Westbrook, G.L. (1987) The physiology of excitatory amino acids in the vertebrate central nervous system. *Prog. Neurobiol.*, 28: 197–276.

Muller, D., Joly, M. and Lynch, G. (1988) Contributions of quisqualate and NMDA receptors to the induction and expression of LTP. *Science*, 242: 1694.

Muller, D. and Lynch, G. (1988) Long-term potentiation differentially affects two components of synaptic responses in hippocampus. *Proc. Natl. Acad. Sci. USA*, 85: 9346–9350.

Nicoll, R.A., Kauer, J.A. and Malenka, R.C. (1988) The current excitement in long-term potentiation. *Neuron*, 1: 97–103.

Nishizuka, Y. (1988) The molecular heterogeneity of protein kinase C and its implications for cellular regulation. *Nature*, 334: 661–665.

Redman, S.J. (1990) Quantal analysis of synaptic potentials in neurons of the central nervous system. *Physiol. Rev.*, 70: 165–198.

Regehr, W.G. and Tank, D.W.(1990) Postsynaptic NMDA-receptor mediated calcium accumulation in hippocampal CA1 pyramidal cell dendrites. *Nature*, in press.

Sastry, B.R., Goh, J.W. and Auyeung, A. (1986) Associative induction of posttetanic and long-term potentiation in CA1 neurons of rat hippocampus. *Science*, 232: 988.

Sayer, R.J., Redman, S.J. and Andersen, P. (1989) Amplitude fluctuations in small EPSPs recorded from CA1 pyramidal cells in the guinea-pig hippocampal slice. *J. Neurosci.*, 9: 840–850.

Sayer, R.J., Friedlander, M.J. and Redman, S.J. (1990) The time course and amplitude of EPSPs evoked at synapses between pairs of CA3/CA1 neurons in the hippocampal slice. *J. Neurosci.*, 10: 826–836.

Segal, M. and Patchornik, A. (1989) Modulation of $(Ca)_1$ by a caged EGTA affects neuronal plasticity in the rat hippocampus. *Soc. Neurosci. Abstr.*, 15: 166.

Skrede, K.K. and Malthe-Sorenssen, D. (1981) Increased resting and evoked release of transmitter following repetitive electrical tetanization in hippocampus: a biochemical correlate to long-lasting synaptic potentiation. *Brain Res.*, 208: 436.

Stevens, C.F. (1989) Strengthening the synapses. *Nature*, 338: 460–461.

Wigstrom, H., Gustafsson B., Huang, Y.Y. and Abraham, W.C. (1986) Hippocampal long-term potentiation is induced by pairing single afferent volleys with intracellularly injected depolarizing pulses. *Acta Physiol. Scand.*, 126: 317–19.

Subject Index